Introduction to Environmental Impact Assessment

A Guide to Principles and Practice

Bram F. Noble

Third Edition

OXFORD
UNIVERSITY PRESS

OXFORD
UNIVERSITY PRESS

Oxford University Press is a department of the University of Oxford.
It furthers the University's objective of excellence in research, scholarship,
and education by publishing worldwide. Oxford is a registered trade mark of
Oxford University Press in the UK and in certain other countries.

Published in Canada by
Oxford University Press
8 Sampson Mews, Suite 204,
Don Mills, Ontario M3C 0H5 Canada

www.oupcanada.com

Library and Archives Canada Cataloguing in Publication

Noble, Bram F., 1975–, author
Introduction to environmental impact assessment : a guide to principles
and practice / Bram F. Noble. — Third edition.

Includes bibliographical references and index.
ISBN 978-0-19-900634-2 (pbk.)

1. Environmental impact analysis—Textbooks. 2. Environmental impact
analysis—Canada—Textbooks. I. Title.

TD194.6.N62 2014   333.71'4   C2014-904943-9]

Cover image: Richard Gillard/E+/Getty

Oxford University Press is committed to our environment.
This book is printed on Forest Stewardship Council® certified paper
and comes from responsible sources.

Printed and bound in Canada

5 6 7 — 19 18 17

MIX
Paper from
responsible sources
FSC® C004071

Contents

3 ◉ Tools Supporting EIA Practice   40

Part II | Environmental Impact Assessment Practice   71

4 ◉ Screening Procedures   72

5 ◉ Scoping and Environmental Baseline Assessments 95

6 ◉ Predicting Environmental Impacts 118

7 ◉ Managing Project Impacts 149

8 ◉ Determining Impact Significance   165

9 ◉ Follow-up and Monitoring   194

Boxes, Boxed Features, Figures, and Tables

Boxed Features

Figures

Tables

Preface

Environmental impact assessment (EIA) in Canada has changed considerably since the second edition of this book. In 2012, the government of Canada introduced its federal budget implementation bill, declaring jobs and economic growth as its top national priorities. Perceiving inefficiencies in EIA as a hindrance to economic growth, the new Canadian Environmental Assessment Act, 2012 was introduced. The new act was intended to reduce the scope and number of federal EIAs in Canada, ensure tighter timelines for regulatory decisions, and eliminate small-project EIAs. For some, this change was long overdue. For others, it was a significant setback in the advancement of EIA in Canada.

Recent years have also been characterized by a considerable growth in awareness of the need to improve the practice of cumulative effects assessment (CEA) in Canada, particularly regional and strategic frameworks for CEA. Although there has been little regulatory improvement to advance CEA, much has occurred outside the regulatory process, including: the development of national guidance for regional strategic environmental assessment by the Canadian Council of Ministers of the Environment; community and industry lobbying for regional and cumulative effects assessment in Canada's western Arctic; the rise of several regional cumulative effects–based monitoring programs, driven by industry, community, and environmental non-government organization partnerships; and considerable advances in the science of impact assessment, including land-use modelling, scenario-based methods, and aquatic cumulative effects assessment.

As with the first two editions, this third edition is directed to students of environmental assessment—this includes current and future environmental assessment practitioners and administrators. The main focus of the book is on the fundamental principles and "good" practices of EIA that are, arguably, common across EIA systems. Since EIA is a subject of interest in many disciplines and to persons of varied backgrounds, the book is interdisciplinary in perspective and attempts to balance discussion on physical and human environments. The book may also serve as a useful reference for key EIA principles and requirements in the new federal regulatory environment.

The book is organized into four parts. Each part introduces a number of principles and procedures, along with case study examples and selected methods and techniques. In response to reviewer feedback on the second edition, each chapter now includes a new feature—Environmental Assessment in Action. This new feature provides in-depth examples of environmental assessment, describing case studies, applications, and lessons learned.

Part I introduces the basic principles of EIA and the basic EIA process. Particular attention is given to the new Canadian Environmental Assessment Act, 2012 and supporting regulations. A significantly revised chapter, focused on the practitioner's toolkit, is also included in Part I. Part II focuses on EIA principles and procedures, from determining the need to conduct an assessment to post-decision analysis. Each chapter has been updated to reflect the new federal act and contains new case examples from the field. Part III focuses on advancing principles in impact assessment, including cumulative effects assessment and strategic environmental assessment. Significant changes have been made to the chapters dealing with these subjects to keep pace with advances in both research and practice. A new chapter, on professional practice, has been added to Part III. This chapter addresses the skill requirements of the EIA practitioner and the standards of ethical conduct. Part IV is an EIA postscript, reflecting on the substantive effectiveness of EIA and making a case for a more regional and integrative approach to environmental assessment.

Bram F. Noble
Professor, Environmental Assessment
Department of Geography and Planning and
School of Environment and Sustainability
University of Saskatchewan

Acknowledgements

I would like to acknowledge the staff at Oxford University Press for their assistance throughout the manuscript writing and editing process for this third edition. I am especially grateful to my students and to my colleagues for their critical insight and suggestions for improvement over the second edition.

To Deana, Noelle, and Elias

I

Environmental Impact
Assessment Principles

Introduction

With growing populations, technological advancement, and increasing demand for natural resources, the need for tools and processes to scrutinize development proposals, inform decision-makers, and manage the impacts of development effectively is of ever-increasing importance. Environmental, economic, and social changes are inherent to development. While the goal of development is typically positive change, it can also create significant and long-term adverse environmental and socio-economic consequences. The challenge facing society is to find ways to support development initiatives while enhancing health and well-being and without adversely affecting the environment. Managing the impacts of development activities, as well as creating positive outcomes from such activities, is essential if development is to be recognized as sustainable.

Defining Environmental Impact Assessment

Originating in the US **National Environmental Policy Act (NEPA)** of 1969, which became law in 1970, **environmental impact assessment (EIA)** is among the most widely practised environmental management tools in the world. Currently, 191 of the 193 member-nations of the United Nations either have national legislation or have signed some form of international agreement that refers to the use of EIA (Morgan 2012). Initially conceived to ensure that the biophysical impacts of major development proposals were considered during decision-making, EIA has been followed by the development of many other forms of impact assessment, including social impact assessment, health impact assessment, strategic environmental assessment, and sustainability assessment, to name a few. Morgan (2012) suggests that the emergence and development of EIA as a key component of environmental management over the past 40-plus years has coincided with the increasing recognition of the nature, magnitude, and implications of environmental and socio-economic change triggered by human actions.

There is no single, universally accepted definition of EIA (Box 1.1), and the term is often used interchangeably with "environmental assessment" (EA) or "impact assessment" (IA). The International Association for Impact Assessment (IAIA) and the UK Institute of Environmental Assessment (IEA) define EIA as: "The process of identifying, predicting, evaluating, and mitigating the biophysical, social, and other

---

Box 1.1 Definitions of Environmental Impact Assessment

EIA is an activity designed to identify and predict the impact on human health and well-being of legislative proposals, policies, programs, and operational procedures and to interpret and communicate information about the impacts (Munn 1979).

EIA is the study of the full range of consequences, immediate and long-range, intended and unanticipated, of the introduction of a new technology, project, or program (Rossini and Porter 1983).

EIA is a planning tool whose main purpose is to give the environment its due place in the decision-making process by clearly evaluating the environmental consequences of a proposed activity before action is taken (Gilpin 1995).

EIA is the evaluation of the effects likely to arise from a major project (or other action) significantly affecting the environment. It is a systematic process for considering possible impacts prior to a decision being taken on whether or not a proposal should be given approval to proceed (Jay et al. 2007).

EIA is a process to predict the environmental effects of proposed initiatives before they are carried out (CEAA 2013).

---

relevant effects of development proposals prior to major decisions being taken and commitments made" (IAIA and IEA 1999). This "look before you leap" approach is the basic concept behind EIA. In simplest terms, EIA is an integral component of sound decision-making, serving both an information-gathering and an analytical component, used to inform decision-makers concerning the impacts and management of proposed developments.

EIA is often described as an environmental protection tool, a methodology, and a regulatory requirement. The IAIA (2012) characterizes EIA as having both a technical guise, providing and analyzing information about the impacts of development actions, and a regulatory or institutional guise, focusing on legal and procedural matters as they relate to stakeholder engagement, laws, and project authorizations. Arguably, EIA is all of these, but perhaps most important, it is a "process" designed to aid decision-making through which concerns about the potential environmental consequences of proposed actions, public or private, are incorporated into decisions regarding those actions. In this regard, EIA can also be viewed as a means of strengthening environmental management processes following the consent decision for development (Morrison-Saunders and Bailey 1999). EIA should not be seen merely as a mechanism for preventing development that might generate potentially negative environmental effects. If this were the case, then few developments would actually take place!

EIA is an organized means of gathering information used to identify and understand the effects of proposed projects on the **biophysical environment** (e.g., air, water, land, plants, and animals) as well as on the **human environment** (e.g., culture, health, community sustainability, employment, financial matters) for people potentially affected. As a formal and frequently regulatory-based process, EIA should not be confused with related environmental studies, such as **environmental site assessments** whose purpose is to determine the nature and extent of contaminant levels at a specific site and to identify clean-up and remediation plans.

Environmental Change and Effects

Important to understanding EIA is recognizing the difference between environmental "change" and environmental "effects." Although the terms are often used interchangeably, an environmental change is a temporal measure, and not all change is project-induced; an **environmental effect** is the difference between change conditions. A distinction can thus be made between **environmental change** and environmental effects in impact assessment. Environmental change is the difference in the condition of a particular environmental or socio-economic parameter, usually measurable, over a specified period of time. Environmental change is typically defined in terms of a process, such as soil erosion, that is set in motion by particular project actions, other actions, or natural processes (Figure 1.1). Actions such as road construction or dam construction, for example, contribute to environmental or condition change. An environmental effect is the change difference: the difference in the condition of an environmental parameter under project-induced change versus what that condition might be in the absence of project-induced change.

The term "environmental effect" is often used interchangeably with **environmental impact**; however, some argue that impacts are estimates or judgments of the value that society places on certain environmental effects or concerns about the difference between the quality of the environment with and without the proposed action. Some consider this distinction little more than a matter of semantics; others argue that the distinction between effects and impacts is particularly important from an environmental management perspective. For example, managing environmental "impacts" may be no more than a damage control exercise if one does not successfully

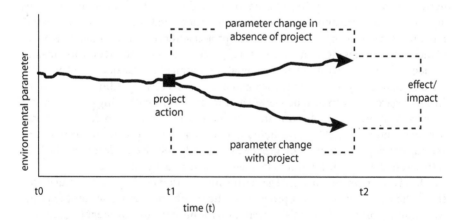

Figure 1.1 ⊚ Conceptualization of an Environmental Effect

identify environmental effects and, perhaps even more important, the project-induced environmental change that is generating the effect or impact from the onset.

EIA Theory and Objectives

The primary purpose of EIA is to facilitate the consideration of environment in planning and decision-making and, ultimately, to make it possible to arrive at decisions and subsequent actions that are more environmentally sustainable. How this is achieved, however, has been the subject of much attention in recent years. Of particular interest has been criticism of the traditional, technocratic model of EIA and the need for more participative and deliberative approaches.

In many respects, the underlying approach to EIA can be traced back to the rationalist model of planning and decision-making that emerged in the 1960s, requiring a technical evaluation as the basis for objective decision-making (Owens, Rayner, and Bina 2004). This rational, comprehensive approach is based on the elements of: (i) defining the problem; (ii) setting goals and objectives; (iii) identifying options; (iv) assessing the options; (v) implementing a preferred solution; and (vi) monitoring and evaluation. Core to this approach is the assumption of a well-defined problem characterized by a range of possible options, complete information, and objective decision-makers. Although the rationalist approach has been criticized in recent years (information in EIA is rarely value-free, often incomplete, and frequently constrained or shaped by political factors and societal interests), it continues to be a valid representation of practice and "represents the framework within which EIA is often used as a tool for planning and decision-making" (Hanna 2005, 7). Others, however, disagree and conceptualize EIA, in its normative form, as a participative and deliberative process designed to facilitate public debate about development priorities and to provide a means for the public to influence government decisions.

The objectives of EIA depend on the lens through which it is viewed. Members of a local community about to be affected by a development, for example, may see EIA as no more than a public relations tool used by developers and politicians to justify decisions, or they may see it as a way to ensure the accountability of developers and provide a means of negotiating consensus solutions such that local concerns are taken into account in the development process. Non-government environmental organizations may also view EIA as a tool to improve stakeholder involvement in development decision-making, or they may see it as a means of preventing development from proceeding but as a "rubber stamp" when it is unsuccessful in doing so. A proponent may view EIA as a time-consuming and expensive regulatory hurdle that must be overcome in order to receive development approval or as a means of improving project design and earning a social licence to operate.

Given the diversity of views and expectations about EIA, Cashmore (2004) argues that it is more relevant to represent the reality of extant EIA as a series of nebulous models, operating along a broad spectrum of philosophies and values. These models concern the role of science in EIA and broadly reflect the range of purposes and objectives of the EIA process (Table 1.1). At one end of this spectrum is the belief that the scientific method provides the basis for EIA theory and practice. In order

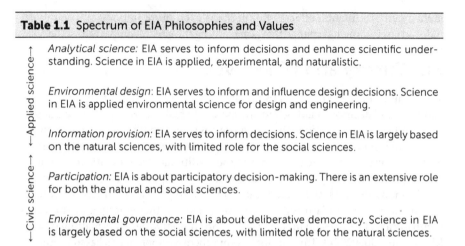

Table 1.1 Spectrum of EIA Philosophies and Values

Analytical science: EIA serves to inform decisions and enhance scientific understanding. Science in EIA is applied, experimental, and naturalistic.

Environmental design: EIA serves to inform and influence design decisions. Science in EIA is applied environmental science for design and engineering.

Information provision: EIA serves to inform decisions. Science in EIA is largely based on the natural sciences, with limited role for the social sciences.

Participation: EIA is about participatory decision-making. There is an extensive role for both the natural and social sciences.

Environmental governance: EIA is about deliberative democracy. Science in EIA is largely based on the social sciences, with limited role for the natural sciences.

Source: Based on Cashmore 2004

to be credible, the EIA process must be based on scientific objectives, modelling and experimentation, quantified impact predictions, and hypotheses-testing. At the other end of the spectrum is the belief that EIA is a decision tool used to empower stakeholders and promote an egalitarian society, with a strong green interpretation of sustainability. In this regard, EIA must be deliberative, promote social justice, and help to realize community self-governance. In Canadian practice, EIA typically reflects neither of these norms but tends to operate on the border of information provision and participation, swaying from one to the other depending on the particular political, project, or local context of application. There is, however, no "best" model, and practitioners should treat these as reference points rather than recipes (Morgan 2012) and adopt an approach to EIA that is best suited to the context at hand.

EIA Output and Outcome Objectives

Because of extant EIA theory and expectations about EIA, the specific objectives of EIA application can vary from one context to the next. However, there are a number of substantive objectives of EIA that, arguably, transcend context. These substantive objectives can be separated into "output objectives" and "outcome objectives."

Output objectives are the immediate, short-term objectives of the EIA process. Output objectives can often be audited, measured, and attributed directly to the EIA process. Many of these objectives are realized during the EIA process itself and are often considered characteristics of "good" EIA. The output objectives of EIA include:

- ensuring that environmental factors are explicitly addressed in decision-making processes concerning proposed developments;
- improving the design of the proposed development;
- anticipating, avoiding, minimizing, and offsetting **adverse effects** relevant to development proposals on the human and biophysical environment;

- ensuring proponent accountability and compliance with relevant laws and regulations respecting development;
- providing a means for public participation in the development process;
- ensuring that information about the potential impacts of the project is made available to the public and decision-makers.

Outcome objectives are the longer-term objectives that are the products of consistent and rigorous, long-term EIA application. Such objectives are typically realized from a broader environmental and societal perspective. Outcome objectives are difficult to measure, often subtle and difficult to associate directly with the EIA process itself. They are often realized only long after completion of an EIA, or series of assessments, and the longer-term implications of development actions and management measures can be observed. The outcome objectives of EIA include:

- protecting the productivity and capacity of human and natural systems and ecological functions;
- facilitating individual, organizational, and social learning;
- increasing environmental awareness;
- promoting sustainability.

The Origins and Development of EIA

EIA Origins

Throughout North America and western Europe, the 1960s was characterized by a sudden growth in awareness of the relationship between an expanding industrial economy and local environmental change. While many characterize the 1960s as an era of environmental "idealism," triggered by a number of environmental challenges and sparked by such works as Rachel Carson's *Silent Spring* (1962), the decade did lead to increasing environmental awareness and public demand and pressure on central governments that environmental factors be explicitly considered in development decision-making. As a result of such pressures, the 1960s and early 1970s witnessed the passage of legislation concerning resource protection, hazardous waste management, and control of water and air pollution. However, perhaps the most significant piece of legislation at the time was the National Environmental Policy Act (NEPA) in the United States, which came into effect in January 1970 after its introduction in 1969.

The term "environmental impact assessment" is actually derived from NEPA, which for the first time required by law that those proposing to undertake certain development projects had to demonstrate that the projects would not adversely affect the environment. To do so, project proponents had to include in their proposal an **environmental impact statement (EIS)** describing the proposed development, the affected environment, likely impacts, and proposed actions to manage or monitor those impacts. During the first 10 years of NEPA, approximately 1000 EISs were prepared annually. Currently, it is estimated that approximately 30,000 to 50,000 EISs are prepared annually in the United States.

Prior to the 1970s, development projects were assessed, but assessment was limited to technical feasibility studies and, in particular, **cost-benefit analysis**. Cost-benefit analysis (CBA) is an approach to project assessment that expresses impacts in monetary terms. Large-scale development projects such as the 114-metre-high Aswan High Dam in Egypt, for example, constructed during 1960 to 1970 and financed by the United States and the United Kingdom, were assessed using CBA techniques. The US Corps of Engineers used CBA for many years to assess and justify large-scale water resource development projects and dam construction in the United States, including the Glen Canyon Dam in Arizona (completed in 1964) and the Oroville Dam in California (completed in 1968). While still commonly used today, obvious drawbacks to CBA include the inability to allocate meaningful dollar values to environmental and human intangibles and the limited scope of fiscal impacts traditionally addressed by CBA. As an "add-on process" to CBA, EIA was initially intended to incorporate all the potential impacts that were excluded from traditional CBA.

EIA Development

The US NEPA is generally recognized as the pioneer of contemporary environmental assessment. Since its beginnings, EIA has gone through a number of evolutionary phases in the United States and Canada, a pattern repeated to varying degrees throughout the world (see Wood 1995; Sadler 1996).

Initial Development

During the early 1970s, immediately following the implementation of NEPA, EIA was characterized by casual and disjointed observations of the biophysical environment, particularly within the local project area. Wider project impacts and potential impact interactions among physical and human environmental components were largely ignored. During this phase, EIA was primarily criticized as a tool used to justify project decisions already made when the process was initiated. The ethos that predominated in this era was "develop now, minimize associated costs and, if forced to, clean up later" (Barrow 1997). In many cases, as illustrated in Chapter 2, lands were leased and project construction was well underway before an EIA started. The end result was that by the time EIA commenced, many of the opportunities to integrate environmental concerns into project planning and to identify more environmentally sustainable courses of action were foreclosed.

Broadening Scope and Techniques

By the mid-1970s and up to the early 1980s, EIA efforts became much more highly organized and technically oriented, reflecting the interests of professionals in the natural sciences. Significant attention was devoted to collecting large environmental inventories—i.e., comprehensive descriptions of the biophysical environment in local project areas. The result was impact statements that were thousands of pages in length, often consisting of little more than a compilation of biophysical environmental facts. It is perhaps not surprising then that during this time, project **scoping** (see Chapter 5) was first introduced in an attempt to identify the important issues and data requirements for project assessment. Greater attention turned toward managing adverse project impacts and the risks associated with particular actions, as opposed to creating large environmental inventories. Public review of project proposals and impact assessment processes also emerged as common practice in EIA, and several new, innovative impact prediction and assessment techniques were introduced. Perhaps more important, EIA advanced beyond the biophysical environment of the local project in recognition that broader regional and social impacts were also important in evaluating the impacts of proposed project actions on the environment.

Institutional Support and Integration

The early 1980s to 2000 witnessed rapid growth in EIA, attributed to, among other things, a number of international events, such as the 1987 World Commission on Environment and Development, the 1991 Convention on Environmental Impact Assessment in a Transboundary Context, the 1992 and 1997 Earth Summits, and the 1998 Convention on Access to Information, Public Participation in Decision-Making, all of which fostered greater international awareness of EIA. Of significant importance in the growth and development of EIA was formation of the International Association for Impact Assessment (IAIA), the leading global network on best practice in the use of impact assessment (Box 1.2).

Box 1.2 The International Association for Impact Assessment

The International Association for Impact Assessment (IAIA) was formed in 1980 to bring together practitioners, researchers, and other users of impact assessment from across the world. Currently, the IAIA has more than 1600 members from over 120 countries, with affiliate branches established in several countries, including Canada (western and northern, Quebec, Ontario), Germany, New Zealand, Portugal, Zambia, Mozambique, and Ghana. The IAIA membership is diverse and includes planners, engineers, social and natural scientists from various disciplines, corporate managers, public interest advocates and environmental non-government organizations, government regulators and senior administrators, private consultants, and educators and students.

The IAIA's mission is to provide an international forum for advancing innovation and communication of best practice in all forms of impact assessment so as to further the development of local, regional, and global capacity in impact assessment. This is achieved, in part, through annual international conferences where research and practice experience is shared and future directions for impact assessment discussed, a formal journal (*Impact Assessment and Project Appraisal*) to disseminate research focused on advancing impact assessment, and a variety of professional networking opportunities. The IAIA's website also contains a variety of resources on different forms of impact assessment, many of which are available to non-members, including special publications on the principles of best practice in environmental, social, and health impact assessment and standards for impact assessment professionals. The website can be accessed at www.iaia.org.

Source: Based on www.iaia.org.

During this period, EIA practice also turned increasing attention toward physical and social interrelationships associated with project development. **Environment** within the context of EIA was defined as inclusive of not only the biophysical environment but also components of the social and economic environments (e.g., labour markets, demography, housing structure, education, health, values, lifestyles) at multiple spatial and temporal scales. Not all current EIA systems, however, adopt such a liberal interpretation of "environment." At the Canadian federal level, for example, "environment" is interpreted as including the physical environment and interacting natural systems, whereas in some Canadian provincial jurisdictions, "environment" in EIA is interpreted much more broadly to include social, cultural, and other human dimensions as well (Box 1.3).

EIA emerged in the 1990s as a system-oriented or multi-dimensional approach and involved application of both qualitative and quantitative models in impact

---

Box 1.3   Scope of "Environment" in Canadian EIA

At the Canadian federal level, the primary focus of EIA is on the adverse effects of a project or activity on the biophysical environment. "Environment," in the context of the **Canadian Environmental Assessment Act, 2012**, sec 2.1, means "components of the Earth, and includes: (a) land, water and air, including all layers of the atmosphere; (b) all organic and inorganic matter and living organisms; and (c) the interacting natural systems that include components referred to in paragraphs (a) and (b)." While de facto, socio-economic issues are, in their own right, invariably part of the assessment process, the scope of environment under Canadian federal EIA legislation is concerned primarily with the physical environment and, specifically, those components of the environment that are within the legislative authority of Parliament (see sec 5(1) of the Canadian Environmental Assessment Act, 2012). What constitutes an "environmental effect" under the Canadian federal EIA system is discussed further in Chapter 4. Interpretations of "environment" under provincial EIA legislation, including that of Newfoundland and Labrador, Ontario, and Saskatchewan, involve what might be considered a broader understanding of environment to encompass direct impacts on communities and on social and cultural systems as well.

Province	EIA Legislation	Scope of Environment
Alberta	Environmental Protection and Enhancement Act	"Environment" means the components of the Earth and includes: (i) air, land, and water, (ii) all layers of the atmosphere, (iii) all organic and inorganic matter and living organisms, and (iv) the interacting natural systems that include components referred to in subclauses (i) to (iii).
Manitoba	Environment Act	"Environment" means (a) air, land, and water or (b) plant and animal life, including humans.

New Brunswick	Clean Environment Act	"Environment" means (a) air, water, or soil, (b) plant and animal life, including human life, and (c) the social, economic, cultural, and aesthetic conditions that influence the life of humans or of a community insofar as they are related to the matters described in paragraph (a) or (b).
Newfoundland and Labrador	Environmental Protection Act	"Environment" includes (i) air, land, and water, (ii) plant and animal life, including human life, (iii) the social, economic, recreational, cultural, and aesthetic conditions and factors that influence the life of humans or a community, (iv) a building, structure, machine, or other device or thing made by humans, (v) a solid, liquid, gas, odour, heat, sound, vibration, or radiation resulting directly or indirectly from the activities of humans, or (vi) a part or a combination of those things referred to in subparagraphs (i) to (v) and the interrelationships between two or more of them.
Nova Scotia	Environment Act	"Environment" means the components of the Earth and includes (i) air, land, and water, (ii) the layers of the atmosphere, (iii) organic and inorganic matter and living organisms, (iv) the interacting systems that include components referred to in subclauses (i) to (iii), and (v) for the purpose of Part IV, the socio-economic, environmental health, cultural, and other items referred to in the definition of environmental effect.
Prince Edward Island	Environmental Protection Act	"Environment" includes (i) air, land, and water, (ii) plant and animal, including human, life, any feature, part, component, resources, or element thereof.
Quebec	Environmental Quality Act	"Environment": the water, atmosphere, and soil or a combination of any of them or, generally, the ambient milieu with which living species have dynamic relations.
Yukon	Environmental and Socioeconomic Assessment Act	"Environment" means the components of the Earth and includes (a) land, water, and air, including all layers of the atmosphere, (b) all organic and inorganic matter and living organisms, and (c) the interacting natural systems that include components referred to in paragraphs (a) and (b).
Saskatchewan	Environmental Assessment Act	"Environment" means (i) air, land, and water, (ii) plant and animal life, including man, and (iii) the social, economic, and cultural conditions that influence the life of man or a community insofar as they are related to the matters described in subclauses (i) and (ii).

prediction and analysis. Emphasis increasingly turned to the assessment of "total **environmental impacts**" and the inclusion of environmental considerations early in the project planning cycle before irreversible decisions were made. This gave way to greater attention to the links between EIA and impact management, social effects assessment, health impact assessment, and adaptive environmental management and led to increasing awareness of the need for better monitoring of environmental effects after project implementation. Originally conceived in reaction to growing environmental awareness and to the voice of environmental lobby groups during the late 1960s, EIA emerged throughout the 1990s as an increasingly important environmental management tool.

Environmental Assessment beyond the Individual Project

In recognition that many environmental impacts are the result of decisions made at the level of policies, plans, and programs (PPP) before individual development projects become a reality, increasing attention has been given to the advancement of **strategic environmental assessment (SEA)**. The concept of assessing PPP impacts was first established under the US National Environmental Policy Act, but the term strategic environmental assessment was not introduced until 1989 in a research report to the European Commission, describing SEA as environmental assessments appropriate to policies, plans, and programs of a more strategic nature than those applicable to individual projects (Wood and Djeddour 1989).

The purpose of SEA is to help in the understanding of the development context of a PPP, to identify problems and potentials, to address key trends, and to assess viable environmental and sustainable options. In other words, SEA is supposed to promote the development of more environmentally sensitive PPPs, help situate EIA and project decisions in a broader environmental framework, and address non-project impacts and cumulative environmental effects. SEA was formally established in Canada in 1990 by way of a federal Cabinet directive and as a separate process from EIA, "making it the first of the new generation of SEA systems" (Dalal-Clayton and Sadler 2005). Current requirements for SEA in Canada exist under the 2010 *Cabinet Directive on the Environmental Assessment of Policy, Plan and Program Proposals*. The principles and practice of SEA are the focus of discussion in Chapter 12

Sustainability Initiatives

In principle, EIA should serve as an integrated planning tool for decision-making, characterized by integrating cumulative and global environmental effects, empowering the public, recognizing uncertainties, favouring a precautionary and adaptive approach, and making a positive contribution to sustainability (Gibson 2002). Although one of the stated purposes of federal EIA in Canada under the Canadian Environmental Assessment Act, 2012 is to "encourage federal authorities to take actions that promote sustainable development" (section 4(1)(h)), the focus of EIA is largely on minimizing adverse effects as opposed to ensuring overall, substantive gains. While some Canadian EIAs have been set within the context of sustainability, such as the Voisey's Bay nickel mine-mill project (Box 1.4), true integration of sustainability in EIA remains limited in practice. In some countries, EIA is still not conducted prior to the design and approval of projects. As a result, there has emerged considerable interest in the practice of **sustainability assessment**, whether as a separate assessment process or integrated as a seamless part of both EIA and SEA.

Sheate (2009) argues that all environmental assessment tools, including EIA, have sustainability as an underlying purpose. Others, however, including Gibson (2012), argue that regardless of their stated purposes, sustainability imperatives have not been met by traditional approaches to impact assessment systems, including EIA. Of particular concern in EIA is the long-standing focus on making project impacts less severe rather than encouraging positive steps toward greater community and ecological sustainability (Gibson 2006). Gibson suggests eight core principles or characteristics of sustainability-focused assessment, namely, a focus on: long-term socio-ecological system integrity; livelihood sufficiency and opportunity for everyone; intra-generational equity; intergenerational equity; resource maintenance and efficiency; socio-ecological civility and democratic governance; precaution and

---

Box 1.4   Voisey's Bay Mine-Mill Project

In 1993, a rich nickel-copper-cobalt deposit was discovered at Voisey's Bay in northern coastal Labrador. The deposit has surface dimensions of approximately 800 metres by 350 metres, extends to depths of about 125 metres, and will be mined using open-pit methods. The deposit contains estimated proven and probable mineral reserves of 30 million tonnes, grading 2.85 per cent nickel, 1.68 per cent copper, and 0.14 per cent cobalt. There is an additional estimated 54 million tonnes of mineral wealth at Voisey's Bay, grading 1.53 per cent nickel, 0.70 per cent copper, and 0.09 per cent cobalt, as well as 16 million tonnes of inferred mineral resource, grading 1.6 per cent nickel, 0.8 per cent copper, and 0.1 per cent cobalt.

The project proponent, Voisey's Bay Nickel Company Limited, a subsidiary of Inco Limited, submitted a proposal in 1997 for the development of a mine-mill complex and related infrastructure to produce mineral concentrates at Voisey's Bay. In the absence of Aboriginal land claims agreements, the Voisey's Bay project was subject to review as set out under Canadian federal and Newfoundland provincial environmental assessment processes and pursuant to a memorandum of understanding between the provincial and federal governments, the Labrador Inuit Association, and the Innu Nation.

The public review panel for the Voisey's Bay project issued guidelines for review in which the proponent was required to discuss explicitly the extent to which the project would "make a positive overall contribution towards the attainment of ecological and community sustainability, both at the local and regional levels" (Voisey's Bay Panel 1997). This was the first major resource development project in Canada for which the impact statement guidelines for the project proponent explicitly identified the sustainability criterion, noting that EIA should go beyond seeking to minimize damage to requiring that an undertaking maximize long-term, durable net gains to the community and the region. Construction of the mine-mill project commenced in 2002. See www.vbnc.com for additional project information.

adaptation; and immediate and long-term integration. The application of these core principles in EIA to help evaluate the significance of potential impacts is discussed in Chapter 7. However, Bond, Morrison-Saunders, and Pope (2012) note that the point has not yet been reached at which there is universal consensus as to what sustainability assessment is or how it should be applied.

International Status of EIA

Since NEPA, EIA has diffused throughout the world (Box 1.5). While NEPA provided the initial basis for EIA, every nation's EIA system is quite distinct, and the enabling legislation, policy directives, and guidelines for EIA vary considerably from one nation to the next. It is not possible to review the status of EIA in all nations, so only a few national EIA systems are highlighted here. Although EIA is expanding internationally, it is still relatively new, since many nations have less than 10 years of EIA experience.

Canada

Canada was first to follow the US NEPA beginnings, formally implementing an EIA system in 1973 as a guidelines order through the federal Environmental Assessment and Review Process. It was not until 1995, however, that EIA became entrenched in Canadian law, and subsequent amendments have since taken place. In 2012, during a period of global economic crisis, Parliament passed Bill C-38, the Jobs, Growth and Long-term Prosperity Act. Included in this budget implementation bill, and prompted by perceived inefficiencies in the EIA process as a hindrance to economic development, was provision for the new Canadian Environment Assessment Act, 2012. The bill passed in June 2012, and the new environmental assessment act came into effect. Currently, all Canadian provinces and territories have their own EIA systems, with additional EIA systems for Aboriginal land-claim areas. The current Canadian EIA system is discussed in greater detail in Chapter 2.

Australia

Australia was next to formally adopt a system of EIA through its Environmental Protection Act in 1974. The act was formally implemented in 1975. As with Canada's provinces, most states in Australia have opted for their own form of EIA legislation in which projects of national significance are assessed under a joint state and federal system. In Western Australia, four environmental assessment processes operate under state legislation. The first of these is compulsory for project-level EIA and includes any action that is likely to create significant environmental impacts. Between 40 and 50 of these assessments are completed each year under the guidance of Western Australia's Environmental Protection Agency. A second compulsory process is the referral of land-use plans for environmental assessment by local government and by the state planning agency. This assessment process is seldom triggered—perhaps only two or three times per year. However, hundreds of informal assessments take place whereby local governments and state agencies

---

Box 1.5 Selected National EIA Systems

Country	Implemented*	Website to Current System
Australia	1974	www.ea.gov.au
Brazil	1981	www.mma.gov.br
Bulgaria	1991	www.moew.government.bg
Canada	1973	www.ceaa.gc.ca
Chile	1993	www.sea.gob.cl/
China	1989	www.zhb.gov.cn
Colombia	1974	www.minambiente.gov.co
France	1976	www.environnement.gouv.fr
Germany	1985	www.bmu.de
Guyana	1996	www.epaguyana.org
Israel	1982	www.sviva.gov.il
Japan	1984	www.env.go.jp
Madagascar	1995	www.madonline.com
New Zealand	1974	www.mfe.govt.nz
Pakistan	1983	www.punjab.gov.pk
South Africa	1984	www.environment.gov.za
Thailand	1975	www.thaigov.go.th
United States	1969	www.epa.gov

* Many nations have undergone revisions in their EIA systems since first introduced, including moving from guidelines or policy directives to formal legislated systems. Canada, for example, introduced EIA in 1973, but it was not until 1995 that EIA was formally legislated. A new Canadian Environmental Assessment Act was introduced in 2012.

---

refer plans for assessment, which are then screened and passed without requiring a full-scale impact assessment. There are also two voluntary processes, one of which is specifically for strategic proposals such as land-use planning strategies, drilling programs, or satellite mining developments. The second concerns the Environmental Protection Agency's authority to report on environmental matters generally. Essentially, proponents outline their future plans and ask for the agency's viewpoint. There is no legally binding outcome, but the process does allow a proponent to pursue an endorsed development option that is more likely to receive approval during formal project impact assessment.

European Union

It was not until 1985 that EIA was formally adopted in Europe through European Union Directive 85/337/EEC, which established a legal basis for individual member states' EIA regulations. The directive was amended in 1997 to align with the UN ECE Espoo Convention on EIA in a Transboundary Context, again in 2003 to align with principles for public participation included under the Aarhus Convention on Public Participation in Decision-Making, and in 2009 to include projects related to the transport, capture, and storage of carbon dioxide. The initial directive of 1985 and its three amendments were codified by Directive 2011/92/EU in December 2011.

There are two categories of EIA under Directive 2011/92/EU, mandatory EIA and discretionary EIA. Some projects, such as long-distance railway lines and hazardous-waste disposal sites, require mandatory EIA. Projects listed in Annex I to the directive are considered as having significant effects on the environment and always require an EIA. For most other projects, listed in Annex II to the directive, EIA is at the discretion of the member states, and the need for assessment is determined on a case-by-case examination of the potential impacts of the project or by screening the project against specified criteria or thresholds. The environmental assessment of public plans or programs, SEA, is required under Directive 2001/42/EC, known as the SEA Directive.

China

EIA was formally introduced to China in the 1970s by way of the Environmental Protection Law of the People's Republic of China (on trial), proclaimed in 1989. Article 6 of the law required that all enterprises and institutions pay adequate attention to the prevention of pollution and damage to the environment when selecting their sites and designing, constructing, and planning production and must prepare a report on the potential environmental effects of the project for approval. In 2003, following a decade of practice, reform, and development of technical guidance and regulations, China introduced the Law of the People's Republic of China on Environmental Impact Assessment, which strengthened its commitment to EIA and also provided for the application of impact assessment to strategic development plans. Under the EIA law, EIA is defined as a system for: analyzing, forecasting, and assessing the potential impact on the environment after implementation of planning and construction projects; establishing strategies and measures to prevent or alleviate adverse impacts on the environment; and implementing follow-up reviews and monitoring. According to Buckley (2013), however, the lack of broad public involvement in China's EIA process, because of insufficient or blocked access to information and the lack of overall public awareness about the EIA process, poses significant challenges to its effectiveness.

Developing Nations and Development Agencies

In developing countries, EIA is practised primarily for two reasons. The first is to comply with EIA provisions in the country; the second is to meet the EIA requirements of development aid agencies (Noble 2011). EIA was introduced early in some developing countries, including Colombia (1974) and the Philippines (1977). However, it was not until well into the 1980s (e.g., Brazil, Indonesia, Mexico, Algeria, and Turkey) and particularly in the 1990s (e.g., Belize, Bolivia, Gambia, Mongolia, and Tunisia) that many developing countries established formal legislative bases for EIA or introduced EIA provisions into their existing environmental legislative frameworks. The World Bank first introduced EIA requirements in 1989 for evaluating projects it was financing; the Asian Development Bank followed in 1993 with similar requirements (Harrop and Nixon 1999). The Canadian International Development Agency also has EIA requirements for international investment and development projects and recently adopted a system of environmental assessment to address the potential environmental impacts associated with policy and program decisions.

EIA legislation in many developing countries is of high quality, much like legislation in developed countries. The main differences lie in implementation. EIA capacity in developing countries needs strengthening, especially within government agencies.

The Basic EIA Process

Notwithstanding the diversity of EIA definitions and its adaptations to different international contexts and circumstances, the core elements of EIA are widely agreed upon (Jay et al. 2007). From an applied perspective, EIA can be thought of as a *process* that systematically examines the potential environmental implications of development actions prior to project approval. In short, EIA is simply responsible and proactive planning. The structure of an EIA process, however, is dictated by the specific issues it attempts to address and by the regulatory and legislative requirements within which it operates. While not all EIA systems contain the same elements or specific design procedure, the broad EIA process emanating from the original NEPA and subsequently diffused throughout the world can be thought of as a series of systematic steps (Box 1.6). Although presented here as a linear framework, in practice EIA is an iterative process in which discussions with stakeholders, public review, scoping processes, and post-project evaluations continue to refine impact predictions and management actions. The various stages of the EIA process are explored in greater detail throughout this book.

Operating Principles

How the EIA process unfolds in practice and the quality of its application vary considerably from one EIA system to the next. However, there are several *operating principles* of EIA that define how the EIA process should unfold. Specifically, EIA should be applied:

- as early as possible in the planning and decision-making stages;
- to all proposals that may generate significant adverse effects or about which there is significant public concern;
- to all biophysical and human factors potentially affected by development, including health, gender, and culture, and cumulative effects;

- consistently with existing policies, plans, and programs and the principles of sustainable development;
- in a manner that allows involvement of affected and interested parties in the decision-making process;
- in accordance with local, regional, national, or international standards and regulatory requirements (IAIA and IEA 1999).

Box 1.6   Basic EIA Process

Project description	Description of the proposed action, including its alternatives, and details sufficient for an assessment	
Screening	Determination of whether the action is subject to an EIA under the regulations or guidelines present, and if so, who is responsible and what type or level of assessment is required	
Scoping and baseline assessment	Delineation of the key issues and the spatial and temporal boundaries to be considered in the assessment, including analysis of the baseline conditions and trends	
Impact assessment	Prediction and evaluation of the potential environmental effects of the project, including cumulative effects	
Impact management	Identification of impact management, mitigation, and enhancement strategies and development of environmental management or protection plans	Public participation
Significance determination	Determination of the significance of potential adverse environmental effects, taking into consideration the effectiveness of proposed impact management actions	
Submission and review of the EIS	Preparation and submission of the EIS and related technical documents and reports for technical and public review	
Recommendations and decision	Recommendations and decision as to whether the proposed action should proceed and, if so, under what conditions or be outright rejected	
Implementation and follow-up	Implementation of the project, management measures, and recommendations (if any); continuous data collection to monitor compliance with conditions and regulations; monitoring the effectiveness of impact management measures and the accuracy of impact predictions	

Opportunities for and Challenges to EIA

In addition to managing the impacts of development on the environment, EIA generates a number of direct benefits, including:

- improvements to project planning and design;
- cost savings for proponents through early identification of potentially unforeseen impacts;
- reduction in the role of the legal system by ensuring early compliance with environmental standards;
- increase in public acceptance through participation and demonstrated environmental and socio-economic responsibility.

However, in an IAIA newsletter, Fuggle (2005) commented on disillusionment with EIA and scepticism about whether EIA is in fact contributing to better decisions. Fuggle warned that too much might be expected of EIA and that there are perhaps too many different ideas as to what EIA can accomplish. EIA is not a "magic bullet"—while it complements broader environmental planning, management, and decision-making, it does not replace them.

That said, there are a number of emerging and enduring challenges to EIA that must be addressed if EIA is to fulfill its role as a process toward improved environmental management. These challenges are certainly not the *only* issues facing EIA, but how and whether they are addressed will in large part determine the next phase of EIA evolution and the role of EIA as either a tool for reactive management or one that facilitates the sustainable development of the environment, the economy, and society.

Ensuring a Quality EIA Process

Many scholars have argued for the need to focus more attention on the substantive effectiveness of EIA. Although process effectiveness has received much more attention than the substantive outcomes or long-term influence of EIA, this is not to suggest that there is no need for improvement in the current practice of EIA. EIA processes have improved considerably since the early 1970s, and there is a wealth of guidance available to practitioners and decision-makers on "good practice," but there is still a need to ensure a quality EIA process and considerable room for improvement, especially in such areas as scoping (Chapter 5), public participation (Chapter 10), follow-up (Chapter 9), and the assessment of cumulative environmental effects (Chapter 11). EIA can be quite ineffective, indeed damaging, if not used properly: (1) if it is not applied to projects with the potential to cause significant environmental effects or to projects with minor effects that may be cumulatively significant over space or time; (2) if proponents undertake EIA late in the project design process; (3) if significant issues of concern are not included in the assessment; (4) if predictions of impacts are not accurate or not verifiable; (5) if decision-makers fail to use EIA results in making development decisions; or (6) if the implementation of projects subject to EIA fails to follow through effectively with the sound environmental management plans developed through the EIA process (Noble 2011).

Balancing Efficiency and Efficacy

The need for greater efficiency in the EIA process is simply a reality of the current political and economic climate, both in Canada and internationally. In his 2012 review of the "state-of-the-art" EIA, Morgan suggests that as governments look to stimulate economic growth and create employment in response to the global financial crisis, many are promoting rapid expansion of physical infrastructure and the exploitation of natural resources in frontier areas and generally seeking to expedite EIA and approval processes for development undertaking. The result, in some instances, is the weakening of provisions for environmental protection (see Gibson 2012) and increased efficiencies in EIA perhaps at the cost of its efficacy. Efficiency can be thought of as "doing things right," whereas efficacy refers to "doing the right things." Efficiency is indeed important to the future of EIA. As Leonard (2012) argues, if EIA is not able to adapt to rapid economic and social change and improve its efficiency, it is in danger of being more and more marginal to the decision-making process. But has this concern with process efficiency come at the expense of effective EIA? At no point has the need for "effective" EIA been so important.

In their 2011 Brief to the House of Commons Committee on the Environment and Sustainable Development on the Canadian Environmental Assessment Act, Hanna and Noble (2011) define the effectiveness of EIA simply as the extent to which it identifies, assesses, and finds ways to mitigate or eliminate the potential negative impacts of development and, importantly, how well EIA helps to improve environmental management and ultimately the state of the environment. They go on to suggest several key characteristics of effective EIA, namely, that EIA:

- ensures stakeholder confidence in the process and decisions taken;
- is integrative and linked to decision-making;
- promotes betterment and longer-term and substantive gains in environmental quality;
- is comprehensive of physical and human environments and the range of actions, processes, and alternatives that affect environmental quality;
- is evidence-based;
- ensures accountability through independent review, clearly defined roles, and open and accessible information;
- is participatory and supportive of participatory processes;
- has a legal basis to ensure compliance and enforce action;
- is sufficiently resourced and embraces innovation and creativity in approaches to assessment and evaluation.

A major challenge for the future of impact assessment is thus how to ensure a degree of efficiency in process such that EIA does not cause undue delay and cost for proponents but at the same time to ensure that the credibility, integrity, and effectiveness of EIA are maintained.

Improved EIA Governance

Finally, good governance is important to good EIA. "Government" refers to the institutions that have the formal power to exercise authority in EIA. "Governance" refers to the function of governing and is inclusive of government and non-government agencies and institutions. In other words, the governance of EIA is about how the EIA process is managed, including who has responsibility for what, determining the nature and extent of collaborative approaches, the choice of methods and how they are applied, and who is involved in EIA and in what capacity (Meuleman et al. 2013). At the same time, EIA is part of an overall governance approach to planning and decision-making about development. In order to ensure effective EIA, Meuleman et al. identify four key areas in the governance of EIA that must be addressed:

1. The destruction of trust: Short consultation deadlines, discrediting evidence from stakeholders, and other hierarchical biases can cause decision-makers to act in ways that negatively affect the trust of stakeholders and the public in the EIA process.
2. Disregarding complexity: Preference for clear rules and problem definition and acting based on "the standard for practice" can result in the fragmentation of projects to avoid addressing complex, cumulative effects problems or in attempts to avoid the problem by invoking urgency in the decision process.
3. A bias for economic efficiency: During periods of economic downturn, assessment and decision-making processes tend to lead to overestimating the value of efficiency in the EIA process.
4. Never-ending talks: The quest for consensus-building and inclusivity may result in EIA becoming an interminable exercise, sometimes even losing sight of its objective to ensure informed decision-making.

Key Terms

biophysical environment
Canadian Environmental
  Assessment Act
Canadian Environmental
  Assessment Act, 2012
cost-benefit analysis
environment
environmental change
environmental effect
environmental impact assessment (EIA)

environmental impacts
environmental impact statement (EIS)
environmental site assessment
human environment
National Environmental Policy
  Act (NEPA)
scoping
strategic environmental
  assessment (SEA)
sustainability assessment

Review Questions and Exercises

1. What are the potential benefits of conducting an EIA?
   a) For communities affected by development?
   b) For proponents undertaking development?
2. How can EIA contribute to sustainable development?
3. Obtain a small sample of EISs from your local library or government registry, or access them online. Examine the table of contents of each, and create a general list of common elements or issues addressed in each assessment. Are the contents similar in terms of the types of issues and topics addressed?
4. Why was there a sudden interest in EIA in North America in the 1960s?
5. Given how EIA has evolved over the years, what direction might you expect it to take in the future?
6. What requirements currently exist in your country or province for conducting EIA? When were these requirements implemented, and how have they evolved over time? Is the environment defined to include both physical and human systems? Who is responsible for conducting EIA? How many assessments have been completed to date? Is there a particular sector or type of EIA that dominates? Why might this be so? Do you feel that EIA is meeting sustainability objectives?
7. Visit some of the websites listed in Box 1.5, and compare the requirements and provisions for EIA to those of your own nation or province.
8. Review some of the challenges to EIA discussed in this chapter. Can you identify one (or more) recent EIA case in which these challenges were evident? How might they be addressed?
9. As noted in this chapter, EIA has evolved to become one of the most widely applied environmental management tools in the world. Are there certain ethical, cultural, or other elements that should be considered when applying EIA in developed versus developing nations? What elements of the process, if any, might vary from one socio-political or cultural context to the next?
10. Richard Fuggle, former president of the IAIA, raised a question as to whether EIA had passed its "sell by" date. What do you suppose this means? Do you agree? Provide evidence to support your claim.

References

Barrow, C.J. 1997. *Environmental and Social Impact Assessment: An Introduction*. London: Arnold.

Bond, A., A. Morrison-Saunders, and J. Pope. 2012. "Sustainability assessment: The state of the art." *Impact Assessment and Project Appraisal* 30 (1): 53–62.

Buckley, L. 2013. "China to strengthen public participation in environmental impact assessments." *Chinawatch* 7 July. Washington: Worldwatch Institute.

Carson, R. 1962. *Silent Spring*. Boston: Houghton Mifflin.

Cashmore, M. 2004. "The role of science in environmental impact assessment: Process and procedure versus purpose in the development of theory." *Environmental Impact Assessment Review* 24: 403–26.

CEAA (Canadian Environmental Assessment Agency). 2013. "Basics of environmental assessment." www.ceaa.gc.ca.

Dalal-Clayton, B., and B. Sadler. 2005. *Strategic Environmental Assessment: A Sourcebook and Reference Guide to International Experience.* London: Earthscan.

Fuggle, R. 2005. "Have impact assessments passed their 'sell by' date?" *Newsletter of the International Association for Impact Assessment* 16 (3): 1, 6.

Gibson, R.B. 2002. "From Wreck Cove to Voisey's Bay: The evolution of federal environmental assessment in Canada." *Impact Assessment and Project Appraisal* 20 (3): 151–9.

———. 2006. "Sustainability assessment: Basic components of a practical approach." *Impact Assessment and Project Appraisal* 24 (3):170–82.

———. 2012. "In full retreat: The Canadian government's new environmental assessment law undoes decades of progress." *Impact Assessment and Project Appraisal* 30 (3): 179–88.

Gilpin, A. 1995. *Environmental Impact Assessment: Cutting Edge for the 21st Century.* Cambridge: Cambridge University Press.

Hanna, K., ed. 2005. *Environmental Impact Assessment: Practice and Participation.* Don Mills, ON: Oxford University Press.

———, and B.F. Noble. 2011. *A Brief to the House of Commons Committee on the Environment and Sustainable Development on the Canadian Environmental Assessment Act (CEA Act).* Available at www.eiaeffectiveness.ca.

Harrop, D.O., and A.J. Nixon. 1999. *Environmental Assessment in Practice.* Routledge Environmental Management Series. London: Routledge.

IAIA. 2012. "Impact assessment." Fastips no. 1. Prepared by M. Partidário, L. den Broeder, P. Croal, R. Fuggle, and W. Ross. Fargo, ND: International Association for Impact Assessment.

———, and IEA. 1999. *Principles of Environmental Impact Assessment Best Practice.* Fargo, ND: IAIA.

Jay, S., et al. 2007. "Environmental impact assessment: Retrospect and prospect." *Environmental Impact Assessment Review* 27: 287–300.

Leonard, P. 2012. "Will impact assessment become extinct, and disappear as the dinosaurs did, or will impact assessment adapt to the new challenges?" IAIA Business and Industry Series no. 1, 8.

Meuleman, L., et al. 2013. "Governance." Fastips no. 4. Fargo, ND: IAIA.

Morgan, R. 2012. "Environmental impact assessment: The state of the art." *Impact Assessment and Project Appraisal* 30 (1): 5–14.

Morrison-Saunders, A., and J. Bailey. 1999. "Exploring the EIA/environmental management relationship." *Environmental Management* 24 (3): 281–95.

Munn, R.E. 1979. *Environmental Impact Assessment: Principles and Procedures.* New York: John Wiley & Sons.

Noble, B.F. 2011. "Environmental Impact Assessment." In *Encyclopedia of Life Sciences.* Chichester: John Wiley & Sons. doi: 10.1002/9780470015902.a0003253.pub2.

Owens, S., T. Rayner, and O. Bina. 2004. "New agendas for appraisal: Reflections on theory, practice, and research." *Environment and Planning A* 36: 1943–59.

Rossini, F.A., and A.L. Porter. 1983. *Integrated Impact Assessment. Social Impact Assessment Series.* New York: Perseus Books.

Sadler, B. 1996. *Environmental Assessment in a Changing World: Evaluating Practice to Improve Performance.* Final report of the International Study of the Effectiveness of Environmental Assessment. Fargo, ND: IAIA.

Sheate, W.R. 2009. "The evolving nature of environmental assessment and management: Linking tools to help deliver sustainability." In W.R. Sheate, ed., *Tools, Techniques and Approaches for Sustainability: Collected Writings in Environmental Assessment Policy and Management*. Singapore: World Scientific.

Voisey's Bay Panel. 1997. *Environmental Impact Statement (EIS) Guidelines for the Review of the Voisey's Bay Mining and Mill Undertaking*. Voisey's Bay Mine and Mill Environmental Assessment Panel.

Wood, C. 1995. *Environmental Impact Assessment: A Comparative Review*. London: Longman Scientific and Technical.

———, and M. Djeddour. 1989. *Environmental Assessment of Policies Plans and Programmes*. Interim Report to the Commission of European Communities. Contract no. B6617-571-572-89. Manchester: EIA Centre, University of Manchester.

EIA in Canada

Canada is recognized internationally as a nation that has contributed significantly to the development and advancement of EIA policy and practice. Since its inception in the early 1970s, thousands of EIAs have been completed in Canada, most of them for relatively small-scale, routine development projects and undertakings. This chapter presents a brief overview of Canadian EIA systems. It is not the intent here to discuss Canadian EIA practice, procedure, and requirements in detail, since these elements are addressed throughout the text. Rather, this chapter briefly outlines Canadian EIA under federal, provincial, and territorial governments and land-claims agreements and illustrates the legal development of EIA in Canada. Particular attention is paid to the new Canadian Environmental Assessment Act, 2012.

Canadian EIA Systems

Environmental impact assessment in Canada is enshrined in the law of the provinces, territories, Aboriginal governments, and the federal government (Figure 2.1). Since Canada is a confederation, EIA at the federal level is not binding on the provinces and territories. Rather, responsibility for EIA is divided among the provinces and territories and the federal government, with laws, regulations, EIA objectives, and procedures varying considerably from one jurisdictional system to the next (Bitter 2008). At the time of Confederation, for example, the four "founding provinces" received authority to make laws governing their natural resources. The Constitution Act, 1930, transferred further responsibility for the development of natural resources, and hence their management, from the Dominion of Canada to the various other provincial governments of the day, including Manitoba, British Columbia, Alberta, and Saskatchewan. The Constitutional Act, 1982, amended the original constitution to give all provinces the authority to make laws and decisions concerning the development and management of their natural resources.

Provincial EIA

Canada's first provincial EIA system, Ontario's Environmental Assessment Act, was enacted in 1975. All provinces and territories now have some form of EIA process, whether EIA requirements under environmental or resource management law, such

Figure 2.1 ◉ Provincial, Territorial, and Land-Claims-Based EIA Systems in Canada

Source: Map produced by J. Bronson, University of Saskatchewan.

as New Brunswick's Clean Environment Act and Newfoundland's Environmental Protection Act, or requirements under EIA-specific law, such as Saskatchewan's Environmental Assessment Act. In 1998, to facilitate co-ordination of EIA applications and responsibilities between the government of Canada and the provinces and territories, the Canadian Council of Ministers of the Environment signed an accord designed to improve cooperation on environmental assessment and protection. The purpose of the accord is to streamline the EIA process and eliminate duplication between federal and provincial or federal and territorial processes in cases where a proposed project could potentially trigger dual EIA systems—referred to as a "one-window approach." The intent of harmonization was not to transfer jurisdictional authority from the federal government to provincial or territorial governments but rather to ensure that federal and provincial or territorial approval processes are co-ordinated such that a project proponent does not need to present the same impact statement twice or in two different formats (Meredith 2004). Harmonization agreements on EIA have been signed between the government of Canada and the governments of Manitoba (2007), Alberta (2005), Saskatchewan (2005), British Columbia

(2005), Quebec (2004), Newfoundland and Labrador (2005), Ontario (2004), and Yukon (2004). Notwithstanding EIA harmonization, co-ordination between federal and provincial or territorial assessment processes and regulatory requirements, as well as inconsistencies between jurisdictional mandates over the role and function of EIA in project assessment and sustainability, remain a critical challenge to EIA effectiveness (see Fitzpatrick and Sinclair 2009).

Northern EIA

North of 60°, EIA is a mixed system of federal jurisdiction, federal–territorial agreements such as the Yukon Territory Environmental and Socio-economic Assessment Act, and regulation under numerous Aboriginal land claims and co-management boards. From the initial Mackenzie Valley Pipeline Inquiry of 1974–7 to the more recent Mackenzie Gas Project, EIA in Canada's North has undergone a number of significant regulatory and legislative changes. In 1973, the government of Canada announced a policy that would permit northern Native groups to seek compensation in the form of a land-claims agreement and to have more control over development activities on their traditional lands. The very first land-claims-based EIA process was initiated shortly thereafter, in 1975, when the governments of Canada and Quebec and the Cree and Inuit of northern Quebec signed the James Bay and Northern Quebec Agreement. Several additional northern EIA agreements have since been established, including the **Nunavut Land Claims Agreement**, the **Mackenzie Valley Resource Management Act** (**MVRMA**), and EIA under the Inuvialuit Land Claim Agreement.

Nunavut Land Claims Agreement

The 1993 Nunavut Land Claims Agreement represents by far Canada's largest Aboriginal land-claims settlement and land-claims-based EIA process. It provided the Inuit with self-government title to approximately 350,000 square kilometres of land in what had been the eastern half of the Northwest Territories. The negotiated settlement area included mineral rights for approximately 35,000 square kilometres of this land. The agreement also included an undertaking for the implementation of a new territory, Nunavut, which was formally established in 1999 under the agreement (Figure 2.1). Included under Article 12.2.2 of the agreement is the establishment of an environmental assessment review board, the **Nunavut Impact Review Board** (**NIRB**), which is the primary authority responsible for EIA activities in the land-claims area. The mandate of the NIRB includes:

- screening proposals to determine whether an environmental review is required;
- gauging and determining the extent of the regional impacts of proposed projects;
- reviewing ecosystemic and socio-economic impacts of proposed projects;
- determining whether proposed projects should proceed and under what conditions;
- monitoring development projects (see www.ainc-inac.gc.ca).

The overall purpose of the NIRB is to protect and promote the future well-being of residents and communities of the Nunavut settlement area, including its ecosystem integrity.

Mackenzie Valley Resource Management Act

For the purposes of the MVRMA, the Mackenzie Valley is defined as including all of the Northwest Territories with the exception of Wood Buffalo National Park and the Inuvialuit Settlement Region (Figure 2.1). Proclaimed in 1998, the MVRMA was implemented by the federal government with the intent of giving northerners increased decision-making authority over natural resource and environmental development matters. A co-management board was also established for the Sahtu and Gwich'in settlement areas of the Northwest Territories to delegate responsibilities for land-use planning and for issuing land-use permits and water-use licences. As part of this establishment, a Valley-wide public board, the **Mackenzie Valley Environmental Impact Review Board** (**MVEIRB**), was created to undertake EIAs and panel reviews within the jurisdiction of the MVRMA. The Canadian Environmental Assessment Act no longer applies in the Mackenzie Valley except for very specific situations involving issues that are transboundary in nature or for which it is deemed necessary by the board and the federal minister of environment. In such cases, a joint MVEIRB–CEAA review panel is established. The MVEIRB is responsible for:

- conducting environmental impact assessments;
- conducting environmental reviews;
- maintaining a public registry of environmental assessments;
- making recommendations to the minister of the Department of Indian and Northern Affairs regarding the approval or rejection of development projects within the region.

Inuvialuit Land Claim Agreement

In 1984, a comprehensive land claim, the Inuvialuit Final Agreement (IFA), was signed between Canada and the Inuit of the western Arctic (Figure 2.1). The IFA encompasses approximately 1,000,000 square kilometres, known as the Inuvialuit Settlement Region, and includes six communities. One of the outcomes of the IFA was the establishment of two co-management agencies, the Environmental Impact Screening Committee and the Environmental Impact Review Board, to deal with environmental assessment screenings and reviews in the Inuvialuit Settlement Region. During the agencies' first 10 years, more than 150 development proposals were screened and two public reviews undertaken for offshore oil and gas exploration programs. In 2000, Canada and the Inuvialuit Environmental Impact Review Board signed an environmental assessment agreement for the Inuvialuit Settlement Region. The agreement concerns how the EIA process under the Environmental Impact Review Board will be substituted for a panel review under the Canadian Environmental Assessment Act and details the process and the steps involved should the Environmental Impact Review Board request such a substitution.

Development of Federal EIA

Environmental assessment over the last 30 years has moved towards being earlier in planning, more open and participative, more comprehensive, more mandatory, more closely monitored, more widely applied, more integrative, more ambitious, and more humble (Gibson 2002, 152).

The current EIA regulatory system in Canada is the product of over 40 years of development, legislative reform, and political controversy (Table 2.1). EIA at the federal level is currently required by means of EIA-specific law, and it is the responsibility of the proponent (the company, government agency or individual proposing the development) to prepare an EIS as per the relevant federal, territorial, or provincial regulations and requirements. When EIA was first formally introduced in Canada in 1973–4, it was never the intent of the federal government that EIA would someday have a legal basis.

EARP Origins

In Canada, EIA originated with the establishment of a task force in 1970 to examine a policy and procedure for assessing the impacts of proposed developments. In 1973, the Cabinet Committee on Science, Culture and Information agreed on the need for a formal process to assess the potential impacts associated with project development. At that point, the first Canadian EIA process, the federal **Environmental Assessment Review Process** (EARP), was born, and in 1984 the **Federal Environmental Assessment Review Office** (FEARO) was created to administer its implementation. Unlike NEPA in the United States, however, the Canadian EARP had no legislative basis and was therefore not legally enforceable; rather, project impact reviews were initially intended to be cooperative and voluntary, not legally binding. The result was that impact assessments under EARP were carried out inconsistently and in some cases not carried out at all. Some of the earliest projects reviewed under EARP included the Point Lepreau nuclear power station (1975), eastern Arctic offshore drilling (1978), and the Arctic Pilot Project (1980). Perhaps the most significant project reviews under EARP, reviews that would change the course of EIA in Canada, were the Oldman River Dam project in Alberta (Box 2.1) and the Rafferty-Alameda Dam in Saskatchewan—both initiated without formal project assessment.

Canadian Environmental Assessment Act

Introduced in Parliament in 1992 by the Conservative government as Bill C-78, the **Canadian Environmental Assessment Act** was created to replace EARP. While the act was intended to make EIA more rigorous and systematic, it also limited the reach of EIA to include only project-level decisions and not broader planning or policy issues. Under EARP, for example, the potential reach of EIA was quite broad, as indicated by the Beaufort Sea hydrocarbon review (1982–4) and Atomic Energy of Canada Limited's nuclear fuel waste management concept (1988–94)—both were area-wide reviews and

Table 2.1 History of Federal EIA in Canada

Period	Development
1973–4	Cabinet makes policy commitment to review the environmental effects of federal decisions. The Environmental Assessment Review Process (EARP) is established to screen projects to "ensure that they do the least possible damage to our natural environment."
1984	EARP is reinforced and codified by the *Environmental Assessment Review Process Guidelines Order*. The Federal Environmental Assessment Review Office (FEARO) is established to administer EARP.
1990	The minister of environment announces a reform package for EIA. The package includes new EIA legislation, an assessment process for policy and program proposals, and a participant funding program to support public participation in EIA.
1992	Consultations on and a parliamentary review of Bill C-13, the Canadian Environmental Assessment Act, take place. Canadian Environmental Assessment Act receives royal assent.
1995	The Canadian Environmental Assessment Act comes into force, providing a legal basis for in EIA Canada. The Canadian Environmental Assessment Agency is established to administer the act, replacing FEARO.
1999	The minister of environment launches a five-year review of the Canadian Environmental Assessment Act. The review is a requirement of the provisions of the act.
2001	The minister of environment introduces Bill C-19 to amend the Canadian Environmental Assessment Act.
2003	Bill C-9, formerly Bill C-19, receives royal assent, and the amended Canadian Environmental Assessment Act comes into force.
2010	The Canadian Environmental Assessment Act is amended to strengthen the role of the Canadian Environmental Assessment Agency and improve the timeliness of EIA.
2012	The 2012 Economic Action Plan on Jobs, Growth and Long-term Prosperity is presented in the House of Commons and includes a commitment to reforming the regulatory system in Canada's natural resource sector by shortening timelines for EIA, reducing duplication and regulatory burden on development, and enhancing Aboriginal consultation. The Jobs, Growth and Long-term Prosperity Act introduces a new Canadian Environmental Assessment Act, 2012 and repeals the former act.

Source: Based on the Canadian Environmental Assessment Agency's (2012) "Introduction to the Canadian Environmental Assessment Act, 2012" training manual. http://publications.gc.ca/site/eng/429080/publication.html.

Box 2.1 Reforming EIA: Oldman River Dam Project, Alberta

Throughout the 1960s and into the 1970s, agriculture was of growing importance in Alberta. Periodic droughts, characteristic of the region's semi-arid continental climate, meant considerable uncertainties in crop production and sustainability of local water supplies. Combined, agricultural growth and climatic unpredictability reinforced earlier initiatives for the development of large-scale irrigation systems. In 1976, plans were announced to construct an earth- and rock-filled dam at the Three Rivers site in Alberta (Oldman, Castle, and Crowsnest rivers). The proposed dam was to be approximately 75 metres in height and more than 3000 metres long, with a reservoir capable of storing nearly 500 million cubic metres of water (Oldman River Dam Project 1992). Several sites for the dam were considered, but the Three Rivers site was favoured from the outset, notwithstanding studies that identified potentially significant adverse effects on fish and wildlife as well as impacts on ranchers whose properties would be partially divided and flooded.

The controversy over the proposed project occurred over two periods. First, between 1976 and 1980, landowners whose lands would be flooded opposed the dam, and the Peigan Indian Reserve argued that the project was intruding on lands within the Blackfoot Nation's territory. This first controversy in part ended in 1980 when the final project site was selected. The second, and more significant, controversy emerged in 1987 when an environmental group, Friends of the Oldman River, challenged the project through the provincial courts. The environmental group was initially unsuccessful, and based on assessments under Alberta's environmental assessment process and with the approval of the federal government under the Navigable Waters Protection Act, project construction commenced in 1988. No federal assessment under the EARP Guidelines Order was initiated, notwithstanding the involvement of a federal authority. The Oldman Dam project was nearly 40 per cent complete when Friends of the Oldman River once again challenged the project in 1989 but this time in the federal court. The environmental group lost the court case, but in an interesting turn of events, the court of appeals in 1992 reversed the initial decision and quashed approval for the Oldman Dam project under the Navigable Waters Protection Act. The court of appeals ordered that the EARP did in fact have the force of law, and the federal government was compelled to comply with the Guidelines Order. While the decision was too late to reverse any damage that had already been created by the dam project, it did pave the way for a legal requirement for EIA in Canada, which would ensure that EIA would be implemented before development took place.

concept-based assessments. With the introduction of the Canadian Environmental Assessment Act, broader regional and concept-based reviews became divorced from the formal EIA process. The act also set out responsibilities and procedures for the environmental assessment of projects involving federal authorities and established a process to ensure that impact assessment was applied early in the planning stages of proposed project developments. When the act applied to a proposed project, the relevant federal government department was designated as the "responsible authority" and was to ensure that EIA unfolded according to the principles of the act.

In 1994, the **Canadian Environmental Assessment Agency** was created to oversee the act and replaced FEARO. The act itself was proclaimed and came into force in 1995. Among its main objectives were:

- to ensure that the environmental effects of projects were reviewed before federal authorities took action so that potentially significant adverse impacts could be ameliorated;
- to encourage federal authorities to take actions that would promote sustainable development;
- to promote cooperation and co-ordinated action between federal and provincial governments on environmental assessments;
- to promote communication and cooperation between federal authorities and Aboriginal peoples;
- to ensure that development in Canada or on federal lands did not cause significant adverse environmental effects in areas surrounding the project;
- to ensure that there was an opportunity for public participation in the environmental assessment process.

When the act was initially proclaimed, section 72 required that a review of the act be undertaken every five years. Accordingly, in 1999 Bill C-13, An Act to Amend the Canadian Environmental Assessment Act, was introduced, and the Standing Committee on the Environment and Sustainable Development organized public consultations on EIA reform. Following consultation, the standing committee reintroduced the bill to Parliament as Bill C-9, which received royal assent on 11 June 2003, and the revised Canadian Environmental Assessment Act became law on 30 October 2003. Among the commitments under the new act were:

1. Promotion of high-quality assessments. The revisions would potentially contribute to better decision-making in support of sustainable development by improving compliance with the act, strengthening the role of follow-up in EIA, providing the minister of environment with the option of requesting follow-up studies or additional information before making a project decision, and improving the consideration of cumulative effects.
2. Assessment of appropriate projects. The revised act was intended to focus the EIA process on projects likely to have significant adverse environmental effects while reducing the need for and time commitment to assessing smaller, routine projects. This involved making use of a more streamlined assessment process for routine projects and exempting certain smaller projects from EIA requirements provided that specified environmental conditions were met.
3. Fair and consistent application of the act. EIA was extended to reserve lands, federal lands, Crown corporations, and the assessment of

transboundary effects that might occur on park reserve lands and areas of land claims.

4. Better co-ordination among participants. To minimize duplication, the position of federal environmental assessment co-ordinator was established to facilitate EIAs that involved more than one jurisdictional authority, such as federal–provincial EIA agreements.

5. Increased certainty in the process. In addition to clarification of key terms and procedures under the act, the process was to make greater use of mediation and dispute resolution to improve EIA efficiency. The Canadian Environmental Assessment Agency committed to playing a greater role in building consensus and co-ordinating the dissemination of information through early involvement in the assessment process.

6. Improvement of public participation. The nature and role of public participation was to be enhanced through improved access to information, including a federal registry of impact statements and reports, expanded opportunities for public input, and better incorporation of Aboriginal perspectives and values into the decision-making process.

Canadian Environmental Assessment Act, 2012

In 2008, in the midst a global economic crisis, the Conservative government of Canada declared its top priority as being "to support jobs and growth and to sustain Canada's economy" (House of Commons Standing Committee on Finance, Sub-committee on Bill C-38, 2012). Perceiving inefficiencies in the current environmental assessment process as a hindrance to economic development, the government included provisions in its federal budget implementation bill (Bill C-38, the Jobs, Growth and Long-Term Prosperity Act) to replace the federal environmental assessment process (the Canadian Environmental Assessment Act) with the **Canadian Environmental Assessment Act, 2012** (Becklumb and Williams 2012). Ensuring an expedited review process and removing barriers to economically attractive resource development ventures, such as the highly contested Enbridge Northern Gateway project, were among the primary drivers for the new act. Among other actions, the bill included elimination of the National Round Table on the Economy and the Environment.

Gibson (2012) notes that, in contrast to the establishment of the former act, the Canadian Environmental Assessment Act, 2012 was not preceded by any preliminary proposals or public consultations. The briskness with which the new act came into force, preceding development of the new regulations needed to implement it, was explained by government representatives as a response to the urgency for changes to an assessment process that was hindering progress. However, Gibson (2012) reports that the new legislation was actually initially discussed in 2009, but because of its controversial nature, the government waited until 2012, when it was in a majority position, to push it quickly through Parliament in an effort to cushion Canada from a global economic crisis by fostering resource development.

The Canadian Environmental Assessment Act, 2012 is now the legal basis for federal EIA in Canada, replacing the former Canadian Environmental Assessment

Act. It consists of 129 sections and three schedules. Most sections are sequential, meaning that they follow the order in which the process is to be carried out. Included among the purposes of the act (section 4) are:

- to protect the components of the environment that are within the legislative authority of Parliament from significant adverse environmental effects caused by a designated project;
- to promote cooperation and coordinated action between federal and provincial governments with respect to environmental assessments;
- to promote communication and cooperation with aboriginal peoples with respect to environmental assessments;
- to ensure that opportunities are provided for meaningful public participation during an environmental assessment;
- to ensure that an environmental assessment is completed in a timely manner;
- to encourage federal authorities to take actions that promote sustainable development in order to achieve or maintain a healthy environment and a healthy economy; and
- to encourage the study of the cumulative effects of physical activities in a region and the consideration of those study results in environmental assessments.

Major Changes under the Canadian Environmental Assessment Act, 2012

Application of the Canadian Environmental Assessment Act, 2012 and determining when the act applies to a particular project is discussed in detail in Chapter 4. But first, it is important to understand how the transition from the former act to the Canadian Environmental Assessment Act, 2012 has changed the nature, application, and scope of federal EIA. Some of these major changes are:

1. Fewer federal authorities involved in EIA. Under the former act, any number of federal departments and agencies could potentially be responsible for carrying out EIA (e.g., Fisheries and Oceans Canada, Parks Canada, Aboriginal Affairs and Northern Development Canada, Transport Canada, Environment Canada), depending on the nature of the proposed development, primary responsibilities for permitting, and various other factors. Under the new act, there are only three responsible authorities for EIA. The Canadian Environmental Assessment Agency is responsible for the majority of assessments, unless the project falls under the purview of the National Energy Board or the Canadian Nuclear Safety Commission.

2. Elimination of EIA for "small projects." The former act contained three levels of assessment: **screening EIA**, **comprehensive study EIA**, and **review panel** EIA—reflecting the increasing complexity of projects and

their potential for significant adverse environmental effects. Screening assessments, which comprised more than 90 per cent of federal EIAs, were eliminated from the new act, leaving the focus instead only on "designated" projects deemed most likely to cause significant adverse environmental effects on matters of direct federal responsibility. The standard level of assessment under the new act is roughly equivalent to the former comprehensive study assessment. Thus, the new act is intended to result in far fewer federal EIAs. Whereas the former act adopted an "all in unless excluded" approach to determining the need or EIA, the current act is much more restrictive in determining which projects may require an assessment.

3.   Delegation, substitution, and equivalency of EIA. In an effort to reduce duplication between federal EIA and assessments carried out under provincial, territorial, or Aboriginal land-claims jurisdiction, the new act expands on existing mechanisms under the old act, namely, the ability to delegate part of the EIA process to another jurisdiction and the ability to substitute a provincial or Aboriginal EIA process for federal assessment by means of a provision for equivalency. Under the new act, the minister can recommend to the governor in council that a designated project be exempt from federal EIA and assessed instead under provincial, territorial, or Aboriginal land claims EIA should the process be deemed "equivalent" for the project at hand.

4.   Established time limits. Under the new act, there are time limits on the EIA process for both standard assessments and panel reviews. Standard assessment must now adhere to a 365-day time limit and panel reviews to a two-year time limit (Figure 2.2). It is important to note, however, that this does not mean that the entire EIA process, including a decision on a project, must be completed within these timelines. The timelines

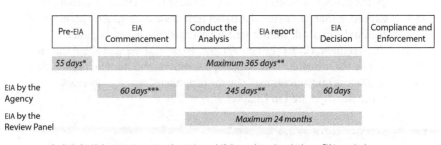

* Includes 10 days to review project description and 45 days to determine whether an EIA is required.
** Government timeline only; the clock is paused while the proponent conducts the analysis and when the proponent is asked to supply new information to support the EIA report.
*** Time limit for the Canadian Environmental Assessment Agency to refer the project to an EIA by review panel.

Figure 2.2 ◉ Timelines under the Canadian Environmental Assessment Act, 2012 for Assessment by the Canadian Environmental Assessment Agency and by Review Panel

apply only to the regulatory process. For example, once a project application is submitted to the Canadian Environmental Assessment Agency, the agency has 55 days (of the 365-day limit) to review the project and determine whether an EIA is required. When the proponent receives the guidelines for conducting the EIA, the clock is suspended while the proponent collects the information required to undertake the assessment. The clock is also suspended after the EIS is submitted should the proponent be asked to prepare and submit any additional or supplementary information.

5. Changes to the definition of "environmental effect." Under the former act, an environmental effect included any change that the project could cause in the environment, as well as the social and economic effects of such change. Under the new act, section 5, an environmental effect includes a change to a component of the environment that is under federal jurisdiction, such as fish or migratory birds; change that may be caused on federal lands or outside the province where the project is to be carried out; and effects on Aboriginal peoples of a change to the environment. The narrower definition constrains the minister in the types of projects that may be designated for an EIA. Paragraph 19(1)(a) of the act adds that these environmental effects include environmental effects of malfunctions or accidents that may occur in connection with the project and any cumulative environmental effects that are likely to result from the project in combination with other physical activities that have been or will be carried out.

6. Focus on "interested parties." In the event of panel review assessments, the former act permitted any member of the public to appear before a review panel to present their views on a proposed project. Under the new act, participation in public hearings is restricted to interested parties, which comprise members of the public who are directly affected by the proposed project as well as those with relevant information or expertise that can assist the panel in its assessment. Any member of the public can still submit written comment to a review panel.

7. Enforcement. Under the new act, the decision statement about the project by the minister, including any conditions about impact mitigation or other commitments made by the proponent, is now enforceable. This is further discussed in Chapter 9.

"Higher-Order" Assessment

Among the most significant developments in Canadian EIA has been the establishment of environmental assessment above the project level—at the level of policies, plans, and programs. In Canada, as discussed above, the environmental assessment of policies was included in the former EARP Guidelines Order, which defined "proposal" as any initiative for which the federal government had decision-making responsibility. With the introduction of the 1992 Canadian Environmental Assessment Act, however, environmental assessment was restricted by law to "projects" or "physical undertakings," and policy, plan, and program assessment was required only by means of federal Cabinet directive.

In 1990, Canada announced a reform package for environmental assessment that included new legislation and a new assessment process for decisions and actions above the project level. The commitment to strategic environmental assessment was strengthened in 1999 with release of the Cabinet directive on the Environmental Assessment of Policy, Plan and Program Proposals. Under the directive, an environmental assessment of a policy, plan, or program is required when a proposal is submitted to a minister or to Cabinet for approval and implementation of that proposal may result in either positive or negative environmental effects. The overall purpose of this strategic approach to environmental assessment is to allow for more informed decisions in support of sustainability initiatives.

The directive was updated in 2004, requiring that Canadian federal departments and agencies prepare a public statement whenever a full strategic environmental assessment had been completed. Federal guidelines for implementation of the directive were most recently updated in 2010. Outside the federal process, however, strategic environmental assessment is still practised, largely on an ad hoc basis. Further, notwithstanding constant and consistent messages about the need to link environmental assessment for policies, plans, and programs with that of projects, there is no mention of strategic environmental assessment in the Canadian Environmental Assessment Act, 2012. Chapter 12 of this book discusses strategic environmental assessment in greater detail.

EIA Directions

Initiated as a policy with little grip, EIA has since evolved to become a rigorous regulatory and legal framework. How this evolution has unfolded over the past 40 years, however, might be described either as the dislocation of an established order in order to improve regulatory requirements, as the natural transition from one system to another, or as a series of unnatural breaks triggered by forces of social and legal change. However one chooses to characterize the history of EIA in Canada, Gibson (2002, 156) captures it best in suggesting that "Canadian environmental assessment policies and laws have evolved slowly . . . this evolution has been hesitant and uneven, though overall it has been positive." What people will say five years from now looking back on the Canadian Environmental Assessment Act, 2012 remains to be seen. Future directions and enduring and emerging issues facing EIA in Canada are discussed in Chapter 14.

Key Terms

Canadian Environmental Assessment
    Agency
comprehensive study EIA

Environmental Assessment Review
    Process
Federal Environmental Assessment
    Review Office (FEARO)

Mackenzie Valley Environmental
   Impact Review Board
Mackenzie Valley Resource
   Management Act

Nunavut Impact Review Board
Nunavut Land Claims Agreement
review panel EIA
screening EIA

Review Questions and Exercises

1. What are some of the critical events that have shaped the course of Canadian EIA at the federal, provincial, and territorial levels?
2. When was EIA enacted in your province (or territory)? What is the name of the legislation that requires EIA? Describe some of the main features of your province or territory's EIA legislation, such as its purpose, how it defines "environment," and when an EIA is required. Compare this to the EIA legislation of another Canadian province or territory.
3. It has been suggested that all of Canada should be governed by a single EIA regulatory framework and assessment process. Do you agree? What might be the advantages and challenges to removing EIA responsibility and authority from provinces and territories and transferring it to a single, composite federal EIA system?
4. What are the major differences between the Canadian Environmental Assessment Act and the Canadian Environmental Assessment Act, 2012? Do you think these changes are positive or negative for strengthening EIA in Canada?

References

Becklumb, P., and T. Williams. 2012. "Canada's new federal environmental assessment process: Background paper." Publication no. 2012-36-E. Ottawa: Library of Parliament.

Bitter, B. 2008. Personal communication (8 October 2008). Environmental Assessment Branch, Ministry of Environment, Government of Saskatchewan.

CEAA (Canadian Environmental Assessment Agency). 2012. "Introduction to the Canadian Environmental Assessment Act, 2012 participant manual." Ottawa: CEAA.

Fitzpatrick, P., and A.J. Sinclair. 2009. "Multi-jurisdictional environmental impact assessment: Canadian experiences." *Environmental Impact Assessment Review* 29: 252–60.

Gibson, R.B. 2002. "From Wreck Cove to Voisey's Bay: The evolution of federal environmental assessment in Canada." *Impact Assessment and Project Appraisal* 20 (3): 151–9.

———. 2012. "In full retreat: The Canadian government's new environmental assessment law undoes decades of progress." *Impact Assessment and Project Appraisal* 30 (3): 179–88.

House of Commons Standing Committee on Finance, Sub-committee on Bill C-38. 2012. *Bill C-38, Part 3. Responsible Resource Development.* Ottawa: House of Commons.

Meredith, T. 2004. "Assessing environmental impacts in Canada." In B. Mitchell, ed., *Resource and Environmental Management in Canada*, 3rd edn, 467–96. Toronto: Oxford University Press.

Oldman River Dam Project. 1992. *Report of the Environmental Assessment Panel*. Report no. 42. Hull, QC: FEARO.

The Practitioner's Toolkit

The practitioner's toolkit consists of a variety of methods and techniques. **Methods** in EIA, such as assessment matrices or checklists, are concerned with the characterization of impacts and the organization, classification, and communication of information—they are not used to "predict" environmental impacts (Barrow 1997; Canter 1996). **Techniques**, on the other hand, are often predictive tools to provide data that are then collated, arranged, interpreted, and presented according to the organizational principles of the methods being used (Bisset 1988). A technique, such as a **Gaussian dispersion model**, provides information on some parameter, such as the anticipated dispersion of air pollutants from a specific industrial development; those data are then organized and serve as input to a particular assessment method, such as an impact matrix, by which the practitioner evaluates, interprets, and presents the data to aid decision-making. Assessment techniques are much more selective than assessment methods and often discipline-specific. Many of the techniques adopted for project- and program-level impact assessments, such as pollutant dispersion models or species population models, are not appropriate at the policy level, where the issues are by nature much broader. Methods, which are concerned with the various aspects of assessment, are often applicable to all levels of assessment.

There are no tools specifically designed for EIA practice, but the choice of tools used in EIA does have significant bearing on the quality of information available to support decision-making. A variety of tools are available for use in EIA, each offering different possibilities and used for different purposes (Table 3.1). Each also has its own costs, data requirements, and other operational characteristics that must be considered when determining which is best at any particular stage of the EIA process and for any particular EIA application. The choice of tools varies depending on the nature of the particular problem, the local socio-economic context, the availability of time and resources, the skill of the practitioner, and the goals and objectives of the assessment. "Good-practice" is about knowing the best tools for deriving and communicating the information needed to support an informed decision. It is not possible to discuss all of the tools used in EIA. This chapter presents only a suite of common tools used in EIA to (i) organize information for the purpose of impact identification, interpretation, and communication; and (ii) predict environmental impacts (see Baker and Rapaport 2005). Many tools serve both functions. Other examples of tools that support EIA practice are discussed throughout this book.

Table 3.1 Select Tools That Support EIA Practice

Tool Type or Focus	Examples of Methods and Techniques
Analogue approaches	Examination of similar projects; literature and case reviews; document analysis; personal communication; synthesis of existing data bases
Spatial analysis	Geographic Information Systems; remote sensing; land-use partitioning analysis; constraints mapping; vulnerability analysis; air photo analysis; participatory mapping
Economic valuation	Cost-benefit analysis; input-output analysis; Monte Carlo analysis; hedonic pricing; contingent valuation; conjoint choice
Weighting and scoring	Multi-criteria analysis; ranking; concordance analysis; Sondheim method; Battelle method
Risk evaluation	Risk perception assessment; quantitative risk assessment; SWOT analysis; cost of environmental degradation analysis
Judgment	Workshops; Delphi technique; reasoned argumentation; causal chain analysis; expert consultation; traditional knowledge
Trends and associations	Network and system diagrams; Markov chain analysis; backcasting; statistical correlations; regression trees
Matrices	Leopold matrices; component interaction matrices; magnitude matrices; weighted magnitude matrices; interaction matrices; weighted interaction matrices; optimum pathway matrices; policy compatibility matrices; goals achievement matrices; Peterson matrix
Scenarios	Simulation modelling; futures workshops; decision trees; impact networks; options appraisal; intent surveys; scenario analysis
Systems modelling	Mechanistic models; balance models; simulation modelling; heuristic modelling; spatial models; stochastic models; deterministic models; statistical models
Checklists	Descriptive matrices; templates and forms; program theory evaluation; summary reports and briefings; programmed-text checklist; questionnaire checklist
Participation	Virtual learning communities; public hearings; public surveys; interviews; advisory committees; town halls; design charrettes; focus groups; community-based monitoring

Tools for Impact Identification, Interpretation, Organization, and Communication

Tools for identifying, interpreting, organizing, and communicating impacts and issues are useful and systematic aids to the EIA process. One of the most important reasons for using such tools is that they provide a means for the synthesis of information and for the evaluation of alternatives on a common basis (Canter 1996). A variety of tools support impact and issues identification and organization of information in EIA. Some of the more common ones are presented here, based on Canter (1992), Shopley and Fuggle (1984), UNEP (2002), and Aura (2009). Many of these tools are revisited in other chapters.

Checklists

Perhaps the simplest systematic tool used in EIA is the checklist. **Checklists** are comprehensive lists of environmental effects or indicators of environmental impacts designed to stimulate thinking about the possible consequences of a proposed action. They are typically used in conjunction with other tools to ensure that a prescribed list of items, regulations, or potential actions and effects is considered in the EIA and to screen particular project actions. Checklists are also commonly used to assist development proponents in complying with EIA requirements. Sometimes called screening checklists or initial environmental evaluations, such checklists help a proponent to describe their project and determine whether it will be subject to an EIA and, if so, what level of assessment.

Programmed-Text Checklists

One type of checklist used in EIA is the **programmed-text checklist**, which is similar to a **questionnaire checklist** and typically contains a series of questions to be answered. For example, a typical question for a bridge construction project might ask "Is there a risk of riverbank erosion?" If the answer is "yes," then the user may be directed to a subset of questions, such as "Is the erosion likely to be severe enough to cause harm to fish habitat?" Programmed-text checklists are best suited for relatively routine undertakings, such as bridge modifications, road realignments, municipal land-use developments, or forest access road construction (Box 3.1). They are useful when the projects involve impacts that are known and for which there are standard impact mitigation measures. The primary advantages of using checklists are that they promote thinking about the range of potential issues and impacts emerging from a project in a systematic way and are relatively efficient and easy to use. Among the disadvantages, checklists:

- are typically not comprehensive of all potential impacts, and as a result not all impacts are considered;
- are often impractical to use if they are comprehensive;
- are often too general and not tailored to specific project environments;
- do not evaluate effects either quantitatively or qualitatively;

- are subjective and qualitative, meaning that different assessors may reach different conclusions using the same checklist;
- do not consider underlying environmental systems or cause-and-effect relationships and hence provide no conceptual understanding of impacts.

Box 3.1 Programmed-Text or Questionnaire Checklists

Propose a set of questions that must be answered when considering the potential effects of a proposed development.

A. Will the project cause pollution of air, water, or soil?	Yes	No	Unsure
i. Will there be a discharge of solid or dissolved substances to waste water?	☐	☐	☐
ii. Is there a risk of discharge of gases that are damaging to health or environment?	☐	☐	☐
iii. Is there risk of a potential impact on drinking water?	☐	☐	☐
iv. Will the activity cause discharge of dust to the atmosphere?	☐	☐	☐
B. Will the project cause waste problems?	**Yes**	**No**	**Unsure**
i. Will waste be created during operations that is hazardous to human health?	☐	☐	☐
ii. Is there a risk that tailings may contaminate local water and soil resources?	☐	☐	☐
iii. Have the long-term environmental impacts of mine waste been considered?	☐	☐	☐
iv. Is the proposed management of hazardous waste in compliance with standards?	☐	☐	☐

Matrices

While specific design and format often vary from project to project, **matrices** are essentially two-dimensional checklists that consist of project activities on one axis and potentially affected environmental components on the other (Figure 3.1). Matrices are perhaps the most commonly used tools for impact identification and communication and are particularly useful for identifying first-order relationships between specific activities and impacts, as well as for providing a visual aid for impact summaries. The disadvantages of matrices are the same as those of checklists. EIA matrices can often be large and difficult to complete, and the volume of information

Forest Management Activity		Moose	Woodland Caribou	White-Tailed Deer	Fur-Bearers	Bats	Rodents	Raptors
Infrastructure development	All-weather access road	I	M	I	I	I	I	I
	Dry-weather access roads	I	M	I	I			
	Winter road		I					
	Camps, timber and fuel-storage sites		I		I			
Harvesting	Logging	M / *	M	M / *	I-M / *	M	I	I-M / *
	Slashing and woody-debris management						I / *	I / *
	Timber storage							
Forest renewal	Site preparation		*		I		I	
	Tree establishment		*		*			
	Mechanical stand tending							
	Chemical stand tending	I	*				I	

Matrix legend:
Blank cell = not applicable or no impact; "I" = insignificant/mitigable; "M" = significant/mitigable; "N" = significant/non-mitigable; * = positive

Figure 3.1 ◉ Section of an Impact Identification Matrix from the Tembec, Pine Falls Operations, Manitoba, Forest Stewardship Plan, 2010–29

Source: Select data from "Environmental Impact Statement of the Forest Management Licence 01, 2010–2029 Forest Stewardship Plan, Executive Summary," p. 3 https://www.gov.mb.ca/conservation/eal/registries/4572tembec/tembec_eis/_exe_sum.pdf

can make them difficult to understand. There are several types of EIA matrices, but the two most basic types, upon which more sophisticated matrices are often developed, are **magnitude matrices** and **interaction matrices**.

Magnitude Matrices

The magnitude matrix goes beyond simple impact identification to provide a summary of impacts according to their magnitude, importance, or time frame (Glasson, Therivel, and Chadwick 1999). The best-known and most comprehensive magnitude matrix is the traditional **Leopold matrix**, originally developed for the US Geological Survey by Leopold et al. (1971). The Leopold matrix consists of a grid of 100 possible project actions along a horizontal axis and 88 environmental considerations along a vertical axis, for a total of 8800 possible first-order project–component interactions. Each cell of the matrix consists of two values: a quantification of the magnitude of the impact and a measure of impact significance. Where an impact is anticipated, the matrix cell is marked with a diagonal line in the appropriate row and column. The magnitude of the impact is then indicated within the top diagonal of each cell, typically on a scale from −10 to +10, and the importance of the impact or interaction is indicated in the bottom half of the diagonal (Box 3.2). An advantage of the Leopold matrix is that it can easily be expanded or contracted based on the specific context of the project and environment being assessed. Although the Leopold matrix is perhaps the most comprehensive of EIA matrices, its sheer magnitude may be a drawback in itself. At the same time, it identifies only direct project impacts, has a primarily biophysical emphasis, and does not directly provide a framework for classifying the timing or duration of impacts. Given that the Leopold matrix was developed for use on many different types of projects, it tends to generate an unwieldy amount of information for any single project application (Glasson, Therivel, and Chadwick 1999). Moreover, it does not provide for direct weighting of the affected components to reflect their relative significance.

In an attempt to provide some indication of the "relative importance" of identified project impacts, **weighted magnitude matrices** assign some measure of importance to each of the affected environmental components. In this way, values or judgments assigned to represent the potential impacts of a particular project action on an environmental component are multiplied by a weight to represent the relative importance of that component. The relative importance of the affected component may be a reflection of its current conditions, sensitivity to change, value to society, or importance in ecological functioning. Box 3.3 presents an example of a simple weighted magnitude matrix in which individual project action–component interactions are multiplied by the component weight and summed across the rows to determine the total impact on each component. Weighted magnitude matrices are particularly useful for comparing the relative impact of project actions across environmental components or for comparing alternative project locations. In the absence of a standardized system for assigning the impacts or the measures of component importance, magnitude matrices remain subjective and, like most other EIA methods, are only as good as the person assigning the values.

Box 3.2 Illustration of a Typical Section of the Leopold Matrix

Matrix instructions:	Components and actions: modification of regime							
1. Identify all actions across the top that are part of the proposed project. 2. Under each action, place a diagonal slash in the cell at the intersection of each component on the side of the matrix where an impact is possible. 3. Indicate the magnitude of the impact with a value from 1 to 10 in the upper left of each cell, where 1 is a low and 10 is a high magnitude. Indicate + for a positive impact or − for a negative impact. In the lower right, indicate a value from 1 to 10 for the importance of the impact.	a) exotic flora or fauna introduction	b) biological controls	c) modification of habitat	d) alteration of ground cover	e) alteration of groundwater hydrology	f) alteration of drainage	g) river control and flow modification	h) noise and vibration

A. CHEMICAL CHARACTERISTICS	1. Earth	a. mineral resources								
		b. construction material								
		c. soils								
		d. land form								
		e. force fields and radiation								
		f. unique features								
	2. Water	a. surface								
		b. ocean								
		c. underground								
		d. quality								
		e. temperature								

For example:

−10 ⟵———— Magnitude (strong negative impact)

1 ⟵———— Importance (minor, perhaps locally contained)

Source: Based on Leopold et al. 1971.

Interaction Matrices

A shortcoming of both the Leopold matrix and the weighted magnitude matrix is that neither goes beyond direct component interactions and impacts. Interaction matrices use the multiplicative properties of simple matrices to generate a quantitative impact score of the proposed project on interacting environmental components.

Box 3.3 Example of a Simple Weighted Magnitude Matrix

Affected Environmental Components		Weight (importance)	blasting	side cleaning	dredging	road construction	waste disposal	equipment transport	Total impact
					Project Actions				
	air quality	0.26	−1			−1	−1		−0.78
	water quantity	0.10	−2	−3	−3				−0.80
	water quality	0.22	−2	−4	−2				−1.76*
	noise	0.04	−2		−1	−2		−2	−0.28
	habitat	0.08		−5		−3			−0.64
	wildlife	0.08	−2	−4		−2			−0.64
	human health	0.22	−2			+3	−3		−0.44

+ = positive impact   No impact =   Moderate impact = 3
− = adverse impact   Neglible impact = 1   Major impact (irreversible or long term) = 4
            Minor impact = 2   Severe impact (permanent) = 5

*Total impact (water quality) = (0.22)(−2) + (0.22)(−4) + (0.22)(−2) = −1.76

In the above matrix, the weights are distributed across the affected environmental components such that the total of all weights is "1", where the larger the weight the more important the component. In this way, all components can be given equal weight, but to increase the importance of one component requires that a trade-off be made and the importance of another component or components be decreased.

There are two general types of interaction matrices: **component interaction matrices** and **weighted impact interaction matrices**.

Component interaction matrices are intended to identify first-, second-, and higher-order interactions and dependencies between environmental components so that indirect impacts resulting from project actions may be better understood. Box 3.4 presents an example of a component interaction matrix. A limitation of component interaction matrices is that for large numbers of components, the data may become quite cumbersome and require computer-based assistance. Moreover, as we reveal more and more indirect linkages, we often have magnitude and less understanding of the nature of those linkages (for example, amplifying or offsetting tendencies) and less control over their impacts. Most impact management strategies can deal effectively with only primary and secondary impacts. This raises the question of how far down the impact chain we should go when identifying project-induced impact interactions. Is there an advantage to identifying potential third-, fourth-, fifth-, or higher-order linkages and interactions? Is it practical to do so?

Similar to weighted magnitude matrices in that impacts are multiplied by the relative importance of the affected environmental components, weighted impact

Box 3.4 Example of a Component Interaction Matrix for an Aquatic Environment

Step 1: Initial matrix of component linkages. Where a direct link exists between those dependent components on the side axis and the supporting components across the top, a value of "1" is entered in the cell; where there is no direct link, a "0" is entered. In other words, fish (row 4) are dependent upon vegetation (column 1), so a value of "1" is entered in the cell. Vegetation (row 1) is not directly dependent upon fish (column 4), so a value of "0" is entered.

<table>
<tr><td rowspan="2" style="writing-mode: vertical-lr">Dependent aquatic components</td><td colspan="6">Supporting aquatic components</td></tr>
<tr><td></td><td>vegetation</td><td>plant detritus</td><td>benthic fauna</td><td>fish</td><td>avifauna</td></tr>
<tr><td></td><td>vegetation</td><td>0</td><td>1</td><td>0</td><td>0</td><td>0</td></tr>
<tr><td></td><td>plant detritus</td><td>1</td><td>0</td><td>0</td><td>0</td><td>0</td></tr>
<tr><td></td><td>benthic fauna</td><td>0</td><td>1</td><td>0</td><td>0</td><td>0</td></tr>
<tr><td></td><td>Fish</td><td>1</td><td>1</td><td>1</td><td>0</td><td>0</td></tr>
<tr><td></td><td>avifauna</td><td>1</td><td>0</td><td>1</td><td>1</td><td>0</td></tr>
</table>

Step 2: Transfer all "1s" in the matrix above, indicating direct or first-order linkages, to the new matrix (below).

Step 3: Multiply the initial matrix (above) by itself to solve only for those cells that do not contain a "1" or direct link. If the result for any cell is greater than "0," then enter a value of "2" in the new matrix below to indicate a "second-order" link. If the result is "0," then enter a value of "0" in the matrix.

<table>
<tr><td rowspan="2" style="writing-mode: vertical-lr">Dependent aquatic components</td><td colspan="6">Supporting aquatic components</td></tr>
<tr><td></td><td>vegetation</td><td>plant detritus</td><td>benthic fauna</td><td>fish</td><td>avifauna</td></tr>
<tr><td></td><td>vegetation</td><td>2</td><td>1</td><td>0</td><td>0</td><td>0</td></tr>
<tr><td></td><td>plant detritus</td><td>1</td><td>2</td><td>0</td><td>0</td><td>0</td></tr>
<tr><td></td><td>benthic fauna</td><td>2</td><td>1</td><td>0</td><td>0</td><td>0</td></tr>
<tr><td></td><td>fish</td><td>1</td><td>1</td><td>1</td><td>0</td><td>0</td></tr>
<tr><td></td><td>avifauna</td><td>1</td><td>2</td><td>1</td><td>1</td><td>0</td></tr>
</table>

The output is a component interaction matrix where a value of "1" indicates a direct link between aquatic components and a value of "2" suggests that the components are linked indirectly through another, common component.

Step 4: To identify possible third-order linkages, the procedure is repeated by multiplying the second matrix above by the initial matrix for all cells with no first- or second-order link to create a new matrix. If the result for any cell is greater than "0," then enter a value of "3" to indicate third-order component linkages, and transfer all "1s" and "2s" to this new matrix. To exhaust all possible linkages, the procedure can be repeated by multiplying the third matrix by the original, and so on.

interaction matrices explicitly incorporate second-order or indirect impacts. One of the more common examples of a weighted impact interaction matrix is the **Peterson matrix** (Peterson, Gemmel, and Shofer 1974) (Box 3.5). The Peterson matrix consists of three individual component matrices: a matrix depicting the impacts of project actions or causal factors on environmental components; a matrix depicting the impacts of the resultant environmental change on the human environment; and a vector of weights or relative importance of those human components. The initial project–environment interaction matrix is multiplied by a matrix depicting secondary human–component impacts resulting from project-induced environmental change. The result is multiplied by the relative importance of each of the human components to generate an overall impact score. The advantage of the Peterson matrix lies in the multiplicative properties of matrices and the ease of manipulation. That said, its mathematical properties are also the primary limitation of the Peterson matrix in that two negative impacts, when multiplied, generate an overall positive impact. As discussed in the previous chapter, few impacts have offsetting tendencies.

Box 3.5 The Peterson Matrix for Route Selection

The following Peterson matrix was developed to score alternative routing options for a highway realignment project to bypass a local community. The project assessment scores for impacts on the biophysical environment were derived using expert judgment and based on experiences with similar highway realignment projects. The effects of the resulting biophysical change on the human environment, including the importance of the affected human components, were derived through community forums and focus groups in the potentially affected community.

The impact data are scaled in such a way that a "minor" adverse impact is indexed as "1" and a "major" adverse impact is indexed as "4." The expert group classified a minor impact simply as one that can be fully mitigated with standard, known impact management actions and, based on previous experiences, is unlikely to cause long-term irreversible effects to the local biophysical environment. The expert group classified a major impact as one that was perceived as having a long-term social or economic effect on the livelihood or well-being of the community at large.

[A] Project routing options and affected biophysical effects

| | | Affected biophysical valued environmental components (VEC_b) | | | | |
		VEC_{b1}	VEC_{b2}	VEC_{b3}	VEC_{b4}	VEC_{b5}
Project routing options	(i)	3	2	4	1	4
	(ii)	1	3	2	4	1
	(iii)	3	2	4	3	2
	(iv)	1	1	3	2	1
	(v)	4	2	1	1	1

continued

[B] Affected biophysical VECs and resulting effects on community

Affected biophysical VCs		Affected community valued environmental components (VEC$_c$)				
		VEC$_{c1}$	VEC$_{c2}$	VEC$_{c3}$	VEC$_{c4}$	VEC$_{c5}$
	VEC$_{b1}$	2	2	1	1	3
	VEC$_{b2}$	1	1	2	2	3
	VEC$_{b3}$	2	2	1	3	2
	VEC$_{b4}$	1	2	4	3	2
	VEC$_{b5}$	2	2	3	2	3

[C] Importance (vector of weights) of community vecs

Community VCs		VEC weight
	VEC$_{c1}$	0.80
	VEC$_{c2}$	0.25
	VEC$_{c3}$	0.60
	VEC$_{c4}$	0.40
	VEC$_{c5}$	0.20

To calculate the Peterson matrix results, multiply the primary biophysical effects matrix [A] by the secondary community effects matrix [B], and multiply the resulting matrix by the vector of community VEC weights [C].

The result is an index for each project routing option as follows:

Project routing option:
   i.   62
   ii. 50
   iii. 62
   iv. 36
   v.   37

The higher index is indicative of a higher overall effect and therefore a less preferred routing option. Each option can now be scaled or standardized to generate a relative ranking of routing options by using the following equation:

$$\frac{i_{max} - i}{i_{max} - i_{min}}$$

Where: i_{max} = the value of the routing option with the highest index
        i_{min} = the value of the routing option with the lowest index
        i = the value for each option, (i) through (v)

This result will yield a preferred option that is always 1, a least preferred option that is always 0, with all other scores falling somewhere in between. In this example, the scaled results are as follows:

    i.   0.00
    ii. 0.46
   iii. 0.00
   iv. 1.00
    v. 0.96

Routing options (iv) and (v) are clearly the preferred options and more than twice as preferred as the next competing option (ii). Options (i) and (iii) are clearly the least preferred routing options.

Trends and Associations

Understanding trends and associations is an important part of EIA baseline assessment and fundamental to the prediction and analysis of a project's potential environmental effects. There are a variety of tools available for identifying trends and associations, ranging from simple statistical correlations and regression trees to more complex Markov chain analysis. However, identifying trends and associations need not always involve quantitative approaches. Sometimes simple, conceptual diagrams, such as network or system diagrams, are quite valuable for scoping potential relationships between project actions and affected environment components or for simplifying complex system relationships.

Network or System Diagrams
Network or system diagrams are based on the notion that links and pathways of interaction exist between individual components of the environment such that when one component is affected, the result will be an effect on other components that interact with it (Cooper 2012). The focus of network analysis is thus on identifying the pathway of an impact through a series of linkages or **network diagrams**. Network diagrams are particularly useful for depicting sequential cause–effect linkages between project actions and multiple environmental components and can be used to describe how project activities could potentially lead to environmental changes that may affect certain components of the environment (Box 3.6). Network diagrams can be constructed based on mathematical representation of dynamic environmental processes or based on more simplified and conceptual understandings of relationships.

Box 3.6 Terrestrial Ecosystem Pathways of Effects for the Proposed Keeyask Hydroelectric Generating Station, Manitoba

In June 2012, Manitoba Hydro and four Cree Nations together submitted an environmental impact statement (EIS) for the development of the Keeyask project—a 695-megawatt hydroelectric generating station at Gull Rapids on the lower Nelson River, Manitoba. The Nelson River and its surrounding

continued

environment have been altered over the past 50 years because of hydroelectric development. The proposed Keeyask project would consist of a power-
house complex, spillway, dams and dykes, and a reservoir. Subject to regulatory
approval, construction of the project is planned for 2014. Once complete, the
project will generate enough electricity on an annual basis to power approximately 400,000 homes. The project was subject to an EIA under the Canadian
Environmental Assessment Act and the Environment Act (Manitoba).

The EIS set out to adopt an ecosystem-based approach to assessment,
focusing on relationships between project actions, disturbances, and responses
in various components of the terrestrial and aquatic ecosystem. As part of the
EIS effects assessment, the project's effects on the terrestrial environment were
examined by constructing several network or system diagrams (see below) to
consider the linkages between the terrestrial environment and changes caused
by the project, both direct change and indirect change. Potential pathways of
effects were identified as well as the expected changes to various components
of the terrestrial ecosystem. This included, for example, identifying linkages
between site-clearing and excavation, changes in habitat intactness, and subsequent changes to terrestrial habitat function. Several approaches were used in the
assessment to support the identification of potential impact pathways and linkages, drawing on a combination of scientific knowledge concerning the causal
relationships between ecosystem components (e.g., how soils are affected by
flooding) and on specific field-based data collection and models. Similar network
or linkage diagrams were also developed to identify aquatic ecosystem pathways
of effects for the project. The project has received all required regulatory and
environmental approvals and licences, and construction is underway. Additional
information on the Keeyask project can be found at www.keeyask.com.

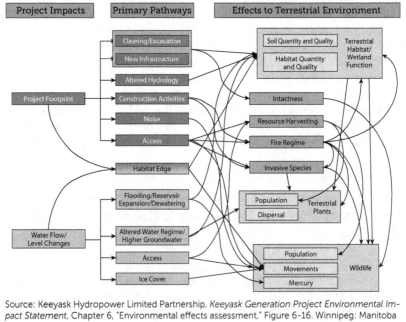

Source: Keeyask Hydropower Limited Partnership, *Keeyask Generation Project Environmental Impact Statement*, Chapter 6, "Environmental effects assessment," Figure 6-16. Winnipeg: Manitoba
Hydro 2012.

Spatial Analysis

Spatial analysis broadly refers to the study of phenomena based on their topological, geometric, or geographic properties. From simple mapping of project features to more complex spatial modelling of impact interactions, spatial analysis and the depiction of spatial information plays a critical role in EIA. Although a variety of techniques are available, most often spatial analysis in EIA is operationalized using Geographic Information Systems.

Geographic Information Systems
The use of **Geographic Information Systems (GIS)** is almost common practice in EIA. GIS are computer-based methods of recording, analyzing, combining, and displaying geographic information such as roads, streams, forest types, human settlement patterns, or any other feature that can be mapped on the ground. In its simplest EIA applications, maps depicting certain environmental or human components are overlaid with project activities to identify areas of potential impact or concern (Figure 3.2). The components may be represented by their actual geographic area, such as would be the case for a protected space or housing area, or represented by map cells and assigned various colours or numerical values according to their importance or the significance of the impact. For example, a map may be created for the impacts of a particular project activity on an aquatic grassland environment whereby the significance of the potential impact, or the vulnerability or sensitivity of the receiving environment, or both, is identified by a numerical value. The same could then be applied to other project activities and the maps overlaid so that when the layers are summed, a high value indicates an area of a significant total impact or concern. Such an approach is visually informative and allows spatial representation of impacts. It is also particularly useful when assessing land-use projects to identify potentially conflicting land-use patterns or optimal locations, such as the routing

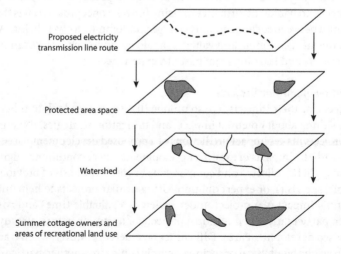

Proposed electricity transmission line route

Protected area space

Watershed

Summer cottage owners and areas of recreational land use

Figure 3.2 ◉ Illustration of Environmental Features Mapping for Impact Identification

of an electrical transmission line or identification of marine shipping patterns so as to detect potential conflicts with fishing activities.

Although costly to operate and maintain, GIS are capable of handling very large data sets for each environmental parameter, and, effectively, an unlimited number of parameters and data points can be processed. GIS also require hardware and software, technical skills, and data collection and conversion to digital format when such data does not already exist. Such capacity issues were major constraints to GIS use in the 1990s and early 2000s and remain concerns in many developing countries today. GIS are now a common tool in North America and used widely in EIA practice, particularly for baseline assessment but also for developing and presenting different scenarios of development, providing multiple options for decision-makers.

Tools for Impact Prediction

The choice of predictive tools depends on the nature of the impact indicator in question and data availability. For example, when predicting the severity of an impact, the approach may be qualitative and based on expert judgment. When predicting the concentration of a particular pollutant, more quantitative-based modelling tools might be used. Tools used for impact prediction cover a wide spectrum of economic, social, ecological, and physical factors. Morris and Therivel (2001) provide a detailed treatment of predictive tools. There is no single comprehensive list of tools used for prediction in EIA, and many tools used to identify, interpret, and communicate impacts can also be applied for predictive purposes. Some of the more common tools are discussed below.

Analogue Approaches

The starting point for impact prediction in almost any EIA is the examination of existing information, referred to as **analogue approaches**. This may involve, for example, drawing on known theories in ecology to predict the response of a particular species to habitat disturbance or using proven conceptual models that depict relationships between socio-economic change and community well-being. Arguably, however, the most common, and valuable, analogue approach is the examination of similar projects and learning from past EIA experiences.

Examination of Similar Projects
Few projects, or their impacts, are so unique that we cannot look to other cases to learn something about potential impacts and mitigation strategies. Examination of similar projects makes impact predictions of a proposed development based on comparisons with the impacts of existing developments where conditions might be similar. Data might be collected and comparisons made on the basis of previous impact statements, site visits, or expert opinion. Using similar projects to help understand the potential impact of a project under review is a valuable time- and cost-saving approach, particularly when multiple projects of a similar nature already exist in the proposed development area. Difficulties do emerge, however, when transferring lessons from one biophysical or socio-economic context to another, particularly with

regard to ensuring that the project environments are sufficiently similar to justify a transfer of findings. In many cases, experiences are not directly transferable from one project environment or one socio-economic setting to the next. Regardless of the limitations, however, examination of similar projects is almost always a first step when evaluating proposed developments.

Judgment

Judgment underlies many EIA tools and is arguably the most commonly used tool in EIA for impact prediction. As suggested by the United Nations Environment Programme (UNEP 2002), the successful application of many EIA methods relies heavily on the nature and quality of expert judgment. Bailey, Hobbs, and Saunders (1992), for example, reviewed artificial waterway developments in Western Australia and found that nearly two-thirds of all impact predictions were based on knowledge and experience. There is no standard by which judgment is integrated in EIA, and techniques for doing so range from unstructured and ad hoc, such as roundtable discussions, to highly structured and organized techniques, such as **intention surveys** or the **Delphi technique**.

Delphi Technique

The term "Delphi technique" was coined by Kaplan, a philosopher working with the Rand Corporation who in 1948 headed a research effort directed at improving the use of expert predictions in policy-making (Woudenberg 1991). The name itself refers to the ancient Greek oracle at Delphi, where people seeking advice were offered visions of the future. The Delphi technique, developed in its contemporary form during the 1950s and early 1960s, is an iterative survey-type questionnaire that solicits the advice of a group of experts, provides feedback to all participants on the statistical summaries of the responses, and gives each expert an opportunity to revise his or her judgment (Figure 3.3). The Delphi typically undergoes a number of distinct phases or survey rounds. In the first survey round, an "open-ended" questionnaire is sent to the panellists, who can then contribute additional information that they may feel is pertinent to the issue. In some cases, when the Delphi facilitator is seeking the opinion or the expert judgment of each individual regarding the issue(s) under consideration, a more structured questionnaire is used rather than an open-ended one. Responses are collected and summarized statistically and often incorporated into a number of analytical-based matrices. With this information at hand, the Delphi facilitator then compiles the round two questionnaires. The second-round and subsequent questionnaires typically include a reiteration of questions from the first questionnaire that engendered considerable discrepancies (e.g., conflict/non-consensus) and new questions based on issues and options raised by respondents during the first round. The process is repeated as many times as necessary until a desired level of consensus is reached. The typical Delphi procedure requires just three survey rounds, since consensus increases strongly over the first two survey rounds.

The Delphi technique is designed for use in situations in which the problem does not lend itself to precise analytical techniques or for which large data requirements

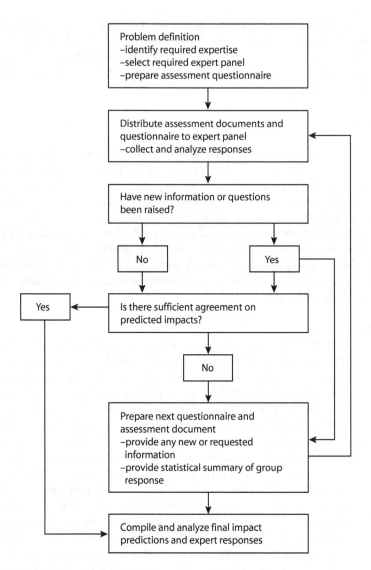

Figure 3.3 ⊚ The Delphi Process

and the lack of available quantitative data prohibit the application of traditional analytical approaches but that can benefit from the subjective judgments of experts on a collective basis (Dalkey 1969). The technique has three key features: anonymity, iteration, and feedback, all of which are key to effective group decision-making. First, the Delphi technique allows the exchange of opinion among the experts while avoiding the pitfalls of face-to-face interaction, particularly conflict and individual dominance. Second, multiple iterations allow the "least informed" participants to change their minds after hearing the group's response. Finally, feedback gives the experts who find the composite group judgments or the judgments of deviating experts more compelling than their own an opportunity to modify their decisions. For these reasons, the Delphi technique is extremely efficient in achieving consensus. However, whether consensus achieved through the Delphi process is "oneness of mind" or simply the illusion of oneness of mind is the focus of much debate. Group pressure to conform does not reflect genuine agreement. In impact assessment, the Delphi's goal should not be to arrive at a consensus but simply to obtain high-quality responses and opinions on a given issue in order to enhance decision-making.

Although there is no standard procedure for determining expertise, several guidelines have emerged from recent EIA practices for identifying expert composition and selecting suitable experts (Box 3.7). There are also three points to keep in mind regarding the use of expert judgment in EIA (Noble 2004):

- No clear linear relationship exists between expertise and the "quality" of impact prediction.
- Contrary to common practice, consensus should never be the primary goal when relying on expert judgment, since the same lack of knowledge that required an EIA, which relies on expert judgment in the first place, is likely to guarantee that a group of diverse experts will disagree. Any consensus that is reached may be a forced consensus and therefore highly suspect. Thus, the dissenting perspectives become of particular importance for further exploration when relying on expert judgment.
- In light of the limitations of consensus, emphasis should be placed on consistency when relying on expert judgment. **Consistency** simply refers to a measure of the extent to which expert judgments were purposefully made or reflect random decisions; in essence, it is a measure of the degree of understanding of the problem at hand.

Box 3.7 Using Expert Judgment in EIA

Criteria for determining the desired number of experts and expert composition:

    a) Sufficient representation of those affected by project and EIA decisions

- affected publics and interest groups
- affected sectors, government departments, and industries

    b) Sufficient representation of those who affect project and EIA decisions

- public administrators
- planners and policy-makers
- scientists and researchers

    c) Appropriate geographic representation

continued

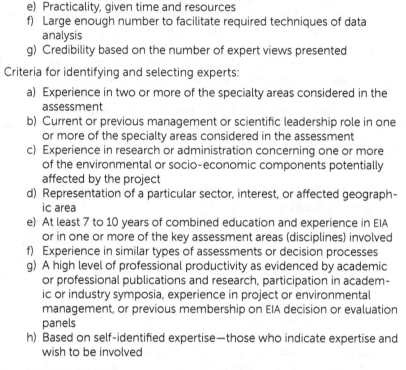

d) Inclusion of necessary expertise and experience
e) Practicality, given time and resources
f) Large enough number to facilitate required techniques of data analysis
g) Credibility based on the number of expert views presented

Criteria for identifying and selecting experts:

a) Experience in two or more of the specialty areas considered in the assessment
b) Current or previous management or scientific leadership role in one or more of the specialty areas considered in the assessment
c) Experience in research or administration concerning one or more of the environmental or socio-economic components potentially affected by the project
d) Representation of a particular sector, interest, or affected geographic area
e) At least 7 to 10 years of combined education and experience in EIA or in one or more of the key assessment areas (disciplines) involved
f) Experience in similar types of assessments or decision processes
g) A high level of professional productivity as evidenced by academic or professional publications and research, participation in academic or industry symposia, experience in project or environmental management, or previous membership on EIA decision or evaluation panels
h) Based on self-identified expertise—those who indicate expertise and wish to be involved

Source: Based on Noble 2004.

Systems Modelling

Models are simplifications of real-world **environmental systems**; they can range from box-and-arrow diagrams to sophisticated mathematical representation based on computer simulation. A wide range of models is available, from those that deal with single issues such as air or water quality to complex models for evaluating ecosystems and interactions. Modelling in EIA is most often used for predicting the future state of environmental and economic variables when sufficient data are available to support such tools. Modelling creates simplified representations of the system under investigation, including causal mechanisms, and is based on explicit assumptions about the behaviour of the system. A significant challenge to modelling approaches, however, is that they are typically very "data demanding." The required input to support reliable modelling approaches is not always readily available from environmental baseline studies, and data may require several field seasons to collect.

Mechanistic, Deterministic, and Stochastic Models

Mechanistic models describe cause–effect relationships in the project environment using flow diagrams or mathematical equations. Examples include Gaussian dispersion models for predicting rates and extents of pollution fallout, traffic noise generation models, population growth models, and economic impact models. Quantitative models used in EIA can be divided into two broad categories—deterministic and stochastic. **Deterministic models** depend on a fixed relationship between environmental components and provide a single solution for the state variables. For example, the **gravity model** of spatial interaction, used to predict population flows, depends on a fixed and inverse relationship between mass (population size) and distance. The system under consideration is thus at any time entirely defined by the initial conditions and model inputs chosen—uncertainty and variability are not explicitly modelled. With a given starting point, therefore, the outcome of the model's response is necessarily the same and is determined by the mathematical relationships incorporated in the model. **Stochastic models** are probabilistic—that is, they give an indication of the *probability* of an event or events within specified spatial and temporal scales and take into consideration the presence of randomness in one or more variables. In other words, a stochastic model includes variability in the model parameters as a function on spatial and temporal aggregation, random variability, and changing environmental conditions. Thus, rather than giving a single-point estimate, stochastic models provide a probability distribution of possible estimates or outcomes—such as the likelihood of a flood as a result of river control systems.

Balance Models

Mass **balance models**, or mass balance equations, are another type of modelling tool used in EIA. Balance models identify inputs and outputs for specified environmental components, such as soil moisture volume or surface runoff, and are essentially composed of inputs, storage, and outputs for a particular environmental system. For example, inputs to a particular system might include water and energy, whereas outputs include waste water and outflow. Balance models, such as surface hydrological models, are particularly useful for predicting physical changes in environmental phenomena when predicted changes equal the sum of the total inputs minus the sum of the total outputs (Glasson, Therivel, and Chadwick 1999). A well-known example is a site water budget, which predicts the impacts of a proposed development on the input, storage, and output of water available in a particular sub-basin.

Statistical Models

Statistical models are often used for extrapolation, or trends analysis, and are based on assumptions about fundamental relationships or correlations underlying an observed phenomenon in order to project beyond the range of available data and information. Such models often include simple linear or multivariate regression and fitting data to mathematical functions. For example, based on hypothesized or proven statistical associations between neighbourhood housing density and quality of life, statistical models can be used to extrapolate changes in the quality of life resulting from a proposed large-scale residential housing development program. Statistical models are also used to determine whether a statistically significant difference exists between the predicted changes in impact indicators due to project influence as opposed to natural changes that would occur in the absence of project development.

Spatial Models

Some models are explicitly spatial in nature. **Spatial models** used in EIA for impact prediction often draw heavily on Geographic Information Systems and range from simple overlay approaches whereby potential land-use conflicts can be "predicted" to more complex spatial modelling and simulation exercises in which the distribution of impacts or phenomena are modelled over space— such as ALCES or MARXAN models (Box 3.8). Spatial tools for impact prediction have the advantage of presenting a visual depiction of potential project effects and are particularly useful in supporting determinations of impact significance based on spatial extent. However, models for spatial simulation require significant data input and computational resources.

Box 3.8 ALCES and MARXAN to Aid Regional Impact Assessment and Planning Practices

There are several modelling and simulation tools available to support impact assessment and regional land-use development planning. Such models are particularly useful for examining the contribution of development projects or land uses to broader-scale, regional disturbance and cumulative change.

ALCES

ALCES (A Landscape Cumulative Effects Simulator) is a system dynamics simulation tool for exploring the behaviour or response of resource systems to disturbances. ALCES was developed by Dr Brad Stelfox, founder of Forem Technologies in 1995 and an adjunct professor at the University of Calgary. It is used for exploring and quantifying the relationships between different land-use sectors and natural disturbances and their potential environmental and socio-economic consequences. As a simulation model, it does not search for optimal solutions, such as an optimal land-use configuration or spatial patters; rather, it is designed to forecast and explore alternative outcomes and conditions. ALCES is a spatially stratified model. It allows users to examine the area and length of different land uses or disturbances, such as roads or forest cut blocks, within each landscape unit. ALCES is not a spatially explicit model. It does not provide descriptions of the spatial arrangement of, or spatial relationships between, landscape features. However, with a GIS, users of ALCES can generate maps that illustrate the plausible location and extent of simulated land-use features and landscape types, including footprint types and disturbed area, and other user-defined indicators such as water use or wildlife population parameters. ALCES has been applied in numerous resources and land-use planning and assessment exercises in western Canada, including the Upper Bow Basin Cumulative Effects Study, State of the Baptiste Lake Watershed report, the Terasen Jasper National Park–Mt Robson Pipeline assessment. For additional information on ALCES, including the modelling tool, applications, and training opportunities, see http://www.alces.ca.

Scenario Analysis

A scenario is broadly defined as "a hypothetical sequence of events constructed for the purpose of focusing attention on causal processes and decision points" (Kahn and Wiener 1967, 6). In other words, a scenario is a plausible but unverifiable account of change in a set of conditions over a defined period of time, depicting what could be if particular trends and rates of change unfold. By comparing multiple, alternative scenarios, decision-makers are able to obtain a vivid picture of the possible consequences of different actions. Scenarios usually serve one of two functions: risk management, where strategies and decisions can be tested against possible futures; and creativity and sparking new ideas through exploratory analysis (Greig and Duinker 2007).

MARXAN

MARXAN (Marine Spatially Explicit Annealing) is a landscape and marine optimization tool. It was developed by Drs Ian Ball and Hugh Possingham at the University of Adelaide and later integrated with ARCVIEW (GIS) through a Nature Conservancy–funded project. MARXAN uses a site-selection algorithm to explore options for conservation and regional biodiversity protection. It divides a landscape into small parcels termed planning units and then attempts to minimize the "cost" of conservation while maximizing attainment of conservation goals. The objective is to select the smallest overall area needed to meet set conservation goals. It does this by using a simulated annealing with an iterative improvement algorithm to select areas of high value for conservation, changing the planning units selected and re-evaluating the cost through multiple iterations. MARXAN uses data on ecosystems, habitats, and other relevant conservation features to find efficient solutions to the problem of selecting a set of conservation areas. In this sense, MARXAN is a useful tool for exploring alternative land-use and conservation scenarios and management options and for facilitating discussion among stakeholders. However, MARXAN is limited in its ability to consider social and economic systems and dynamics. It has been used in several conservation planning initiatives, including rezoning of the Great Barrier Reef and the North American Wildlands Project, and by the World Wildlife Fund to define a global network of marine protected areas. MARXAN has not been widely used for applications in impact assessment, but the Great Sand Hills Regional Environmental Study in Saskatchewan (2004–7) demonstrated the utility of MARXAN as a tool in strategic environmental assessments to support scenario-based analysis and cumulative effects assessment of human-induced disturbances due to cattle ranching, oil and gas development, and linear developments (see Chapter 12). For additional information on MARXAN, including the modelling tool, applications, and training opportunities, see http://www.uq.edu.au/marxan.

Scenario analysis allows impacts to be assessed based not only on what has happened in the past but also on potential future trends, which may include a number of surprises (Noble 2008). Several methods are available for developing scenarios, including environmental scanning, futures workshops, simulation modelling, expert opinion, and Delphi surveys. Greig, Pawley, and Duinker (2002) provide a comprehensive review of futures methods, including scenario development, in their report to the Canadian Environmental Assessment Agency's Research and Development Program (see www.ceaa.gc.ca). Although scenario analysis is not new, it has been underutilized in EIA for dealing with the often uncertain futures associated with project development. When used, scenario analysis is often associated with large-scale development projects, such as the Mackenzie Gas Pipeline Project (see Chapter 11); however, scenarios are also useful for smaller projects and can assist decision-makers in better understanding potential outcomes from alternative project actions and in the consideration of the most appropriate impact mitigation measures (see Environmental Assessment in Action: Scenario-Driven Impact and Mitigation Analysis of the Louis Riel Trail Highway Twinning Project, Saskatchewan).

Environmental Assessment in Action

Scenario-Driven Impact and Mitigation Analysis of the Louis Riel Trail Highway Twinning Project, Saskatchewan

Wetlands provide a number of important ecological services, including habitat provision, nutrient cycling, carbon sequestration, and flood control. However, wetlands are under considerable threat due to human-induced surface disturbances. Next to agriculture, road development is among the most significant sources of wetland degradation in the Canadian prairies. In 2007, the Saskatchewan Ministry of Highways and Infrastructure received approval for the construction of an approximately 110-kilometre highway-twinning project between Saskatoon and Prince Albert, the Louis Riel Trail. The highway is one of the busiest in the province. The project application was approved subject to the conditions of an aquatic habitat protection permit and standard upland wetland mitigation strategies. The approval was based on the conclusion that impacts to wetlands occurring within the provincially regulated 31-metre highway right-of-way would be subject to standard mitigation practices and no other significant adverse effects to wetlands were likely to occur. The 31-metre right-of-way is measured outward from the centreline of a proposed highway and includes the highway itself and the roadside ditches. No EIA was required for the project under the Saskatchewan Environmental Assessment Act. However, because of concerns about the potential for cumulative wetland loss along the entire length of the project, an independent assessment was carried out by researchers at the University of Saskatchewan in partnership with Ducks Unlimited Canada, funded by the Natural Sciences and Engineering Research Council of Canada..

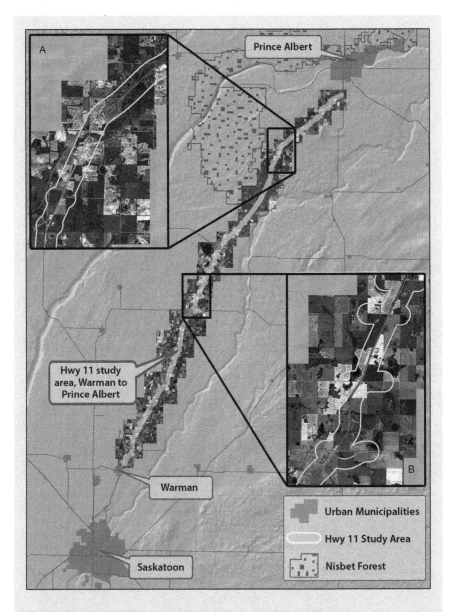

The assessment was based on a one-kilometre-wide corridor, centred on the proposed highway centreline. Notwithstanding a provincial decision of no significant effects based on a 31-metre right-of-way, previous science has shown that adverse effects to wetland functions can occur up to distances of 500 metres from highways (see Houlahan et al. 2006). As such, baseline assessment was completed to identify the total number and area of wetlands

continued

within a 500-metre buffer zone on either side of the centreline of the proposed highway. Project disturbance, wetland numbers, and wetland area were inventoried using aerial photos and panchromatic satellite imagery, ground-truthed for accuracy using a Trimble GeoXM field computer, and mapped using a GIS. Given the time and resources available for the assessment versus the complexity of the science involved, it was not feasible to establish an experimental design to test wetland connectivity and impacts to wetland functions. Consequently, "wetland area" was used as a proxy to assess potential effects to both wetland habitat and function (see Dahl and Watmough 2007; Government of Alberta 2007).

Using a GIS, mapped wetlands within the 500-metre zone were classified based on their potential to be affected by project activities, either directly

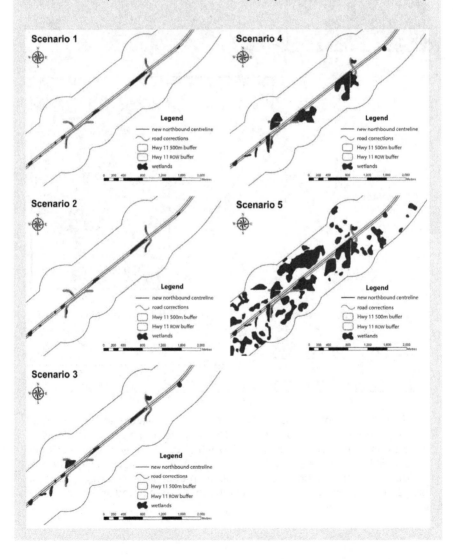

or indirectly. Potentially affected wetlands were determined by delineating "impact zones" of increasing distance from the proposed highway and based on visual surface connectivity of wetlands. The 31-metre right-of-way was considered the "zone of direct impact," and wetlands contained in this area were considered to be at high risk of a complete loss of both habitat and function. Wetlands with more than 50 per cent of their area overlapping the zone of direct impact were also considered to be at high risk. The distance from the edge of the 31-metre right-of-way up to 500 metres was classified as the "zone of indirect impact." Wetlands in this zone were considered at risk because of functional degradation as a result of surface connectivity to directly affected wetlands and due to potentially induced impacts. Five impact and mitigation scenarios were then developed, each describing a scenario of affected wetlands and more or less spatially ambitious mitigation strategies. A section of the highway depicting each of the scenarios is shown below.

Multi-criteria analysis, a structured approach to decision analysis, was used to evaluate alternative impact mitigation options based on the input of a panel of government and non-government experts in wetland conservation, hydrology, transportation engineering, and EIA. Each of the impact and mitigation scenarios was assessed based on the financial cost of implementation, ensuring compliance with a federal policy of no-net-loss of wetland functions, technical feasibility with regard to implementation, the potential to capture cumulative effects, administrative requirements in terms of permitting needs and land access for implementation, and public acceptability.

A total of 458 wetlands (1115 hectares) were identified to be "at risk" within a 500-metre buffer on either side of the proposed highway. Approximately 50 hectares of wetland habitat was located within the zone of direct impact, denoted by the 31-metre highway right-of-way, and at high risk of complete loss. An additional 1065 hectares of wetlands was determined to be at risk of some level of functional degradation. Under the current approach to wetland mitigation practice for highway developments (see Golder 2006) in the province, only 50 hectares of wetlands would be subject to mitigation—the wetland area completely inside the regulatory 31-metre highway right-of-way. Of the 458 wetlands identified inside the 500-metre buffer, more than half were wetlands less than 1 hectare in size, including many seasonal wetlands that are often not captured in wetland mitigation plans.

The assessment identified the need to consider a much more ambitious approach to wetland impact mitigation than the conditions attached to the project permit. The expert panel's preferred mitigation scenario was the most ambitious in comparison to current practice, demonstrating a decision outcome based on the importance of wetland conservation and dissatisfaction with current practices. The distribution of wetland size in the study area is typical of a prairie landscape and indicative of the need for an approach to mitigation that does not overlook the cumulative loss of many small and seasonal wetlands.

Source: Based on Nielsen, Noble, and Hill 2012

Complementary to scenario-based approaches is the use of **backcasting**. Instead of forecasting or painting alternative future scenarios based on possible events, backcasting focuses on identifying desirable and attainable future conditions. The approach involves working backwards from a particular future condition judged to be desirable and then determining the feasibility of achieving that condition, given project actions and changing environmental conditions. Mitchell (2002) explains that backcasting is another name for "critical path analysis."

Tool Selection

There is no single best method or technique that can accomplish all that is required for EIA, and each method and technique has its own strengths and limitations. Horberry (1984) examined 140 EIAs and found 150 different methods and techniques being used! Multiple methods and techniques are necessary in any single assessment. The objective is to select the methods and techniques best suited to the issue at hand while minimizing the overall limitations. When selecting tools for use in impact assessment, particular attention should be given to:

- *Objectives*. What is tool application expected to accomplish?
- *Time availability*. How much time is available to collect data and apply the tool?
- *Resource availability*. What resources are available for tool development and implementation, including human and financial resources?
- *Data availability*. What type and amount of data is required and available, including time series availability and availability of data for a range of environmental components?
- *Robustness*. Has the tool been proven in previous applications?
- *Nature of the project*. Is the tool appropriate, given the size and scale of the assessment and its potential impacts?
- *Selectivity*. Does the tool help to identify the impacts that are most important and critical to project decision-making?
- *Comparability*. Is the tool applicable across projects and over time?
- *Transparency*. Can sources of bias or uncertainty be identified and communicated?

The best EIA tools are those that are capable of organizing large amounts of data and information, providing an overall summary of information in a way that is easily understood, allowing aggregation and disaggregation of data and information without losing information that is important to understanding and decision-making. The general rule of thumb is to use what is available, given the nature and quality of the data and the desired resolution, to get the job done.

Key Terms

ALCES
analogue approaches
backcasting
balance models
checklists
component interaction matrices
consistency
Delphi technique
deterministic models
environmental systems
Gaussian dispersion model
Geographic Information Systems (GIS)
gravity model
intention surveys
interaction matrices
Leopold matrix
magnitude matrices

MARXAN
matrices
mechanistic models
methods
models
multi-criteria analysis
network diagrams
Peterson matrix
programmed-text checklist
questionnaire checklist
scenario analysis
spatial models
statistical models
stochastic models
techniques
weighted impact interaction matrices
weighted magnitude matrices

Review Questions and Exercises

1. In small groups, develop a simple questionnaire checklist to be used for assessing the construction phase of hydroelectric dam projects. Exchange your questionnaire checklist with others, and critically examine other groups' checklists. Are certain items not considered in the checklist that should be included? Compile an aggregate checklist of all groups' submissions, and discuss, among all groups, the advantages and limitations of a checklist approach to impact scoping.

2. Using Box 3.3 as a guide, construct a simple matrix identifying project actions and affected environmental components for a proposed highway development. Once agreement is reached as to the project actions and affected components, break into small groups, and assign weights and assessment scores to the components and impacts. Calculate the total impact score for each affected component, and sum up the results to generate an overall project impact score. Compare your results with those of other groups. Are there noticeable differences in the results? Why might this be so? Discuss the advantages and limitations of the magnitude matrix method. How might these limitations be addressed?

3. Use the example presented in Box 3.4 to identify potential third- and fourth-order linkages between the dependent and supporting aquatic components. Discuss the usefulness and practicality of identifying linkages beyond the first and second orders for project and impact management.

4. Complete an example of the Peterson matrix identified in Box 3.5. Work in groups to identify and add project actions and affected components and to assign impact values on a scale from "1" to "5" where "1" indicates a minor impact

and "5" indicates a major impact. Assign negative values where impacts are adverse. Assign weights to the human components on a scale of "1" to "5" where "1" indicates "unimportant" and "5" indicates "very important," and calculate the overall impact score. Discuss the usefulness of this value. What happened when the "negative" impact scores were multiplied? Discuss the advantages and limitations of the Peterson matrix.

5.   Sketch a simple network diagram for a proposed development that involves the clearing of a forest area in the upper regions of a watershed. Provide a statement explaining the nature of each linkage in the network. Discuss the advantages and disadvantages of the network method.

6.   Obtain a completed project EIS from your local library or government registry, or access one online. Scan the assessment, and identify the types of predicting tools used for: (i) impacts on the biophysical environment; (ii) impacts on the human environment. Compare your results with those of others. Do certain types of tools appear to be common among most impact statements?

References

Aura Environmental Research and Development Ltd. 2009. *Strategic Environmental Assessment Toolkit*. Ottawa: Canadian Environmental Assessment Agency.

Bailey, J., V. Hobbs, and A. Saunders. 1992. "Environmental auditing: Artificial waterway developments in Western Australia." *Journal of Environmental Management* 34: 1–13.

Baker, D., and E. Rapaport. 2005. "The science of assessment: Identifying and predicting environmental impacts." In K. Hanna, ed., *Environmental Impact Assessment: Practice and Participation*. Toronto: Oxford University Press.

Barrow, C.J. 1997. *Environmental and Social Impact Assessment: An Introduction*. London: Arnold.

Bisset, R. 1988. "Developments in EIA methods." In P. Wathern, ed., *Environmental Impact Assessment: Theory and Practice*, 47–61. London: Unwin Hyman.

Canter, L. 1992. "Advanced environmental impact assessment methods." Paper presented at the 13th International Seminar on Environmental Assessment and Management, Centre for Environmental Management and Planning, Aberdeen.

———. 1996. *Environmental Impact Assessment*. 2nd edn. New York: McGraw-Hill.

Cooper, L. 2012. "Network analysis in CEA, ecosystem services assessment and green space planning." *Impact Assessment and Project Appraisal* 28 (4): 269–78.

Dahl, T.E., and M.D. Watmough. 2007. "Current approaches to wetland status and trends monitoring in prairie Canada and the continental United States of America." *Canadian Journal of Remote Sensing* 33 (1): 17–27.

Dalkey, N. 1969. "An experimental study of group opinion: The Delphi method." *Futures* 1: 408–20.

Glasson, J., R. Therivel, and A. Chadwick. 1999. *Introduction to Environmental Impact Assessment: Principles and Procedures, Process, Practice and Prospects*. 2nd edn. London: University College London Press.

Golder Associates Ltd. 2006. *Wetland and Grassland Habitat Compensation Plan for the Trans Canada East Twinning Project and Highway No. 16 Twinning Project*. Saskatoon: Golder Associates Ltd.

Government of Alberta. 2007. *Provincial Wetland Restoration/Compensation Guide*. Edmonton: Alberta Environment.

Greig, L., and P. Duinker. 2007. *Scenarios of Future Developments in Cumulative Effects Assessment: Approaches for the Mackenzie Gas Project*. Report prepared by ESSA Technologies Ltd, Richmond Hill, ON, for the Joint Review Panel for the Mackenzie Gas Project.

Greig, L., K. Pawley, and P. Duinker. 2002. *Alternative Scenarios for Future Development: An Aid to Cumulative Effects Assessment*. Gatineau, QC: Canadian Environmental Assessment Agency.

Horberry, J. 1984. "Development assistance and the environment: A question of accountability." (Massachusetts Institute of Technology, PhD thesis).

Houlahan, J., P.A. Keddy, K. Makkay, and C.S. Findlay. 2006. "The effects of adjacent land use on wetland species richness and community composition." *Wetlands* 26 (1): 79–96.

Kahn, H., and A. Wiener. 1967. *The Year 2000*. New York: MacMillan.

Keeyask Hydropower Limited Partnership. 2012. *Keeyask Generation Project Environmental Impact Statement*. Winnipeg: Manitoba Hydro.

Leopold, L.B., et al. 1971. *A Procedure for Evaluating Environmental Impact*. Geological Survey Circular no. 645. Washington: United States Geological Survey.

Mitchell, B. 2002. *Resource and Environmental Management*. 2nd edn. New York: Prentice-Hall.

Morris, P., and R. Therivel, eds. 2001. *Methods of Environmental Impact Assessment*. 2nd edn. London: Taylor and Francis Group.

Nielsen, J., B.F. Noble, and M. Hill. 2012. "Wetland assessment and impact mitigation decision support framework for linear development projects: The Louis Riel Trail, Highway 11 North Project, Saskatchewan, Canada." *The Canadian Geographer* 56 (1): 117–39.

Noble, B.F. 2004. "Strategic environmental assessment quality assurance: Evaluating and improving the consistency of judgments in assessment panels." *Environmental Impact Assessment Review* 24: 3–25.

——. 2008. "Strategic approaches to regional cumulative effects assessment: A case study of the Great Sand Hills, Canada." *Impact Assessment and Project Appraisal* 26 (2): 78–90.

Peterson, G.L., R.S. Gemmel, and J.L. Shofer. 1974. "Assessment of environmental impacts: Multiple disciplinary judgments of large-scale projects." *Ekistics* 218: 23–30.

Shopley, J., and R. Fuggle. 1984. "A comprehensive review of current environmental impact assessment methods and techniques." *Journal of Environmental Management* 6: 27–42.

UNEP (United Nations Environment Programme). Economics and Trade Programme. 2002. *Environmental Impact Assessment Training Manual*. 2nd edn. New York: UNEP.

Woudenberg, F. 1991. "An evaluation of Delphi." *Technological Forecasting and Social Change* 40: 131–50.

Environmental Impact Assessment Practice

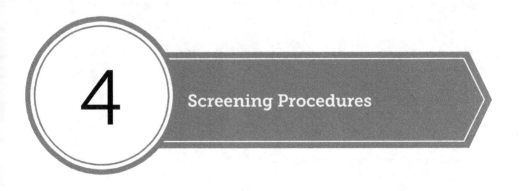

Screening

The number of projects that could potentially be subject to EIA is quite large; thus, the first stage in any EIA process is to determine whether the project requires an assessment and, if so, to what extent. Screening is most often the responsibility of the project proponent, but in some cases the leading regulatory EIA authority, also referred to as the **responsible authority**, plays the lead role. **Screening** simply refers to the narrowing of the application of EIA to projects that require assessment because of perceived significant environmental effects or specific regulations. Essentially, screening is the "trigger" for EIA and asks: "Is an EIA required?" A screening process will normally result in one of the following decisions:

- no EIA is required;
- EIA is required;
- a limited EIA is required, consisting of a preliminary assessment or mitigation plan;
- further study is necessary, an initial environmental evaluation, to determine whether an EIA is required.

The purpose of screening is to ensure that no unnecessary assessments are carried out but that developments warranting assessment are not overlooked. As Barrow (1999) explains, screening caused problems during the first few years of the US NEPA. The primary screening procedure under NEPA was for "major actions *significantly* affecting the environment," but "significantly" was an arbitrary concept—hence the importance of criteria and approaches for project screening.

Screening Approaches

Is an EIA Required?

There are ways to systematize the screening process, thus improving accountability and decision transparency in determining whether a proposal requires an EIA. The most straightforward way is to compare the anticipated impacts of a project with the quality parameters of the environment specified in relevant legislation. However, listing quality parameters of the environment is not a common practice in

EIA legislation. Requirements for screening are highly variable from one EIA system to another. Generally speaking, however, there are three approaches to EIA screening: case-by-case, threshold-based, and list-based.

Case-by-case screening, also referred to as discretionary or criterion-based screening, involves evaluating project characteristics against a checklist of regulations, criteria, or general guidelines as projects are submitted. In Saskatchewan, for example, the need for an EIA is largely determined on the basis of screening criteria identified in the Saskatchewan Environmental Assessment Act. Section 2(d) of the act indicates that any project, operation, or activity, or any alteration or expansion of such, is considered a development and subject to an assessment if it is likely to:

- have an effect on any unique, rare, or endangered feature of the environment;
- substantially utilize a provincial resource and pre-empt the use, or potential use, of that resource for other purposes;
- cause the emission of pollutants or create by-products or residual or waste products that require handling or disposal in a manner that is not regulated by any other Act or regulation;
- cause widespread public concern because of potential environmental changes;
- involve a new technology that is concerned with resource utilization and that may induce significant environmental change; or
- have a significant impact on the environment or necessitate a further development that is likely to have a significant impact on the environment.

Case-by-case screening allows for maximum flexibility to either *screen in* or *screen out* potential developments—it is sensitive to context, is dynamic, and provides for better consideration of the particular features of the local and regional environment in which development is proposed. At the same time, however, case-by-case screening can be time-consuming, inconsistent, and sometimes difficult to defend if the screening criteria are too vague. The process is easily susceptible to challenge.

Also referred to as prescriptive screening, **list-based screening** involves a list of projects for which an EIA is (or is not) required, based on the potential of that project to generate significant effects or based on regulatory requirements and responsibilities. In California, for example, the California Environmental Quality Act lists projects for which a full EIA must always be completed, based on project characteristics, thresholds, and geographic location. As well, a negative list outlines projects for which an EIA is not required. Such lists are typically referred to as inclusion and exclusion project screening lists. **Inclusion lists** include projects that have either mandatory or discretionary requirements for EIA. **Exclusion lists**, on the other hand, include the sorts of projects that would be subject to an EIA unless they were on the exclusion list—for example, projects carried out in response to a national emergency, projects carried out for national security reasons, or routine projects considered of only minor significance. Inclusion and exclusion lists were used in Canada under the former Canadian Environmental Assessment Act.

Threshold-based screening involves placing proposed projects in categories and setting thresholds for each type, such as project size, level of emissions generated, or population affected. Threshold-based screening is typically used in conjunction with list-based screening. Under threshold-based screening, thresholds are often based on different types or classes of development or project size or magnitude (e.g., total electrical generation capacity for a hydroelectric generating station) or based on environmental thresholds as established by regulations (e.g., total emission levels or concentrations). While threshold-based approaches are consistent and relatively easy to use, a problem arises when projects lie just below an established threshold. For example, if the threshold for an EIA for an electrical transmission line development on a new right-of-way is set at 50 kilometres in length, transmission lines that are only 40 kilometres in length are not subject to EIA even though they are just as likely to generate similar environmental effects.

The World Bank, for example, has explicit categories of projects that require assessment (Box 4.1) but cautions that project lists should be used flexibly with reference to particular geographic settings and project scales. The Nova Scotia Environmental Assessment Regulations Schedule A—Class I and Class II Undertakings under section 49 of the Nova Scotia Environment Act (1994–5, c. 1) lists specific projects and identifies thresholds for which an EIA is required. For example, an EIA is required for a facility for the incineration of municipal solid waste. An EIA is also required for the construction of a water reservoir, provided the reservoir has a planned storage capacity that exceeds the mean volume of the natural water body by 10 million cubic metres or more. Britain's EC Directive 85/337/EEC First Schedule, Part II 1989 Regulations, similarly identifies a list of the types of projects for which an EIA is required, some of which are defined according to certain thresholds.

Hybrid Screening

List-based and threshold-based approaches to screening are relatively straightforward screening tools, are widely used, and provide for an efficient screening process. However, list-based approaches should be used cautiously and should not be used as the sole screening mechanism. The automatic requirement for assessment based on thresholds, for example, may result in circumvention of the assessment process when proponents propose developments that are just below the threshold. While such concerns are relevant to case-by-case screening as well, the important factor that is missing in list-based approaches is the opportunity to consider context—especially cumulative, compounding, or induced impacts. Additional problems associated with list-based screening, identified by Mayer et al. (2006) in an international study of screening mechanisms, include:

- listed project types, characteristics, or thresholds may be outdated and may not conform with the current state of the art;
- thresholds may be identified even though there is no sound reason for them;
- regional differences in the sensitivity of the environment are not considered;
- specified thresholds may not be appropriate in all situations and contexts;

Box 4.1 World Bank Project Screening Lists

Category A Projects

These projects are likely to have significant adverse environmental impacts that are diverse, are unprecedented, or affect an area beyond the specific project site. A full EIA is required for such projects as:

- large-scale industrial plants
- dams and reservoirs
- port and harbour development
- large-scale irrigation
- river basin development
- hazardous or toxic materials involvement
- reclamation and new land development

- large-scale land clearance
- oil, gas, and mineral development
- large-scale drainage or flood control
- thermal or hydropower development
- manufacture, transport, use of pesticides
- resettlement
- forestry and production projects

Category B Projects

These projects are likely to have adverse impacts but are less significant than Category A projects. Most impacts are reversible, manageable, and site-specific. Projects include:

- electricity transmission
- renewable energy development
- tourism development
- small-scale irrigation and drainage
- rural water supply or sanitation
- watershed management or rehabilitation

- agro-industries
- rural electrification
- small-scale aquaculture
- rural electricity supply
- small project maintenance or upgrading

Category C Projects

These projects are likely to generate only minimal or no significant adverse environmental impacts. No EIA is required for projects concerning:

- education initiatives
- nutrition programming
- institution development
- most human resource projects

- family planning
- health initiatives
- technical assistance

Source: World Bank 1993.

- thresholds are set either too high or too low to properly capture significant impacts.

As summarized by Snell and Cowell (2006), lists have a useful role, but they cannot replace case-by-case considerations. In case-by-case or discretionary screening, determination of the need for assessment depends heavily on the relative weight given to each criterion considered and to the context within which significance decisions about the project's potential impacts are being made. In contrast to list-based approaches, case-by-case evaluations are dynamic and provide for a better consideration of particular features of the environment (e.g., environmental sensitivities, unique landscapes, endangered species) that are critical in significance determination. That said, case-by-case approaches are time-consuming and often inconsistent, and sometimes the decision is difficult to defend.

An alternative approach is to use the above screening mechanisms in combination such that projects above specified thresholds or located in sensitive areas, for example, are subject to mandatory assessment; whereas projects that fall below the threshold or are not located in sensitive areas are screened for the need for EIA on a case-by-case basis. A **hybrid screening** mechanism (Box 4.2) is essentially a threshold-based screening system with allowances for case-by-case consideration. Under this approach, a prescriptive screening mechanism provides for a list-based screening approach to projects and thresholds for which assessment is always required, while a discretionary case-by-case screening mechanism is applied to projects that fall below mandatory thresholds, fall in between inclusion and exclusion thresholds or criteria, or are included in descriptive lists (DETR 1998).

Screening Requirements in Canada

Screening practices vary across Canada. Because it is not possible to consider them all, this chapter focuses on screening under the Canadian Environmental Assessment Act, 2012, with an emphasis on determining the need for EIA for designated projects when the Canadian Environmental Assessment Agency is the responsible authority. The information presented here is based on the Canadian Environmental Assessment Act, 2012, the **Regulations Designating Physical Activities** under the act, and the Canadian Environmental Assessment Agency's "Introduction to the Canadian Environmental Assessment Act, 2012 participant manual."

Requirements for EIA

The Canadian Environmental Assessment Act, 2012 applies only when (Figure 4.1):

i)   a project is a "**designated project**"; or
ii) certain authorities have a decision-making responsibility in relation to a project on federal lands or outside Canada.

Under the Canadian Environmental Assessment Act, 2012, a "designated project" means physical activities that are

i)   carried out in Canada or on federal lands;

Box 4.2 Hybrid Screening Approaches in the UK

EC Directive 85/337/EEC First Schedule, Part II of the 1989 Regulations is illustrative of inclusion and threshold-based screening, identifying projects and thresholds subject to EIA as including:

- urban development projects that would involve a total area greater than 50 hectares for new or extended urban areas and an area greater than 2 hectares within existing urban areas;
- oil and gas pipelines that exceed 80 kilometres in length;
- installations for the manufacture of cement;
- industrial estate development projects that exceed 15 hectares;
- waste-water treatment plants with a capacity greater than 10,000 population or equivalent;
- all fish meal and fish oil factories;
- petroleum extraction, excluding natural gas.

In addition, a case-by-case provision states that at the discretion of a local planning authority, an EIA may be required for projects smaller than the specified thresholds but likely to have a significant effect on the environment.

The nature of the hybrid approach is clearly illustrated by the DETR (1998) proposal, which includes an "exclusive threshold" for projects that are exempt from EIA, an "indicative threshold" for projects requiring case-by-case analyses, and an "inclusive threshold" for projects for which EIA is always required.

EIA always required	**Inclusive threshold**
[EIA more likely to be required, but test remains likelihood of significant environmental effect]	
Case-by-case consideration	**Indicative threshold**
[EIA less likely to be required, but test remains likelihood of significant environmental effect]	
EIA not required	**Exclusive threshold** (except projects in sensitive areas)

Sources: Based on Gilpin 1999; EC Directive 85/337/eec First Schedule Part II Regulations 1989; DETR 1998 (proposed amendment to Directive 97/11).

---

ii) designated by the Regulations Designating Physical Activities or designated in an order by the minister of environment; *and*

iii) linked to the same federal authority as specified in those regulations or that order.

The federal Regulations Designating Physical Activities identifies those activities that constitute "designated projects" that may require an EIA by the Canadian Environmental Assessment Agency, the Canadian Nuclear Safety Commission,

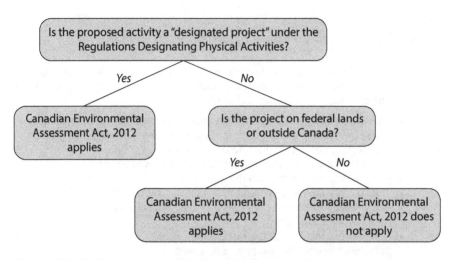

Figure 4.1 ⊚ Determining Whether the Canadian Environmental Assessment Act, 2012 Applies

Source: Based on Canadian Environmental Assessment Act, 2012

or the National Energy Board. The minister of environment can also designate a physical activity not included in the regulations should he or she believe that it has the potential to cause adverse environmental effects or if warranted by public concern. Designated projects in the Regulations Designating Physical Activities are listed and typically defined by specified thresholds—thus using a combination of list- and threshold-based screening.

An activity may not be a designated project under the Regulations Designating Physical Activities, but the Canadian Environmental Assessment Act, 2012 may still apply if the project is on federal lands or outside Canada and a federal authority is carrying out the project or must exercise its authority (e.g., providing access, issuing permits or authorizations). Such projects might include, for example, the expansion or modification of a golf course in a national park or the operation of a waste disposal site on Aboriginal lands (CEAA 2012a). An EIA is not required for any federal actions, authorizations, or other activities that are considered matters of national energy security or carried out in response to a national emergency or other emergency situation where a delay in action would result in damage to property or the environment or pose a risk to health and safety.

Responsible Authorities

There are three responsible authorities for EIA for designated projects under the Canadian Environmental Assessment Act, 2012. The Canadian Environmental Assessment Agency is responsible for most activities designated under the act, with the exception of those regulated under the Nuclear Safety and Control Act, the National Energy Board Act, or the Canada Oil and Gas Operations Act. The Canadian Nuclear Safety Commission is the responsible authority for activities regulated under the Nuclear Safety and Control Act, such as uranium mining operations or the construction or decommissioning of nuclear projects. The National Energy Board is the responsible authority for activities under the National Energy Board Act and the Canada Oil and Gas Operations Act, such as international and interprovincial pipelines and electrical transmission lines. For those projects that are not designated projects but are projects on federal lands or lands outside Canada, federal authority rests with the department or agency responsible for those lands or for exercising a duty in relation to those lands, such as Parks Canada as the federal authority for most projects that occur in national parks.

Table 4.1 provides examples of proposed undertakings, the responsible authority, and their designation under the Canadian Environmental Assessment Act, 2012 Regulations Designating Physical Activities. The specific sections of the act and regulations can be viewed at www.ceaa.gc.ca.

Determining Whether an EIA Is Required

A positive determination that the Canadian Environmental Assessment Act, 2012 applies does not *always* mean that an EIA is required. For projects that are *not* "designated projects" under the Regulations Designating Physical Activities but are located on federal lands, the federal authority responsible for those lands or for the project must first determine whether the project is likely to cause significant adverse environmental effects before it can carry out the project or exercise any power or duty (e.g., issue permits or authorizations), unless it is determined by the Governor-in-Council that the effects are justified under the circumstances. If a project *is* a designated project, then a determination must be made as to whether an EIA is required. A designated project will always require an EIA when either the Canadian Nuclear Safety Commission or the National Energy Board is the responsible authority. Also, if the minister of environment designates a project not described in the regulations, an EIA is always required.

If a designated project is linked to the Canadian Environmental Assessment Agency, an EIA *may* be required. In determining whether an EIA is required, the Canadian Environmental Assessment Agency will consider the information contained in the "description of a designated project" provided by the proponent, the potential for adverse environmental effects, public comments received during the public review period of the project description, and the results of any relevant regional studies that have been conducted in the project's regional environment.

Description of a Designated Project
If a project is covered by the regulations, a proponent must contact one of the three relevant responsible authorities—the Canadian Nuclear Safety Commission, the National Energy Board, or the Canadian Environmental Assessment Agency. In the case of designated projects linked to the Canadian Environmental Assessment Agency, a project proponent must submit a project description to the agency. The project description serves two purposes: first, to determine the need for an environmental assessment; second, to facilitate efficiency and co-ordination of the

Table 4.1 Examples of Projects and Designations under the Canadian Environmental Assessment Act, 2012 and Regulations Designating Physical Activities under the Act

Proposal[1]	Responsible authority or other federal authority	Example of applicable section(s) of the act or designation(s) under regulations	Determination as to whether the act applies
Sawco Inc. has submitted its intent to develop a softwood pulp and paper mill on the Athabasca River. The proponent has applied for federal subsidies to partially support the undertaking.	Canadian Environmental Assessment Agency (responsible authority)	Regulations Designating Physical Activities, sec 18: "The construction, operation, decommissioning and abandonment of a pulp mill or pulp and paper mill"	Yes. Designated project.
The town of Somewhere in southern Ontario is planning to construct a new 32 km public highway to provide an alternate access route to the local municipal park. There are no water crossings.	None	Project falls short of the 50 km threshold identified in sec 28(b) of the Regulations Designating Physical Activities: "The construction, operation, decommissioning and abandonment of an all-season public highway that will be more than 50 km in length and . . ."	No. Not a designated project. Project is not on federal lands or outside Canada.
Electrical Utilities is proposing to construct a new 178 km, 375 kV electrical transmission line on a new right-of-way in Quebec.	Canadian Environmental Assessment Agency (responsible authority)	Regulations Designating Physical Activities, sec 5: "The construction, operation, decommissioning and abandonment of an electrical transmission line with a voltage of 345 kV or more that is 75 km or more in length on a new right of way"	Yes. Designated project.
Parks Canada is constructing a new visitor centre in one of its national parks in Manitoba. Construction will involve construction of buildings, installation of water and sewer facilities, and establishment of a 25 km walking trail starting and ending at the new facility.	Parks Canada (other federal authority)	Project is not a designated project under the Regulations Designating Physical Activities. But sec 67 of the Canadian Environmental Assessment Act, 2012 applies. The project is a physical undertaking on federal lands for which a federal agency is responsible.	Yes. Project on federal lands.

Following the discovery of a new ore deposit, Uranium Inc. plans to construct and operate a new uranium mine and mill in northern Saskatchewan. The proposed operation is outside the boundaries of its currently operating mine sites.	Canadian Nuclear Safety Commission (responsible authority)	Regulations Designating Physical Activities, sec 32: "The construction, operation, decommissioning and abandonment, or an expansion ... of (a) a uranium mine, uranium mill or waste management system any of which is on a site that is not within the boundaries of an existing licensed uranium mine or uranium mill"	Yes. Designated project.
The province of Newfoundland and Labrador is proposing a new waterfront development project in one of its east-coast communities. The project will include a municipal park, restaurant, parking facilities, and a small arts-and-crafts marketplace.	None	The project is not a designated project under the Regulations Designating Physical Activities. The project is not on federal lands.	No
Because of favourable market conditions, Gold Miner Co., British Columbia, is proposing an expansion of their existing mining operations, resulting in an increased ore production capacity of 600 t/d, bringing total production capacity to 1300 t/d.	Canadian Environmental Assessment Agency (responsible authority)	Regulations Designating Physical Activities, sec 16(c): "The expansion of an existing gold mine, other than a placer mine, that would result in an increase in its ore production capacity of 50% or more ..."	Yes. Designated project.

[1] Proponent names and proposals are hypothetical.

Source: The examples used in this table were based in part on a workshop facilitated by the Canadian Environmental Assessment Agency as part of its March 2013 training course on the Canadian Environmental Assessment Act, 2012, held in Ottawa.

environmental assessment. Three tiers of project information are typically required, including:

1. General information: This includes the nature and proposed location of the project; information on consultation with affected parties and concerned authorities; details on other provincial, territorial, and federal EIA legislation or land-claims agreements to which the project might be subject; identification of affected or involved government departments and agencies, including landownership; information concerning required permits and project authorizations.
2. Project information: This includes a description of the project's components and structures; activities and activity scheduling associated with construction, operation, and decommissioning phases; engineering design specifications; resources and material requirements; waste production, discharges, and management.
3. Site-specific information regarding the project: This includes the project location and extent of operations; location and distribution of potentially affected environmental components; likely affected environmental components; and previous and current land-use patterns and impacts.

Project information is normally provided by the project proponent; however, other valuable sources of project information, such as previous land-use patterns and conditions, can be obtained through consultations with the general public, professionals, governments, and special interest groups and can be based as well on experiences elsewhere. The basic requirements of a project description under the Canadian Environmental Assessment Act, 2012 are outlined in Box 4.3.

Describing the "need for" and "purpose of" a proposed project are important parts of a good project description. The "need for" a proposed project is the particular problem or opportunity that the project is intended to address or satisfy. The "purpose of" the project is what is intended to be achieved by actually carrying out the project. The need for and purpose of the project are typically established from the perspective of the project proponent. For example, the need for a proposed electrical generation station may be defined on the basis of the demand for electricity supply or the inability of the current system to supply reliable electricity. The purpose of the project might be to supply cost-efficient, reliable electricity to expanding city residential neighbourhoods in a manner that is profitable to the utility company. The need for a particular project and its purpose are an important context for the identification and evaluation of project alternatives.

Environmental Effects
Under the Canadian Environmental Assessment Act, 2012, an "environmental effect" has a particular meaning and scope. As discussed in Chapter 2, under the former Canadian Environmental Assessment Act, an environmental effect included any change that the project could cause in the environment, as well as the social and economic effects of such change. Under the new act, an environmental effect includes

Box 4.3 Preparing a Description of a Designated Project under the Canadian Environmental Assessment Act, 2012

Proponents of designated projects (other than designated projects regulated by the Canadian Nuclear Safety Commission or the National Energy Board) under the Canadian Environmental Assessment Act, 2012 are required to submit a description of the project to the Canadian Environmental Assessment Agency to inform the decision-maker on whether an EIA of the project is required. Details on what must be included in the project description are set out in the federal Prescribed Information for the Description of a Designated Project Regulations. Upon submission of a complete project description, the Canadian Environmental Assessment Agency has 45 days to review the submission, including a 20-day public comment period, to determine whether an EIA is required. The project description must include the following information:

General information and contacts:	A description of the project and proposed actions; proponent contact information; list of jurisdictions and Aboriginal groups consulted during preparation of the project description; indication of whether the project is subject to assessment or other review under other jurisdictions
Project information:	A description of the project's components and activities; size of operations; nature and sources of emissions, discharges, or wastes; scheduling of construction, operation, decommissioning, and abandonment
Project location:	A description of the specific location, detailing project components and operations; identification of water bodies, linear features, location of Aboriginal groups and settlement areas, federal lands, communities, fishing areas, environmentally sensitive areas, land and water use and project land and water use, and access requirements
Federal involvement:	A description of whether there is any proposed or anticipated federal financial support; use of or access to federal lands; federal licensing or permits required
Environmental effects:	A brief assessment of the environmental interactions of the project, including: a description of the physical and biological setting and potential effects; any changes that may affect fish and fish habitat, aquatic species, or migratory birds; and any effects on other jurisdictions and Aboriginal peoples
Proposed engagement and consultation with Aboriginal groups:	A list of Aboriginal groups that may be interested in or potentially affected by the project; a description of any engagement or consultation activities that have been carried out with Aboriginal groups, including details on the date, nature of the consultation, approach to consultation, and key issues or concerns raised; and a plan for ongoing engagement and consultation
Consultation with the public and other parties	Details of any consultation (as per above) that has occurred or will occur with the public and other parties, other than Aboriginal consultation

Source: Synthesized from CEAA 2012b.

a change to a component of the environment that is under federal jurisdiction, such as fish or migratory birds; change that may be caused on federal lands or outside the province where the project is to be carried out; and effects on Aboriginal peoples of a change to the environment.

The definition of an environmental effect is thus much narrower in scope under the current act. Under section 5(1) of the act, for example, the environmental effects to take into account are:

(a) a change that may be caused to the following components of the environment that are within the legislative authority of Parliament: (i) fish as defined in section 2 of the Fisheries Act and fish habitat; (ii) aquatic species as defined in the Species at Risk Act; (iii) migratory birds as defined in the Migratory Birds Convention Act; and (iv) any other component of the environment that is set out in Schedule 2;

(b) a change that may be caused to the environment that would occur: (i) on federal lands; (ii) in a province other than the one in which the project or designated project is being carried out; or (iii) outside Canada;

(c) with respect to Aboriginal peoples, an effect occurring in Canada of any change that may be caused to the environment on: (i) health and socio-economic conditions; (ii) physical and cultural heritage; (iii) the current use of lands and resources for traditional purposes; or (iv) any structure, site or thing that is of historical, archaeological, paleontological or architectural significance.

In section 5(2) of the act, however, a provision is made for the consideration of other types of effects when a project, activity, or designated project requires federal authority to exercise a power or perform a duty or function (e.g., grant an authorization, issue a permit) under any act of Parliament other than the Canadian Environmental Assessment Act, 2012 (e.g., Navigable Waters Protection Act, Migratory Birds Convention Act). In such cases, the act requires that the following environmental effects are also taken into account:

(a) change, other than those referred to in paragraphs (1)(a) and (b), that may be caused to the environment and that is directly linked or necessarily incidental to a federal authority's exercise of a power or performance of a duty or function that would permit the carrying out of the activity, project or designated project; and

(b) an effect, other than those referred to in paragraph (1)(c), of any changes referred to in paragraph (a) on: (i) health and socio-economic conditions; (ii) physical and cultural heritage; or (iii) any structure, site of thing that is of historical, archaeological, paleontological or architectural significance.

Understanding the definition of an environmental effect is important to understanding whether an environmental assessment is required and what must be considered in that assessment (Box 4.4).

Box 4.4 Environmental Effect under the Canadian Environmental Assessment Act, 2012

Consider the following scenario. The project proponent, a rural municipality, is proposing to develop and operate a sand and gravel quarry operation. The quarry is expected to produce between 1.5 and 2 million tonnes of product per year and is expected to last for at least 15 years. The municipality will use the sand and gravel for their own local purposes, sell product to other rural municipalities and private companies, and supply two local concrete plants. The quarry is adjacent to a river, which drains into a lake about 500 metres from the proposed quarry site. Sand and gravel will be excavated by a dredge, crushed and stockpiled locally, and transported out by truck. Although excavation will be below the water table, no pit dewatering will occur. The project will require the construction of a new access road and a new bridge across the river to provide access for truck traffic. The closest community to the quarry operation is located 20 kilometres away. There are several cottages near the lake, and the lake is a popular local recreational site for camping. There is a rich history of First Nations use in the area. The lake is an important place for fishing, and the lands within the vicinity of the proposed quarry are used by First Nations for seasonal hunting. There is a national park 1.5 kilometres from the proposed quarry site.

The project description, which includes comments from community members, cottage owners, and First Nations, has identified several issues and concerns. Are these issues and concerns within the scope of the definition of an "environmental effect" under the Canadian Environmental Assessment Act, 2012?

Issue of concern	Supporting rationale for an "environmental effect" under the act
Local cottage owners have expressed concern about changes to water quality due to the construction of the new bridge and their ability to continue to swim safely in the river.	A permit or authorization is required for a river crossing under the Navigable Waters Protection Act. Thus, based on sec 5(2) of the act, consideration must be given to changes caused by the bridge on such matters as "health and socio-economic conditions" (sec 5(2)(i)), such as water quality to support safe recreational activity.
The national park authority has raised concern about the potential impacts of noise from quarry operations and disruptions to non-migratory birds in the park.	Sec 5(1)(b)(i) refers to a change that may be caused to the environment that would occur "on federal lands," such as a national park. Sec 5(1)(c)(iii) may also apply if there is reason to believe that the birds are used for traditional purposes by the First Nations.
The project proponent identified the loss of habitat supporting a species listed under the Species at Risk Act due to quarrying operations.	There is no effect based on the definitions under sec 5(1), and there is no federal authority being exercised, sec 5(2), concerning the project with regard to

continued

<table>
<tr><td></td><td>the Species At Risk Act. Any assessment of impacts would thus be required under the Species At Risk Act but not under the Canadian Environmental Assessment Act, 2012.</td></tr>
<tr><td>First Nations have raised concern about the loss of habitat for amphibians and reptiles due to construction of the bridge.</td><td>Sec 5(1)(a) applies, provided it can be demonstrated that the habitat is also an important part of the food supply for fish. Sec 5(1)(c)(iii) may apply, provided it can be demonstrated that the amphibians and reptiles affected are used by the First Nations for traditional purposes. Sec 5(2) is also relevant, since the effects are linked to a federal permit for authorization of the bridge, a river crossing, under the Navigable Waters Protection Act.</td></tr>
</table>

Source: This scenario was based in part on a workshop facilitated by the Canadian Environmental Assessment Agency as part of its March 2013 training course on the Canadian Environmental Assessment Act, 2012, held in Ottawa.

Type of EIA Required

Not all projects must undergo the same level of assessment. For example, a proposed nuclear energy facility is likely to require a considerably larger and more complex assessment than a simple road construction project. Determining the requirements for and level of EIA, however, is not always a straightforward process based solely on case-by-case, threshold-based, or list-based screening criteria. For example, the EIA procedures in Canada and Western Australia incorporate a two-tiered system for determining whether an EIA is required and, if so, what type of EIA.

Under Western Australia's system, a notice of intent is submitted to the Australian Environmental Protection Agency describing the nature of the project, potential environmental effects, alternatives to the project, and proposed impact management measures. This is referred to as an **initial environmental examination (IEE)** or **environmental preview report (EPR)**. This information is then used to determine whether an EIA is needed and what level of assessment is required. Based on Australia's procedures, the decision stemming from an IEE may be that:

- no further assessment is required;
- a full EIA is required;
- an examination by an independent commission of inquiry is necessary; or
- a less comprehensive public environmental report is required.

Under the current Canadian federal EIA system, once it is determined that an EIA is required, there are two main assessment options: EIA by responsible authority or EIA by review panel (Figure 4.2).

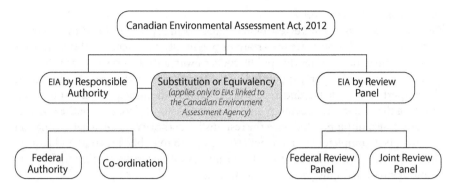

Figure 4.2 ● Types of EIA under the Canadian Environmental Assessment Act, 2012

Source: Based on Canadian Environmental Assessment Act, 2012

Assessment by Responsible Authority

Assessment by responsible authority (Canadian Environmental Assessment Agency, Canadian Nuclear Safety Commission, National Energy Board) applies to designated projects with the potential for adverse environmental effects. Projects involving the expansion of a mining operation, water diversion, or industrial plant would typically be subject to assessment by responsible authority. If deemed appropriate by the federal minister of environment, such projects may be referred to assessment by review panel within the first 60 days of EIA commencement. See Chapter 2, Figure 2.2, for an overview of the timelines for assessment by review panel. Self-assessments are carried out by federal authorities other than responsible authorities, under section 67 of the Canadian Environmental Assessment Act, 2012, for projects that are not designated but are on federal lands or outside Canada. See also Table 4.1, on page 80.

Assessment by responsible authority is in many respects similar to those assessments designated as comprehensive study assessments under the former Canadian Environmental Assessment Act. Under the former act, the majority of EIAs, about 99 per cent, were screening EIAs (see Chapter 2) undertaken for small projects. A **screening EIA** systematically documents the potential environmental effects of a proposed project, identifies mitigation measures, and determines the need to modify the project plan or recommends further assessment to eliminate or minimize those effects. Elimination of EIA for small projects has been met with mixed views. Some perceived the numerous EIAs required for small projects under the former act as a waste of time and resources, arguing that such projects are unlikely to generate significant adverse environment effects if mitigation actions and permitting requirements are followed. Others viewed small-project assessment as an important means of reducing the potential for significant adverse cumulative environmental effects over time, or across space, but argued that to be effective the process needed to be strengthened and EIAs for small-project assessments better co-ordinated. An independent assessment of the impacts of small-scale petroleum and natural gas wells on grasslands in southern Saskatchewan, by Nasen, Noble, and Johnstone (2011), illustrates the importance of assessing the impacts of small projects. See "Environmental Assessment in Action: Screening-out Small Projects."

Assessment by Review Panel

Referral to **assessment by review panel** would normally occur only if the environmental effects of a project are uncertain or potentially significant or if the minister of environment determines that public concern warrants it. A review panel is composed of a group of experts appointed by the minister of environment, selected for their expertise, knowledge, and experience relevant to the anticipated environmental effects of the project. The review panel is responsible for the conduct of the EIA process and is usually appointed once the responsible authority makes a determination that the EIS is complete. The panel will then proceed to public hearings and require the proponent to submit additional information, if necessary, before making its recommendation on the project to the minister.

Substitutions and Equivalency

Under both types of assessment, assessment by responsible authority and assessment by review panel, EIA can be conducted jointly with another jurisdiction through co-ordination of the EIA process. For EIAs by responsible authority, when the responsible authority is the Canadian Environmental Assessment Agency, the minister of environment can apply "**substitution EIA**" or "**equivalency EIA**" whereby another jurisdiction can be delegated responsibility to carry out part of the EIA process or a provincial or Aboriginal EIA process can be used as a substitute for federal assessment. Substitution or equivalency does not apply to EIAs for designated projects for which the Canadian Nuclear Safety Commission or the National Energy Board is the responsible authority.

Environmental Assessment in Action

Screening-out Small Projects: Impacts of Petroleum and Natural Gas Wells on a Grassland Ecosystem

There are more than 30,000 operating petroleum and natural gas (PNG) wells in the prairie ecozone, of which 35 per cent are located in Saskatchewan (CAPP 2007). Approximately 80 per cent of PNG production over the next 10 years will come from currently undrilled wells. This is of particular concern to grassland conservation efforts, given that nearly all of Saskatchewan's unconventional gas deposits are located in grassland ecosystems. A typical PNG site contains one to five wellheads and supporting infrastructure, which can include pumps, separators, and solution tanks for conventional reserves and water injection and disposal facilities for non-conventional reserves.

In Saskatchewan, the assessment and management of the impacts of PNG activities on Crown lands occur largely through the EIA process, administered federally under the Canadian Environmental Assessment Act, 2012 and in Saskatchewan under the Saskatchewan Environmental Assessment Act. The effects of PNG development on grasslands relate predominantly to surface

Typical PNG lease site in southwest Saskatchewan for conventional oil (Photo by Lawrence Nasen)

disturbance associated with well sites and access roads and the disruption of nutrient and water exchange through the soil. The problem, however, is that an application for development of an individual PNG site is rarely considered significant enough to trigger a formal EIA.

In 2010, an independent assessment was conducted to examine the impacts of nearly 50 years of PNG development on southern Saskatchewan grasslands (see Nasen, Noble, and Johnstone 2011). The assessment examined the effects of PNG activity on Prairie Farm Rehabilitation Administration lands, Agriculture and Agri-Food Canada, in the Swift Current–Webb Community Pasture. The pasture is approximately 9882 hectares, and as of 2006, 170 wells had been drilled within the pasture boundaries.

The physical footprint of PNG infrastructure in the pasture was examined using aerial photos and SPOT 5 satellite imagery. Land-use and fragmentation metrics were calculated for five-year periods, from pre-1970 to 2006, including patch density and edge density and the percentage of pasture occupied by roads, trails, and PNG lease sites. Field data were collected from a sample of 31 of the total 170 PNG sites. At each lease site, measurements were made along four transects that extended outwards from the lease-site centre. Environmental and biotic data were collected at sampling locations along the transect, and for every PNG site surveyed, a reference sample of non-PNG pastureland was also sampled. Data collected included percent herbaceous, percent bare ground, percent litter, percent aggregates, percent club moss, percent oil spill/contaminated, soil compaction, pH, conductivity, percent silt, percent sand, percent clay, and range health.

Between 1979 and 2005, the physical footprint of PNG infrastructure was found to have increased from 0.20 per cent to 1.00 per cent of the landscape. During this same time period, lease access roads increased from occupying

continued

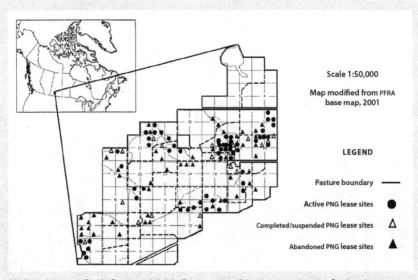

PNG activity on Swift Current-Webb Community Pasture, southwest Saskatchewan.

0.13 per cent to 0.42 per cent of the landscape. Patch density values per 100 hectares increased from 0.11 in 1979 to 0.42 in 2005. On average, over the 50-year period of PNG development, 3 site access roads were associated with each new PNG site constructed. The edge density of the 11 access roads in the study area in 1979 was 3676 metres per hectare compared to 1954 metres per hectare for 41 access roads in 2005. Edge density for the 14 wells in the study area in 1979 was 419.4 metres per hectare, while edge density for 108 wells in 2005 was 595 metres per hectare. In 2006, 33 per cent of the total surface disturbance in the study area was attributed to PNG site access roads, 49 per cent to PNG lease sites, and 18 per cent to other roads and trails.

The 31 PNG lease sites surveyed were found to have lower percent cover of herbaceous plants, club moss, and litter and greater cover of bare ground than reference pasture sites. The upper 20 centimetres of soil at PNG sites had, on average, greater pH, electro-conductivity, and percent clay values than the reference pasture sites. The analysis of plant community data indicated that PNG sites had a greater undesirable species abundance and diversity than reference pastures. Active and suspended PNG sites also had range health values significantly lower than reference pastures.

The effects of PNG activity on grasslands were found to extend well beyond the direct disturbance of the physical infrastructure itself. On average, the impacts of PNG sites extended 25 metres from the PNG wellhead. In 1979, the percentage of the study area occupied by PNG sites and PNG access roads was 0.07 per cent and 0.13 per cent, respectively. When the spatial extent of the ecological effects of PNG activities are considered, the total footprint of PNG sites (25 metres) and access roads (20 metres) increased to 0.31 per cent and 0.67 per cent, respectively. The total footprint of PNG development pre-1979 accounted for 25 per cent of the total disturbed area in the pasture. In

2005, when the ecological effects of PNG sites and PNG access roads are considered, the spatial extent of PNG impacts increased to 5.1 per cent, accounting for 75 per cent of the total disturbance in the area.

The effects of PNG development on grasslands in Saskatchewan is an example of how small disturbances on the landscape, based on their physical footprint, can be quite significant when assessed based on their ecological footprint. Even more important, the assessment showed that a combination of seemingly insignificant projects can result in significant cumulative effects. In the case of the Swift Current–Webb pasture, cumulatively, PNG development now accounts for 75 per cent of all disturbances to grasslands in the area. Moreover, PNG lease sites developed in 1955 have shown no significant improvements in terms of ground cover, species diversity, and range health compared to those developed more recently. Small projects do not always result in small or short-term impacts.

Source: Based on Nasen, Noble, and Johnstone 2011.

General Guidance for Screening

Barrow (1999) suggests that screening criteria for determining the need for an EIA might include, at least in principle, the following:

- Some component of the project will likely reach a specified threshold.
- The site for which development is proposed is sensitive.
- The proposed development involves known or suspected dangers or risks.
- The development will potentially contribute to cumulative impacts.
- There are unattractive input–output considerations, such as excessive labour migration or pollution.

Based on Annex 1 of the European EIA Directive, a number of general criteria are used to determine the need for some level of EIA:

- general condition and character of the receiving environment;
- potential impacts of the proposed development;
- resilience of the affected physical and human systems to cope with potential project-induced change;
- level of confidence associated with likely project impacts;
- consistency or compatibility with existing policy or planning frameworks;
- degree of public interest or perceived effects.

Screening plays a critical role in the EIA process; it is that part of the process when a decision is made as to whether a project requires an EIA and what level of assessment is necessary. While countries with EIA systems are starting to adopt national and formal requirements and guidelines for screening, Barrow (1999) suggests that screening is not widely practised and that in cases where screening is done, authorities often fail to screen adequately. The result is that many projects that may have significant adverse effects are often overlooked and thus proceed with little prior knowledge and understanding of their potential environmental consequences. In this regard, a more precautionary approach to screening may be warranted.

Screening and the Precautionary Principle

EIA can be seen as fostering precaution insofar as it provides a mechanism for anticipating adverse environmental impacts associated with a proposed development (Glasson, Therivel, and Chadwick 1999). Within the context of screening, precaution focuses on determining what is a *sufficiently* sound or credible basis for requiring an EIA. The screening stage is pivotal in this regard in that it ensures that activities that are "likely" to cause significant environmental impacts are given proper treatment and assessment. The problem is that it is often not possible at the screening stage for the competent authority to prove, with scientific certainty, the significance or insignificance of the impacts of a project. The suggested approach is to err on the side of caution; however, as Snell and Cowell (2006) explain, avoiding harm in the absence of evidence does not always sit well with the ethos of EIA in that decisions should be made on the basis of sound information.

The **precautionary principle** suggests that when scientific information is incomplete but there is a threat of adverse impacts, the lack of full certainty should not be used as a reason to preclude or to postpone actions to prevent harm. In other words, when considerable uncertainty exists as to whether a proposed activity is likely to cause adverse environmental effects, the lack of certainty should not be a reason for approving the proposed activity, for not requiring an EIA, or for not requiring rigorous mitigation and monitoring measures (IAIA 2003). Erring on the side of caution, the precautionary principle as a screening guideline places the burden of proof on the proponent, making the proponent responsible for demonstrating "insignificance" (Lawrence 2005).

However, the precautionary principle is not without problems, and the principle itself has been misused to justify everything from minimal change to rejecting any project proposal. Moreover, the interpretation of precaution within EIA is troubled by concerns over the efficiency of the process as well as by the need to ensure that precaution about likely, significant impacts does not unduly constrain development (Snell and Cowell 2006). That said, the underlying objective of adopting a precautionary approach to screening is to provide better assurance that potentially significant effects are captured during the screening process and that an EIA is carried out when needed.

Key Terms

assessment by responsible authority	equivalency EIA
assessment by review panel	exclusion list
case-by-case screening	hybrid screening
designated project	inclusion list
environmental preview report	initial environmental examination

list-based screening
precautionary principle
Regulations Designating Physical
Activities
responsible authority

screening
screening EIA
substitution EIA
threshold-based screening

Review Questions and Exercises

1. What is the purpose of screening in EIA? Should all proposed developments be subject to EIA?
2. Discuss the relative advantages and disadvantages of case-by-case, threshold, and list-based screening approaches. Which type of screening approach is used in your political jurisdiction (e.g., province)?
3. Using the Canadian Environmental Assessment Act, 2012 Regulations Designating Physical Activities (available at www.ceaa.gc.ca), determine (i) whether each of the following is a "designated project" under the act; and (ii) who would be the responsible authority:
    a) the construction of a fossil fuel–fired electrical generating station with a production capacity of 250 MW
    b) the decommissioning of a public highway in a migratory bird sanctuary
    c) the construction of a coal mine with a production capacity of 4250 t/d
    d) the expansion of an existing pulp and paper mill, resulting in an increase in its production capacity by 150 t/d, or 40 per cent
    e) the decommissioning of an old Class 1B nuclear facility for refining uranium that had a production capacity of 180 t/a.
    f) the expansion of an existing petroleum storage facility from its current storage capacity of 500,000m³ to 700,000m³.

References

Barrow, C.J. 1999. *Environmental Management: Principles and Practice*. New York: Routledge.

Canadian Environmental Assessment Act, 2012. S.C. 2012, c. 19, s.52. Assented to 2012-06-29. Ottawa, ON: Minister of Justice CEAA (Canadian Environmental Assessment Agency). 2012a. "Introduction to the Canadian Environmental Assessment Act, 2012 participant manual." Ottawa: CEAA.

———. 2012b. "Guide to preparing a description of a designated project under the Canadian Environmental Assessment Act, 2012." Ottawa: CEAA.

CAPP (Canadian Association of Petroleum Producers). 2007. www.capp.ca/default.asp?V_DOC_ID1/4677.

DETR (UK Department of Environment, Transport, and the Regions). 1998. *Draft Town and Country Planning (Assessment of Environmental Effects) Regulations*. London: DETR EC Directive 85/337/EEC First Schedule Part II Regulations 1989; DETR 1998 (proposed amendment to Directive 97/11).

Gilpin, A. 1999. *Environmental Impact Assessment*. Cambridge: Cambridge University Press.

Glasson, J., R. Therivel, and A. Chadwick. 1999. *Introduction to Environmental Impact Assessment*. London: Spon Press.

IAIA (International Association for Impact Assessment). 2003. "Social impact assessment international principles." Special Publication Series no. 2. Fargo, ND: IAIA.

Lawrence, D. 2005. "Significance criteria determination in sustainability-based environmental assessment." Report to the Mackenzie Gas Project Joint Review Panel. Langley, BC: Lawrence Environmental.

Mayer, S., et al. 2006. *Projects Subject to EIA: D 2.4 Report WP 4*. Vienna: Austrian Institute for Regional Studies and Spatial Planning.

Nasen, L., B.F. Noble, and J. Johnstone. 2011. "Effects of oil and gas lease sites in a grassland ecosystem." *Journal of Environmental Management* 92: 195–204.

Snell, T., and R. Cowell. 2006. "Scoping in environmental impact assessment: Balancing precaution and efficiency?" *Environmental Impact Assessment Review* 26: 359–76.

World Bank. 1993. "Sectoral environmental assessment." *Environmental Assessment Sourcebook Update No. 4*. Washington: World Bank.

Scoping

If the quality of an EIA were measured by how much shelf space (or DVDs) the EIS required, then the state-of-practice must be improving. Complex projects do require a significant volume of information and analysis; however, the number of volumes and maps contained in an EIS and the total page count are not indicators of the quality of an assessment. Good EIA is focused on those issues most important to managing the impacts of the proposed project and to informing the decision at hand. Too often, EIAs are rich in description and weak in analysis.

To consider *all* issues, impacts, and environmental components in any single EIA is neither feasible nor desirable. **Scoping** is about determining the important issues and parameters that should be addressed in an environmental assessment, establishing the spatial and temporal boundaries of the assessment, and focusing the assessment on the relevant issues and concerns. Scoping serves to identify the components of the biophysical and human environment that may be affected by development and for which there is scientific and public concern. This involves determining what elements of the project to assess, what environmental components are likely to be affected, how these environmental components have changed over time and what factors have driven such change, and how these components may be affected by other actions or disturbances in the project environment.

There are two broad types of scoping: closed and open. In **closed scoping**, the content and scope of the EIA are predetermined by law. Any modifications occur through closed consultations between the proponent and the competent authority or regulatory agency. In **open scoping**, the content and scope of the EIA are determined by a transparent process based on consultations with various interests and publics.

No matter what the type of scoping, the scoping process serves a number of important functions in the EIA process:

- ensuring input from potential stakeholders early in the process;
- identifying public and scientific concerns and values;
- evaluating these concerns to focus the assessment and provide a coherent view of the issues;
- ensuring that key issues are identified and given an appropriate degree of attention;

- reducing the volume of unnecessarily comprehensive data and information;
- avoiding a standard inventory format for EIA that may miss key elements or issues;
- defining the spatial, temporal, and other boundaries and limits of the assessment;
- ensuring that the EIA is designed to maximize information quality for decision-making purposes.

Scoping Requisites

Similar to determining the need to carry out an EIA, scoping requirements and practices vary from one EIA system to the next. The Netherlands, for example, has a formal scoping stage as part of a legislated EIA process. Under UK Directive 85/337, scoping is interpreted as in part mandatory and in part discretionary. Under the amended UK Directive 11/97, for example, the competent authority, at the request of the developer, is required to give opinions as to what should be addressed in EIA, but how scoping is carried out in terms of the consideration of alternatives or identification of indirect impacts is largely at the discretion of the proponent.

While the requirements and guidelines for scoping vary, a number of steps should be followed to facilitate good-practice EIA scoping:

- scope project alternatives;
- identify valued environmental components;
- delineate the assessment's spatial and temporal boundaries;
- establish the environmental baseline and trends;
- identify potential impacts and issues of concern.

Identification of Project Alternatives

Many environmental impacts can be prevented before irrevocable project location and design decisions are made by considering options to those proposed (see Chapter 7). The consideration of alternatives is a central element to good-practice EIA and is described by the US Council on Environmental Quality as the heart of the EIA process. Alternatives are defined as options or different courses of action to accomplish a defined end. There are two types of alternatives that should be considered—alternatives to the project and alternative means of carrying out the project (Figure 5.1).

Alternatives to a project are functionally different ways of meeting the need and purpose of the described project. Proposing alternatives to a proposed project involves developing criteria and objectives for environmental, socio-economic, and technical variables and identifying the preferred alternative based on the relative benefits and costs of each option with reference to the identified variables.

In practice, "alternatives to" are rarely considered comprehensively in project EIA. When considered, they often reflect narrow agency goals or are inherently biased toward the proposed project. Alternatives to a proposed project are typically limited to the "no action" alternative, which is interpreted as either "no change" from the

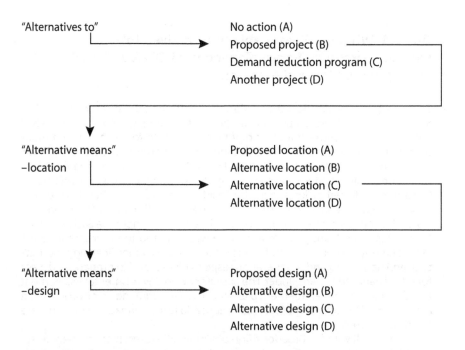

Figure 5.1 ● Hierarchy of Project Alternatives

current or ongoing activity or "no activity," which refers to not proceeding with the proposed development. This bias, to a certain extent, is understandable in that by the time a project is proposed, a proponent has already dedicated considerable resources to it. For example, land for the construction of an industrial complex may already have been leased by the time a proponent submits an application for development. Many alternatives that may be more environmentally sound or socially beneficial have thus already been foreclosed by the time the EIA commences.

In contrast to its predecessor, the Canadian Environmental Assessment Act, 2012 does not require the assessment of "alternatives to" the proposed project. However, it does still permit such considerations (section 19(1)(j)) on a case-by-case basis at the discretion of the minister of environment. Some may interpret this as a significant limiting of the scope of EIA in Canada at the federal level; arguably, however, "alternatives to" are best addressed at the early planning stages through the process of strategic environmental assessment (see Chapter 12) rather than by the proponent at the stage of project application (Box 5.1).

Alternative means can effectively be addressed at the project-scoping stage. "Alternative means" refers to different options for carrying out a project—it has been accepted at this stage that the proposed project is the most suitable alternative. Alternative means typically include alternative ways a project can be implemented, such as technical or engineering design or alternative locations (e.g., alternative routings of pipelines or transmission lines). However, it is not always possible to consider alternative locations for all proposed developments. There are few possible

---

Box 5.1 What Are Reasonable "Alternatives To"?
Pasquia-Porcupine Forest Management 20-Year Forest
Management Plan and Assessment

The Pasquia-Porcupine Forest Management Area is located along the Saskatchewan–Manitoba provincial border. It encompasses approximately 2 million hectares, more than half of which is suitable for commercial timber production. In 1995, a forest harvesting and management partnership was formed between MacMillan Bloedel Limited, one of Canada's largest forestry companies, and a subsidiary of the Saskatchewan Crown Investments Corporation, together known as the Saskfor-MacMillan Limited Partnership (SMLP). An application for development of the Pasquia-Porcupine forest was forwarded to the Saskatchewan government for approval. Under the Saskatchewan Forest Resource Management Act, SMLP was required to enter into a Forest Management Agreement with the province, which outlines licensing, permitting, and harvesting and regeneration responsibilities, and to prepare a 20-year forest management plan. Under The Saskatchewan Environmental Assessment Act, 20-year forest management plans are listed as requiring an environmental assessment. The environmental assessment and forest management plan were endorsed in 1999.

Consistent with the broader vision of sustainable forest management as set out under the province's Forest Resource Management Act and Forest Accord, and based on public consultation, SMLP identified eight objectives for their forest management plan:

- to provide quality products to meet customers' needs and provide a fair return to stakeholders;
- to provide safe and stable jobs;
- to safeguard heritage resources and traditional uses of the forest management area;
- to maintain the diversity of life forms in the area, including species and ecosystems;
- to maintain and enhance the forest ecosystem's long-term health;
- to minimize hazards from forest fires, insect infestations, and diseases;
- to protect primary resources of air, water, and soil;

---

alternative locations for a proposed gold-mining operation other than where the mineral deposit is located. That said, alternative routes for access roads to the mine site might be selected so as to avoid ecologically sensitive habitats or culturally significant areas, and alternative designs for the access road can be identified to minimize collisions with wildlife.

Alternative means should be feasible in terms of meeting the purpose of the project and can be identified through previous or similar experiences elsewhere, expert judgment, public consultation and brainstorming, and/or more complex decision support systems. The number of alternative means can be narrowed by eliminating

- to ensure that forest areas regenerate after harvesting.

Based on these objectives, SMLP identified and evaluated five "alternatives to" as part of their forest management plan and environmental assessment process, namely:

1. no timber harvesting (the current baseline condition);
2. low timber priority, with reforestation based on historical levels, including retention of mature forest stands and hydrological constraints on clear-cutting;
3. intermediate timber priority, with enhanced reforestation, including retention of mature forest stands and hydrological constraints on clear-cutting;
4. high timber priority, with enhanced reforestation, reforestation of all unstocked productive land, no maintenance of old-growth forests, and no hydrological constraints on clear-cutting;
5. the option proposed by SMLP in its application, consisting of a combination of the above, with enhanced reforestation, restoration of insufficiently restocked areas, retention of old-growth forests, and hydrological, species-specific, and ecosystem sensitivity constraints on clear-cutting.

Alternatives were assessed by a multi-disciplinary team of experts using forest land inventories and computer modelling technology. Using a Geographic Information System and a forest simulation model, changes to the forest environment were assessed for each alternative. Socio-economic impacts were similarly assessed by a panel of SMLP-commissioned experts, using an input–output model. The impacts of each alternative were assessed based on a variety of biophysical and socio-economic variables, including direct employment, income projections, tax revenues, demographic change, soil erosion, nutrient depletion, water turbidity, and long-run sustained forest growth. Perhaps not surprisingly, option 5, the SMLP-proposed option, was identified as the preferred one based on the analysis of biophysical and socio-economic trade-offs.

Sources: Noble 2004; SMLP 1997.

those that fail to meet certain minimum project, environmental, or socio-economic requirements. For example, site-screening criteria can be used to identify various environmental criteria or project objectives (Box 5.2).

Candidate alternatives must be systematically compared to identify the preferred option among those that remain for detailed impact assessment. Several methods are available for evaluating alternatives (Table 5.1); the objective is to compare across alternatives using similar criteria or objectives. For example, alternatives can be evaluated using a simple rating or ranking procedure based on potential economic, social, or environmental impacts or contributions (Table 5.2). In cases where the

Box 5.2 "Alternative Means" and the Mackenzie Gas Pipeline Project

In 2003, Imperial Oil Resources Ventures Limited, Shell Canada, ConocoPhillips, ExxonMobil Canada, and the Mackenzie Valley Aboriginal Pipeline Limited Partnership submitted a preliminary information package describing the proposed Mackenzie Gas Project and its related environmental and socio-economic issues.

The proposed Mackenzie Gas Project is one of the largest energy development projects in North America. It involves the development of natural gas reserves on the Mackenzie River Delta in the Northwest Territories, natural gas processing, and the construction of a pipeline for natural gas shipment through the Mackenzie Valley to northern Alberta. At an estimated construction cost of $7 billion, the proposed pipeline will be approximately 1220 kilometres long and will transport 34 million cubic metres of natural gas per day. More than 8000 workers will be employed at peak construction, with an estimated 150 persons employed to operate the pipeline system. Up to 32 communities will be affected by the project, including 26 in the Northwest Territories and 6 in northwestern Alberta.

Given the project's magnitude and northern location, it was subject to EIA under three jurisdictions: the Canadian Environmental Assessment Act, the Mackenzie Valley Resource Management Act, and the Western Arctic Claims Settlement Act. A joint review panel was commissioned to administer the environmental impact review. The project received regulatory approval in 2011. Construction is not expected to commence until 2015 at the earliest, with first gas expected to be produced in 2018.

Various alternatives were considered in the project EIS, including alternatives to the project (functionally different ways of producing, processing, and transporting gas to southern markets) and alternative means (technically and economically feasible ways to implement the proposed project). With regard to "alternatives to," the proponent argued that there were no viable alternatives to the development and transportation of Mackenzie Delta gas to southern markets as advanced as the proposed project. The only true "alternative to" was the no-go alternative, the result of which, according to the proponent, would be the loss of benefits and opportunities for both the project investors and northern residents.

Several "alternative means" were considered by the proponent prior to submitting its EIS and application for development, including alternative locations for facilities and infrastructure sites, route alternatives within the preferred pipeline corridor, and alternative methods for construction and reclamation. Of considerable concern, in terms of both environmental and community impacts, was the route selected for pipeline construction. Examining historical pipeline routes in the region, and on the basis of field surveys, baseline data collection, and discussions with local communities, the preliminary pipeline route was divided into 29 segments, each containing several route alternatives for evaluation.

The objectives of the route and site selection process were to avoid

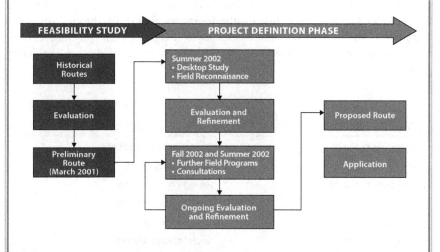

sensitive environmental and cultural areas, reduce disturbance to communities and the landscape, satisfy engineering and construction requirements, and reduce cost. Evaluation criteria for preferred route selection included:

- route placement, including options for reducing the length of the pipeline, potential facility sites, and locating the right-of-way in order to avoid encroaching on existing habitats, be close to existing infrastructure, and be parallel to or use existing linear disturbances;
- reducing the number, complexity, and width of watercourse crossings;
- geotechnical considerations such as avoiding springs, perched aquifers and steep, ice-rich, or unstable slopes, and the distribution of discontinuous permafrost;
- environmental considerations such as land-use plans, socio-economic concerns, and reducing proximity to critical wildlife habitat and important cultural or archaeological sites;
- construction matters such as slopes, muskeg and fen areas, grading, access, ground conditions, crossing linear facilities, and the need for adequate workspace;
- community interests;
- relative costs of the route alternatives.

Alternative route assessment occurred over three phases: first, in August 2002, to scope preliminary route options based on terrain, physical limitations, and traditional land-use patterns; second, in September 2002, based on watercourse crossings to narrow and refine the preferred route options; and third, in August 2003, to confirm the preferred route mapping and to align the final route with proposed pipeline facilities and infrastructure. The preferred route and analysis of alternative routes formed an important part of the proponents EIS.

selection among alternatives is not straightforward, assessment criteria may need to be weighted to determine their relative importance. In short, the environmental and socio-economic effects of alternative means, including technical and economic feasibility, should be identified and clear justification provided for the preferred project alternative.

Table 5.1 Selected Methods for Evaluating Alternatives*

Contingent ranking	Objectives hierarchies
Contingent valuation	Opportunity costs
Cost-benefit analysis	Paired comparisons
Decision trees	Performance measures
Expert systems	Queuing models
Life-cycle assessment	Social choice theory
Linear programming	Spatial decision support systems
Moment estimation methods	Utility functions
Multi-attribute analysis	Weighted scoring
Networks	Willingness to pay
Net present value	

* Many of these are also applicable to impact prediction. See Chapter 6.

Table 5.2 Sample Alternative Site Evaluation Procedure

Candidate sites	Minimize economic cost	Minimize impacts on water quality	Topographic suitability	Ease of access	Minimize community impacts	Minimize wildlife/ habitat impacts	Total score
A	3	4	5	3	2	1	18
B	3	2	2	3	2	1	13
C	4	3	3	5	4	4	23
D	2	4	2	2	2	4	16

Legend: 5 = site meets criterion in full
4 = intermediate decision
3 = site meets criterion in part
2 = intermediate decision
1 = site does not meet criterion

Valued Environmental Components

A large volume of baseline data could be collected to characterize the biophysical (Table 5.3) and human environments (Table 5.4) when conducting an EIA, and a long list of techniques for collecting that data could be employed from sources such as census data, historical records, land-use plans, field surveys, and sampling procedures. However, EIA is not about undertaking data collection for the purpose of a descriptive "environmental study." Scoping serves to establish clear boundaries for spatial and temporal assessment according to the important components of the environmental baseline. This includes limiting the amount of information to be gathered in baseline studies to a manageable level and identifying specific objectives and indicators to guide the assessment. In this way, establishing the assessment boundaries focuses baseline studies, making them more efficient for and relevant to subsequent stages of the assessment process (Figure 5.2).

Identifying Valued Environmental Components

Comprehensive baseline studies can often be a waste of time and resources. Environmental baseline studies must be undertaken within the context of clearly defined scope and objectives; otherwise, too much information is acquired that is often of too little use. While it is important to ensure that all potentially affected environmental components are given consideration, attention should focus on the

Table 5.3 Scoping the Biophysical Environment

Air	Water
• current pollutant concentration • pollutant dispersion • emission levels • emission types • temperatures • windspeeds and directions	• surface quantity • surface water withdrawal • groundwater quantity • groundwater withdrawal • chemical content • turbidity • stream flow • bank stability • levels of eutrophication • current pollution discharges • fish and fish habitat

Soil	Coastal zone
• erosion rates • moisture content • fertility • organic matter • electrical conductivity • chemical composition • stability • soil pollutants	• water temperature • flood frequency • tidal activity • sedimentation • marine resource populations • bank or cliff stability

Terrestrial
• level of fragmentation • widlife populations • vegetation cover and composition • airborne and water-borne pollutants • levels of light pollution • vegetation health

Table 5.4 Scoping the Human Environment

Economics	Housing
• local and non-local employment • labour supply • wage levels • skill and education levels • retail expenditures • material and service suppliers • regional multiplier • tourism	• public and private housing • house prices • homelessness and housing problems • density and crowding
Demographics	**Local services**
• population • population characteristics (family size, income, ethnicity) • settlement patterns	• educational services • health services • community services (police, fire) • transportation services and infrastructure • financial services
Health	**Socio-cultural**
• quality of life (actual and perceived) • medical standards • worker death or injury rates • current disease transmission • mental and physical well-being	• family life • seasonality of employment • culture and belief systems • crime rates, substance abuse, divorce rates • community conflict and cohesion • traditional foodstuffs • community perception • gender relations

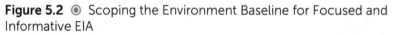

Project charateristics, activities, and environment	+	Baseline scoping, identifing key VECs, objectives, and indicators	=	Focused, timely, efficient, and relevant EIA

spatial and temporal bounding	VEC objectives and indicators	trends and issues identification	impact prediction and evaluation	management and monitoring

Figure 5.2 ◉ Scoping the Environment Baseline for Focused and Informative EIA

valued environmental components (VECs) most likely to be affected. VECs, some-times referred to as simply valued components or VCs, are aspects of the environment, physical and human, that people value and are considered important from scientific or public perspectives, thus warranting detailed consideration in the impact assessment. In most cases, VECs are such components as fish and wildlife species or species groups whose distribution and abundance are of management interest because they are har-vested, at risk, or sensitive to disturbance. However, VECs can be identified in relation to physical attributes (e.g., a wildlife population), ecological processes (e.g., carbon sequestration), and even broad and abstract concepts (e.g., ecological integrity).

In the Mackenzie Gas Project, for example, wildlife VECs in the natural gas pipeline corridor were identified on the basis of their regulatory status, ecological importance, socio-economic importance, and conservation concern. In the Voisey's Bay EIS, 17 key VECs, including water, caribou, plant communities, Aboriginal land use, and family and community, were identified on the basis of social, economic, regulatory, and technical values and concerns. In addition, the EIS identified cer-tain VECs that were also "pathways" for environmental effects that might lead to effects or impacts on other VECs. For example, the VEC "water" was identified as a biophysical effects pathway that could potentially affect other VECs: freshwater fish habitat, waterfowl, and seabirds.

A recent master's thesis by Olagunju (2012) examined the rationale for VEC selec-tion across a sample of Canadian EIAs and found that ecological importance and societal value were common rationales, but other rationales for VEC selection included:

- rarity
- fragility
- scientific value for study or monitoring
- importance to legal compliance
- economic importance
- professional judgment
- biodiversity and conservation value
- medicinal importance
- recreational value
- spiritual importance

There are a variety of information sources that can be used to assist with VEC selection, including project descriptions filed by other proponents in the project's environment, legislation, academic literature, regional environmental studies, trad-itional or local knowledge, and monitoring programs. These sources can be used to understand the current status of knowledge for those issues identified or to identify known regional issues of concern. In their analysis of VEC selection across a sample of 35 federal and provincial EIAs in the South Saskatchewan watershed, Ball, Noble, and Dubé (2012) found that for aquatic VECs, VEC selection was based most often on the need to address a proponent's exposure to liability under the Fisheries Act and the Species at Risk Act. Those VECs protected by punitive federal legislation were used more than twice as often as other aquatic VECS. However, Ball, Noble, and Dubé also observed that the number and diversity of VECs considered was higher in assessments that engaged some form of public consultation in the scoping process.

To ensure that baseline studies are purposeful and not simply a compilation of information about VECs, at least two questions should be asked about each VEC prior to moving forward and collecting detailed baseline environmental data:

1.   Is the VEC likely to be affected, directly or indirectly, by project activities? If not, then there is no need to compile a comprehensive baseline about that VEC. This is not to say that the VEC should be disregarded, since it may still be of significant public concern and warrant some consideration in the impact assessment process.
2.   Is it possible to predict impacts on the VEC or on indicators of that VEC or to relate either quantitatively or qualitatively project-induced change to VEC conditions? If not, the VEC should be given relatively low priority unless it is a VEC of significant public value.

These questions should be reassessed when considering the potential cumulative effects (see Chapter 11) of the project on VECs.

Environmental Assessment in Action

Principles for VEC Selection in the Elk Valley

The Elk Valley is located in the Rocky Mountains in the southeastern region of British Columbia, home to the communities of Elkford, Sparwood, Hosmer, Fernie, Morrissey, and Elko. The Elk River, which is approximately 220 kilometres-long, cuts through the Elk Valley. Its total drainage basin is 4450 square kilometres, providing important habitat for cutthroat trout—a popular freshwater sport fish and important indicator species of general ecosystem health. The valley is also home to grizzly bears, bighorn sheep, American dipper elks, northern goshawks, and bald eagles. Although most know the Elk Valley for its scenery and tourism, the valley plays a significant role in the province's coal-mining industry and contains the largest producing coalfield in British Columbia. There are five surface metallurgical coalmines operating in the valley, all owned by Teck Coal.

In 2012, because of increasing concerns about the potential impacts of human development in the Elk Valley, including tourism, angling, roads and trails, railway operations, and coal mining, a workshop was organized by Teck Coal to initiate a process to develop and implement a cumulative effects assessment and management framework for the Elk Valley. Workshop participants included representatives of Teck Coal, the Ktunaxa Nation Council, various provincial government agencies (e.g., Ministry of Environment, Ministry of Forests, Lands and Natural Resource Operations, Ministry of Aboriginal

Relations and Reconciliation, Ministry of Energy and Mines), federal agencies (e.g., Environment Canada, Fisheries and Oceans Canada), municipal and regional government representatives, and environmental non-government organizations.

The overall purpose of the initiative that emerged from the workshop, referred to as the Elk Valley Cumulative Effects Management Frameworks (CEMF), is to manage the cumulative effects of past, current, and possible future human actions on VECs in the Elk Valley. During the workshop, participants established, by consensus, the spatial and temporal boundaries of the CEMF and compiled a list of values to inform VEC selection. A smaller working group was established from among workshop participants and was commissioned with the responsibility to develop the Elk Valley CEMF, identify potential VECs, and ensure that the CEMF is translated into action.

To help guide VEC selection for the initial development and application of the assessment framework, the working group developed a number of principles:

- VECs will arise from the complete list of values identified by workshop participants at the initial CEMF meeting and will consider potential linkages between values;
- any selected VEC must reflect several values (ecological, social, cultural, and economic);
- there must be reason to believe that the VEC is or will be affected either directly or indirectly by current or future activities in the region;
- the VEC must be sensitive to several important disturbing and supportive ecological and human processes, including interactions among these processes;
- there must be sufficient scientific knowledge about the VEC to allow selection of indicators that will respond in a way that can be readily measured;
- any assessment of the VEC must help to inform regulatory decisions likely to be included in permits and licences for development decisions;
- the VEC must represent traditional knowledge and traditional uses of the land and water resources of the Elk Valley;
- results of the assessment of the VEC must be capable of contributing to more confident decision-making.

At the time this book was written, the Elk Valley CEMF was initiating its baseline analysis, focused on retrospective assessment of changes in VEC condition and indicators. The case is an example of a collaborative and transparent approach to scoping and VEC identification, guided by agreed-upon principles and grounded in the values of those most affected by development and management decisions in the Elk Valley.

Selecting VEC Indicators

Establishing **VEC indicators** is critical to understanding actual change in VEC conditions and, more important, to providing an early warning of potential adverse effects. VECs employ nomenclature used in EIA, but they are not always measurable in their own right (Antoniuk et al. 2009). Often, indicators that are measurable need to be identified to assess or evaluate the status of, or threats to, VECs. Indicators provide a sign or signal that relays a complex message, potentially from numerous sources, in a simplified and useful manner (Jackson, Kurtz, and Fisher 2000). In assessing the VEC "surface water quality," for example, attention may focus on a specific indicator (e.g., phosphorus concentrations, benthic invertebrate abundance) that provides direct, measurable information about the condition or state of the VEC. These are referred to as **condition-based indicators**. In other cases, attention may focus on measurable stress that directly affects the VEC (e.g., amount and distribution of surface disturbance in the watershed or stream crossing density). These are referred to as **stress-based indicators**, or disturbance-based indicators (Table 5.5).

Indicators are the most basic tools for analyzing environmental change—they allow practitioners and decision-makers to gauge environmental change efficiently by avoiding impact "noise" and focusing on parameters that are responsive to change, generate timely feedback, and can be traced effectively over space and time. The choice of indicators is often influenced by data availability (Cairns, McCormick, and Niederlehner 1992). As such, the highest-priority indicators are often those where there is already good scientific or local knowledge of how human activities and natural changes affect the indicator (Antoniuk et al. 2009). Good indicators must

Table 5.5 Examples of Condition- and Stress-Based Indicators for Terrestrial and Aquatic VECs

Examples of condition-based indicators used for terrestrial VECs:	Examples of condition-based indicators used for aquatic VECs:
habitat effectiveness index; habitat suitability index; soil conductivity; range health; soil pH; soil erosion and compactness; presence/absence of invasive plant species; species richness; species population; species distribution; habitat corridor size; core habitat area; habitat patch size	total and dissolved organic carbon; pH; temperature; chloride; suspended sediments; total ammonia; metal indicators (various); nutrients; total dissolved phosphorous; benthic invertebrates (abundance, richness, evenness, index of biological integrity); fish population (relative abundance, catch per unit effort, condition factors, nutritional health, gonad size, fish tissue chemistry); riparian plant community; Hilsenhoff biotic index; ambient concentrations of regulated discharge parameters; quantity/flow
Examples of stress- or disturbance-based indicators used for terrestrial VECs:	**Examples of stress- or disturbance-based indicators used for aquatic VECs:**
industrial footprint; core habitat area loss; habitat patch size; habitat corridor size; linear features density; km roads/km2; total cleared area; % area disturbed; total area burned; wildlife–vehicle collisions; number of hunting licences	stream access density; stream crossing density; number of hung culverts; km roads/km^2; total cleared area; water abstraction rate; % area disturbed by class of activity; disturbed riparian area; % impervious surfaces; total area burned; species harvest rate; number of angling licences

be indicative of the causes or sources of change in, or stress to, a VEC and not only the existence of change. At a minimum, good VEC indicators are:

- measurable, either in a qualitative or quantitative fashion;
- indicative of the VEC of concern;
- sensitive and detectable in terms of project-induced stress;
- appropriate to the spatial scale of the VEC of concern;
- temporally reliable;
- diagnostic to change;
- applicable across different types of development projects;
- cost-effective to collect, measure, or analyze;
- predictable and accurate with an acceptable range of variability;
- understandable by non-scientists;
- useful for informing management actions or decisions.

Establishing VEC Objectives

Establishing objectives, usually based on benchmarks or thresholds, is important to understanding when a significant, acceptable, or unacceptable level of change in the indicator has occurred or is about to occur. A **benchmark** is a standard or point of reference against which change may be compared or assessed, such as the range of natural variability in an environmental phenomenon. A **threshold** is an established limit of change and may include:

- absolute ecological or socio-economic thresholds: carrying capacity, limit of tolerance;
- acceptable limits of change: what is acceptable from a broader societal perspective;
- desired VEC conditions or objectives: desired outcomes.

When environmental effects or conditions change beyond acceptable levels, or when VEC objectives are not met, then project impacts are considered significant, and some form of impact management is necessary. Salmo et al. (2004) suggest a tiered approach to thresholds in impact assessment, consisting of cautionary thresholds, target thresholds, and critical thresholds (Figure 5.3). At a **cautionary threshold**, monitoring efforts should be increased to more closely monitor VEC or indicator conditions and the effectiveness of best-management practices verified to prevent any further adverse change. A **target threshold** is typically politically or socially defined—a margin of safety and a mandatory trigger for management action. A **critical threshold** defines maximum acceptable change, socially or ecologically, beyond which impacts may be long-term or irreparable.

Establish Assessment Boundaries

Spatial Bounding

What are the spatial limits of the assessment? In delineating the spatial limits of assessment, some consideration must be given to scale. João (2002) suggests that scale

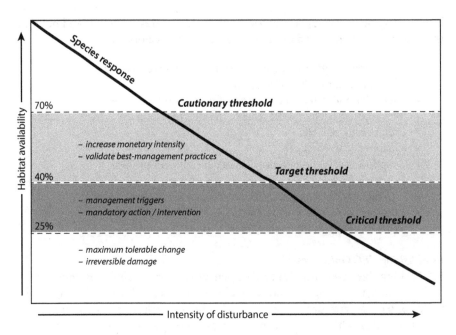

Figure 5.3 ◉ Cautionary, Target, and Critical Thresholds

Source: Based on Salmo et al. 2007

has two interrelated meanings relevant to EIA: first, scale as the *spatial extent* of the assessment; second, scale as the amount of *geographic detail*. When the spatial scale of an EIA covers a very large area, a high degree of geographic detail or granularity is less feasible. In other words, as the geographic boundary of an assessment increases, granularity typically decreases. When considering the impacts of linear developments, such as a pipeline or transmission line project, on caribou, for example, a practitioner may adopt coarse stress- or disturbance-based indicators, such as the density of linear features or various landscape metrics (e.g., habitat intactness, industrial footprint). The spatial scale of an assessment is an important factor in the nature and types of indicators used to understand potential impacts to the VECs of concern.

Since different spatial scales and levels of detail are often required to understand project interactions with different environmental receptors, different environmental receptors should be examined at different scales, and for any given receptor there are various scales at which the receptor can be assessed (Box 5.3). For example, air quality can be measured across a metropolitan area or an airshed or within the confines of a particular administrative space. In other words, factors affecting air quality can be assessed across a variety of **functional scales,** ranging from point source to local, to regional (including regions of various sizes), to national (Chagnon 1986).

It is also important to distinguish between two types of information: the activity information and the receptor information (Irwin and Rodes 1992). **Activity informa-tion** characterizes the types of effects a project might generate, such as habitat frag-mentation or increased use of non-renewable resources. **Receptor information** refers

---

Box 5.3 Basic Principles for Spatial Bounding

- Boundaries must be large enough to include relationships between the proposed project, other existing projects and activities, and the affected environmental components (Cooper 2003).
- The scope of assessment should cross jurisdictions if necessary and allow for interconnections across systems (Shoemaker 1994).
- Natural boundaries should be respected (Beanlands and Duinker 1983).
- Different receptors will require assessment at different scales (Shoemaker 1994).
- Boundaries should be set at the point where effects become insignificant by establishing a maximum detectable zone of influence (Scace, Grifone, and Usher 2002).
- Both local and regional boundaries should be established (Canter 1999).

Geographic boundaries for any particular assessment will vary depending on a number of factors, including the nature of the project itself, sensitivity of the receiving environment, nature of the impacts, extent of transboundary impacts, availability of baseline data, jurisdictional boundaries and cooperation, and natural physical boundaries.

---

to the processes resulting from such effects. Both of these must be taken into consideration when delineating spatial boundaries. For example, the Canadian Environmental Assessment Agency's operational policy for spatial bounding regarding offshore oil and gas exploratory drilling suggests that the spatial area of an assessment should be the "composite" of the spatial area of all affected environmental receptors (Figure 5.4) and should consider uncertainties concerning the precise location of proposed activities and effects, as well as the need to reconsider individual project bounding when multiple projects are being conducted in adjacent resource areas. Several of the methods identified in Chapter 3 can be used to assist in spatial bounding, including Geographic Information Systems, network and systems analysis, matrices, public consultation, quantitative and physical modelling, and expert opinion.

Temporal Bounding

What are the temporal limits of the assessment? The temporal scale should include the past, present, and future. This involves consideration of previous and current activities in the project region affecting environmental conditions as well as potentially foreseeable activities. For example, assessment of a proposed hydroelectric transmission line project should consider previous linear disturbances that have affected the regional baseline environment as well as current and proposed land-use activities that may contribute to additional habitat fragmentation. How far into the past and how far into the future will depend on the quality and quantity of data available, the certainty associated with project environmental conditions, and what the assessment is trying to achieve. Examining past conditions may be as simple as examining land-use maps or census data, and it may be feasible to incorporate

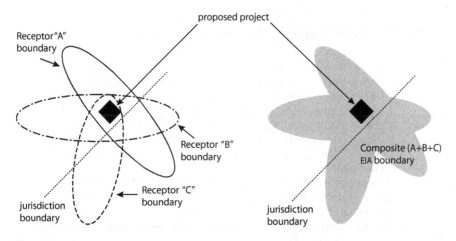

Figure 5.4 ◉ Spatial Bounding Based on Composite Environmental Receptor Information

50 years of historical data if deemed necessary for the proposed project. The most appropriate past temporal bounds for an assessment is a point in time when the VEC of concern was most abundant or before the current drivers of change in the region were most dominant. For example, in the context of the Athabasca River system in Alberta, Seitz, Westbrook, and Noble (2011) suggest that an appropriate past temporal bound for impact assessment would be the late 1960s—prior to rapid agricultural intensification, oil sands development, and population growth in the watershed.

Establishing an appropriate future boundary for assessment may be based on the end of the operational life of a project, after project decommissioning and reclamation, or after affected environments recover. Future boundaries are much less certain, since data are often hypothetical or based on a range of future scenarios. Typically, future boundaries do not exceed a decade beyond project decommissioning; however, this may be too restrictive from a sustainability perspective or when dealing with legacy projects, such as mining operations and contaminated sites. Further attention is given to temporal boundaries in Chapter 11 in the context of assessing potential cumulative environmental effects.

Jurisdictional Bounding and Transboundary Effects

In principle, EIA should be bounded on the basis of ecological units, such as watersheds or eco-regions. In practice, administrative boundaries play an important role, and institutional boundaries and the administrative authority to implement impact management measures almost always temper ecological boundaries. The challenge is that certain, if not most, impacts spread across administrative boundaries and affect environments in other jurisdictions. In other cases, EIA screening may indicate that the nature of the proposed project and its impacts requires the participation of multiple jurisdictions.

At the international level, the Convention on Environmental Impact Assessment in a Transboundary Context, commonly referred to as the Espoo Convention, establishes the responsibilities of nations assessing the transboundary impacts of projects. This convention outlines the obligations of the "host" country to notify, and include participation by, the affected neighbouring country. Members of the European Union and the East African Community have adopted these principles to varying degrees. The focus of attention is largely on consultation responsibilities concerning major developments or land-use changes in one jurisdiction that are likely to cause transboundary impacts in neighbouring countries. Emphasis is placed on "transboundary effects."

Unfortunately, the majority of transboundary provisions for EIA in the Canadian context are associated with physical infrastructure that is of a transboundary nature (e.g., pipelines). This is currently the major focus of transboundary screening and assessment in the Mackenzie Valley of Canada's Northwest Territories, for example, where transboundary is primarily interpreted in relation to a development located partly in the Mackenzie Valley and partly in another area (see MVEIRB 2004). The focus is on the physical activities associated with the project as opposed to the geographic extent of the effects.

There are no specific rules for establishing jurisdictional boundaries in impact assessment or for determining when an adjacent jurisdiction should be involved in the assessment process. At a minimum, however, consideration should be given to:

- what jurisdictions are affected by the project;
- the extent to which adjacent jurisdictions are affected by the project;
- what jurisdiction or jurisdictions have decision-making authority;
- what jurisdiction or jurisdictions are responsible for managing project impacts and over what area;
- the EIA provisions and capacity of affected adjacent jurisdictions to manage potential impacts.

Conduct the Environmental Baseline Study

An **environmental baseline** takes into account the past, present, and likely future state of the environment without the proposed project or activity. A **baseline study** consists of identification and analysis of conditions over space and/or time for the purpose of delineating change, trends, patterns, or limits to assist in impact assessment, impact evaluation, and impact monitoring activities. Three key questions should be considered in a baseline study:

- What do we need to know about the baseline environment to make a decision on the project?
- What are the relevant background conditions that have influenced the current environment?
- What is the likely baseline condition in the future in the absence of the project?

A key component of establishing the environmental baseline is identifying and establishing baseline trends. The purpose of a baseline study is to understand what is happening in the project's regional environment, specifically:

- What did VEC or VEC indicator conditions look like "then"?
- What do VEC or VEC indicator conditions look like "now"?
- What are the trends, rates, patterns, and suspected drivers of change?
- What is the magnitude of change?
- Is the level of change that has occurred in the VEC or VEC indicator to date significant?

Baseline studies are sometimes referred to as retrospective assessments, emphasizing the importance of examining past conditions. Too often, the focus of baseline studies is on a description of current conditions (Box 5.4) rather than on assessing how conditions have changed over time and identifying the suspected drivers of change. Core to a good baseline study is an analysis of what VEC indicators looked like in the past, how they have changed over time, and what types of stressors resulted in or triggered such changes. An attempt should be made to identify relationships between indicators of change in VEC conditions (e.g., caribou population, water-quality indices) and measures of human or natural disturbance so as to determine trends and associations that can be used to predict VEC conditions or responses to future change.

In cases where VECs or VEC indicators are not amenable to interpolating trends, information on past conditions can be used alongside current conditions as benchmarks against which potential project-induced stress can be identified and assessed. If, for example, the trend in the project environment has been toward decreasing physician-to-patient ratios and the health care system is currently at or near capacity, then any induced stress caused by an increase in health care demand will be significant in terms of community health and well-being and thus warrants detailed consideration in impact assessment and mitigation.

Scoping Continuation

Once the project is described and the baseline assessment completed, potential project impacts and areas of concern can be identified. The objective is to highlight for further and more intensive study those issues and impacts that will form an important part of the decision as to whether a project will be approved and under what conditions. Methods and tools supporting impact identification and characterization were discussed in Chapter 3. A good scoping process thus has a positive ripple effect throughout the rest of the EIA process. Scoping ensures that all of the necessary factors are considered in the assessment and at the same time sufficiently narrows the scope of the assessment to focus only on centrally important elements. While scoping is typically identified as a single phase of the EIA process, it is an ongoing activity and should continue in the post-project implementation and monitoring phases. Good-practice scoping is a process by which the environment is continuously scanned to detect signs of adverse environmental changes or variables of importance that may not have been considered during initial project scoping exercises.

Box 5.4 Shifting Baseline Syndrome

Too often, baseline studies are focused only on current conditions rather than also on changes in VEC or indicator conditions from past to present and whether thresholds of concern are being approached or already have been exceeded. This approach defines only the current environment as the baseline for impact assessment. But because the existing environment is a result of the influence of past actions, this approach attributes the effects of past and present actions to the current condition rather than to contributions to cumulative change in VEC or indicator conditions. The magnitude of the effects of past projects are discounted and treated as part of the current baseline condition.

Consider a scenario in which available core area habitat for caribou is already at a target threshold (see Figure 5.3) due to 25 years of linear disturbances caused by highways, pipelines, transmission lines, forestry operations, and mineral lease blocks. It is tempting for a project proponent to use the "current available habitat" as the baseline against which the project's potential environmental effects are assessed. In so doing, a proponent may be able to present a convincing argument that their project is contributing to only a minor loss of caribou habitat, perhaps less than 2 per cent of what is currently available. However, if past conditions are considered and it is observed that caribou core area habitat in the project's regional environment had declined by 60 per cent over the previous 25 years, the project's contribution of an additional 2 per cent loss now appears much more significant.

Current conditions are sometimes adopted as a "new normal" in baseline studies rather than considering current conditions *relative to past conditions* and evaluating the nature and significance of *change* in VECs in the study area. This was the approach adopted in the recent Manitoba Hydro Bipole III project, an approximately 1400-kilometre transmission line project from northern Manitoba, near Gillam, south to Winnipeg. The Manitoba Clean Environment Commission report on public hearings for the project identified several concerns about limited baseline data against which to properly assess project impacts and ignoring the effects of past actions and changes in conditions over time (see Manitoba Clean Environment Commission 2012). The proponent's EIS adopted past effects as the norm, thus precluding possible restoration or other mitigative actions that might be required to improve current conditions. This is poor practice.

Key Terms

activity information
alternative means
alternatives to
baseline study
benchmark
cautionary threshold

closed scoping
condition-based indicator
critical threshold
environmental baseline
functional scale
open scoping

receptor information   threshold
scoping   valued environmental components
stress-based indicator   (VECs)
target threshold   VEC indicators

Review Questions and Exercises

1. What is the purpose of EIA scoping?
2. What are valued ecosystem components?
3. Obtain copies of topographic and political maps of your region. Suppose a new coal-fired electricity generating plant is being proposed to meet a growing demand for electricity:
    a) Identify a "suitable" project location on the map sheet.
    b) What VECs would you identify in the region as important to consider? Provide a statement as to why these environmental components are classified as VECs. In other words, what makes them so important that you would include them in the assessment?
    c) Given current VEC conditions, what objectives would you attach to each EIA?
    d) Identify a list of indicators that you might use to assess and monitor the conditions of each VEC.
    e) What spatial boundary or boundaries would you establish for the assessment? Sketch these boundaries on a copy of the map. Identify the criteria you used to determine these boundaries.
    f) What temporal boundaries might you consider? Why?
    g) Brainstorm a number of feasible alternatives that might be considered in the project assessment, including alternative project locations.
4. Obtain a completed project EIS from your local library or government registry, or access one online. Identify the number and types of alternatives, if any, considered in the assessment. Were these alternatives assessed and, if so, to what extent? What VECs were identified? Is there a rationale for their selection? Compare your results with the findings of others.

References

Antoniuk, T., et al. 2009. *Valued Component Thresholds (Management Objectives) Project.* Report no. 172. Calgary: Environmental Studies Research Funds.

Ball, M., B.F. Noble, and M. Dubé. 2012. "Valued ecosystem components for watershed cumulative effects: An analysis of environmental impact assessments in the South Saskatchewan River watershed, Canada." *Integrated Environmental Assessment and Management* doi: 10.1002/ieam.1333.

Beanlands, G.E., and P.N. Duinker. 1983. "Lessons from a decade of offshore environmental impact assessment." *Ocean Management* 9 (3/4): 157–75.

Cairns, J., Jr, P.V. McCormick, and B.R. Niederlehner. 1992. "A proposed framework for developing indicators of ecosystem health." *Hydrobiologia* 263: 1–44.

Canter, L. 1999. "Cumulative effects assessment." In J. Petts, ed., *Handbook of Environmental Impact Assessment*, vol. 1, *Environmental Impact Assessment: Process, Methods and Potential*. London: Blackwell Science.

Chagnon, S. 1986. "Atmospheric systems: Management perspective." In Canadian Environmental Assessment Research Council (CEARC), ed., *Cumulative Environmental Effects Assessment in Canada: From Concept to Practice*. Calgary: Alberta Society of Professional Biologists.

Cooper, L. 2003. *Draft Guidance on Cumulative Effects Assessment of Plans*. EPMG Occasional Paper 03/LMC/CEA. London: Imperial College.

Imperial Oil Resources Limited. 2004. *Environmental Impact Statement for Mackenzie Gas Project*. Ottawa: National Energy Board and Joint Review Panel.

Irwin, F., and B. Rodes. 1992. *Making Decisions on Cumulative Environmental Impacts: A Conceptual Framework*. Washington: World Wildlife Fund.

Jackson, L.E., J.C. Kurtz, and W.S. Fisher. 2000. *Evaluation Guidelines for Ecological Indicators*. Report no. EPA/620/R-99-005. Washington: Environmental Protection Agency.

João, E. 2002. "How scale affects environmental impact assessment." *Environmental Impact Assessment Review* 22: 289–310.

Manitoba Clean Environment Commission. 2012. *Bipole III Transmission Project: Report on Public Hearing*. Winnipeg: Manitoba Clean Environment Commission.

MVEIRB (Mackenzie Valley Environmental Impact Review Board). 2004. "Environmental impact assessment guidelines." Yellowknife: MVEIRB.

Noble, B.F. 2004. "Integrating strategic environmental assessment with industry planning: A case study of the Pasquia-Porcupine Forest Management Plan, Saskatchewan, Canada." *Environmental Management* 33 (3): 401–11.

Olagunju, A. 2012. "Selecting valued ecosystem components for cumulative effects in federally assessed road infrastructure projects in Canada." (University of Saskatchewan, MES thesis).

Salmo Consulting Inc. 2004. *Deh Cho Cumulative Effects Study. Phase I: Management Indicators and Thresholds*. Report prepared for the Deh Cho Land Use Planning Committee. Fort Providence, NT: Salmo Consulting Inc.

Scace, R., E. Grifone, and R. Usher. 2002. *Ecotourism in Canada*. Ottawa: Canadian Environmental Advisory Council.

Seitz, N., C. Westbrook, and B.F. Noble. 2011. "Bringing science into river systems cumulative effects assessment practice." *Environmental Impact Assessment Review* 31: 172–9.

Shoemaker, D. 1994. *Cumulative Environmental Assessment*. Waterloo, ON: University of Waterloo, Department of Geography.

SMLP (Saskfor-MacMillan Limited Partnership). 2004. *Twenty-Year Forest Management Plan and Environmental Impact Statement for the Pasquia-Porcupine Forest Management Area*. Assessment main document. Regina: SMLP.

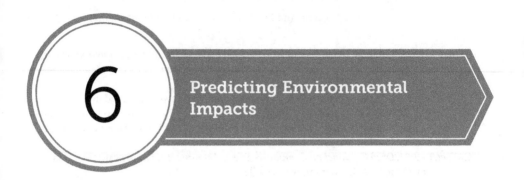

6 Predicting Environmental Impacts

Impact Prediction

Impact assessment requires both forethought and foresight about the potential implications of a proposed development project. A prediction is a statement specifying, without direct measurement, the past, present, or future condition of a particular VEC or VEC indicator, given particular system characteristics (Duinker and Baskerville 1986). Prediction involves identifying potential changes in select *indicators* of environmental receptors (or VECs) identified during EIA scoping and requires:

i)   knowledge of the initial or reference conditions and trends established during the baseline assessment (see Chapter 5);
ii) predicting future conditions, trends, or disturbances in the absence of the proposed development;
iii) predicting future conditions, trends, or disturbances under the project development scenario(s) (Figure 6.1).

Understanding baseline trends is thus important to understanding the nature of a predicted impact. For example, if employment in the area for which the project

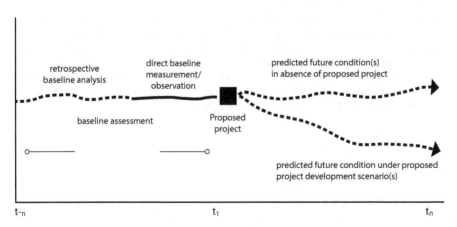

Figure 6.1 ◎ Components of Environmental Impact Prediction

is proposed has been steadily increasing over the previous five years, then this base-line trend should be predicted in the absence of the project in order to allow for a more informed decision concerning the significance of the predicted employment contributions of the proposed project. Similarly, if a project is predicted to have a negative effect on fish populations, then this information should be considered in light of past conditions and predicted trends in fish populations in the area.

Predicting the potential impacts of proposed projects is complex and often a highly uncertain task, since many cause–effect relationships are unknown and physical and human environments are "moving targets." Morris and Therivel (2001) suggest that impact prediction requires several elements, including:

i) sound understanding of the nature of the proposed undertaking;
ii) knowledge of the outcomes of similar projects;
iii) knowledge of past, present, or approved projects whose impacts may interact with the proposed undertaking;
iv) predictions of the project's impacts on other environmental and socio-economic components that may interact with those directly affected by the project;
v) information about environmental and socio-economic receptors and how they have in the past and might in the future respond to change.

Predicting Impacts on the Biophysical Environment

Predicting the effects or impacts of project development on the biophysical environ-ment has been at the heart of EIA since its formal inception under the US NEPA. Most EIAs during the 1970s and well into the 1980s focused almost exclusively on predicting biophysical impacts. Biophysical impact prediction continues to play the dominant role in EIA, but its focus has broadened considerably beyond the direct and immediate project environment to include secondary and ecosystem-based bio-physical change.

What to Predict

Project developments often trigger four broad types of change from baseline condi-tions that affect biophysical systems:

i) biological change
ii) chemical change
iii) physical change
iv) ecological change

Based on these changes, a number of specific effects and impacts often emerge (Table 6.1). There is no single, comprehensive list of biophysical impacts that should be considered in all EIA, since the specific impacts to be predicted depend largely on the nature of the project and the sensitivity of the local environmental setting and are defined by the scoping process.

Several tools are available for predicting environmental impacts. Often the tools are discipline-specific and vary based on the particular issue or problem at hand. Some of the types of tools used for predicting impacts were described in Chapter 3. Numerous tools exist for predicting biophysical impacts, including plume models (Box 6.1), water site budgets, wildlife population modelling, simulation modelling, dose-response factors, hydrographic extrapolation, and expert judgment (Table 6.2). Morris and Therivel (2001) give techniques for predicting biophysical environmental impacts much more comprehensive treatment. A comprehensive discussion of the full range of techniques available for predicting biophysical environmental impacts is beyond the scope of this book

Table 6.1 Selected Biophysical Environmental Impacts: Examples of What to Predict

Air quality impacts	Water impacts
pollutant concentration (various)	surface quantity
pollutant dispersion	surface water withdrawal
emission levels	groundwater quantity
emission types	groundwater withdrawal
temperature	chemical change
changes in wind speed	turbidity
Soil quality impacts	streamflow change
	bank erosion (flood risk)
erosion	biological change (eutrophication, algal
moisture	blooms)
fertility (nutrient change)	pollution discharge rates (assimilative capacity)
chemical change	biological resources (fish and fish habitat)
soil compaction	riparian disturbance
soil pollution	**Coastal zone impacts**
stability	
	water temperature
Terrestrial impacts	flooding
	alterations to tidal activity
fragmentation	sedimentation
wildlife populations	bank or cliff stability
vegetation cover and composition	
airborne and water-borne pollutants	
light pollution	
linear disturbance	
cleared area	

Predicting Impacts on the Human Environment

Predicting the impacts of project development on the human environment is complex, uncertain, and rarely done well in EIA. While attention is often given to impacts over which the proponent has direct control, such as employment or infrastructure development, social impacts, such as quality of life or perceptions of well-being,

Box 6.1 Dispersion Models: An Example of Predicting Plume Rise and Dispersion

Air pollution from industrial emissions can affect both human and physical environmental health. The dispersion of such pollutants is dependent on meteorological conditions (wind speed and patterns, insulation, cloud cover, precipitation patterns) and specific parameters of the pollution stack emissions (temperature, gas molecular structure, gas velocity). Pollutants exit a stack in the form of a "plume," which resembles a moving cloud. The particular shape of the plume depends on local atmospheric conditions, which in turn determine the amount of dispersion. Predicting plume dispersion is critical to determining the ground-level concentration of the pollutants at various distances from the pollution source. Such information will allow project and environmental planners to determine appropriate stack height, severity of ground-level air pollution at various locations, the need for various mitigation measures, and compliance with emissions standards.

Of particular importance is predicting what is commonly referred to as the "worst-case scenario." Several techniques are available to predict plume dispersion and concentration; the most commonly applied is the Gaussian dispersion model. This model assumes that emissions spread outward from the source in an expanding plume according to the prevailing wind direction and that the distribution of pollution concentration decreases with increasing distance from the plume. The Gaussian model is based on a specific mathematical equation that suggests that plume dispersion, pollutant concentration, and pollutant concentrations at the surface are a function of wind speed, wind direction, and atmospheric stability. One particular example of plume analysis for predicting plume height is based on Holland's equation:

continued

$$\Delta h = \frac{V_s D}{u} \left(1.5 + 2.68 \times 10^{-3} \, PD \, \frac{(T_s - T_a)}{T_s} \right)$$

where Δh = plume rise, V_s = stack gas exit velocity, D = stack diameter, u = wind speed, P = atmospheric pressure, T_s = stack gas temperature, and T_a = air temperature. Thus, plume height is determined by Δh plus the stack height.

Many types of Gaussian models are used to predict pollutant dispersion and ground-level concentration, including the Industrial Source Complex model, developed by the US Environmental Protection Agency, and the Atmospheric Dispersion Modelling System, developed by Cambridge Environmental Research Consultants UK.

Sources: Based on Elsom 2001; Harrop and Nixon 1999.

Table 6.2 Sample Tools Used for Prediction in the Biophysical Environment

Air	dispersion modelling
	box models
	air quality indices
	monitoring from analogues
Biological	species population models
	habitat simulation modelling
	ecological risk assessment
	biological assessments
Surface water	waste load allocations
	statistical models
	hydrological models
	water usage and allocation studies
Groundwater	pollution source surveys
	mixing models
	flow and transport models
	soil and groundwater vulnerability indices

Source: Based on Canter 1996.

remain the "orphan" of the EIA process. The role of human impact assessment in EIA is still widely debated; however, as noted at the outset of this book, "environment" is defined here to include both the biophysical and the human environment and the relations and interdependencies between them. Thus, whether human impacts are assessed separately from principal project EIAs or as part of the project EIA process, consideration of them is critical to ensuring sustainability through development.

What to Predict

Project developments often trigger five broad areas of change that affect human systems:

i)   demographic change
ii) cultural change
iii) economic change
iv) health and social change
v)   institutional change

Based on these changes, a number of impacts often emerge that are important to consider in project assessment (Table 6.3). That said, there is no single, comprehensive list of human impacts that should be considered in all EIAs, since the specific impacts to be predicted depend largely on the nature of the project and the local environmental setting and are defined by the scoping process.

As noted above with regard to impacts on the biophysical environment, a comprehensive discussion of the full range of techniques available for predicting human environmental impacts is beyond the scope of this book. There is a long list of techniques for predicting human environmental impacts, including trends extrapolation, population multipliers, intention surveys, gravity models of spatial interaction, gaming, decision trees, traffic modelling, visibility mapping, focus group meetings,

Table 6.3 Selected Human Environmental Impacts: Examples of What to Predict

Direct economic impacts	Housing impacts
local and non-local employment	housing demand
labour supply and training	public and private housing
wage levels	house prices
employment demand by skill group	homelessness and housing problems
	density and crowding

Indirect economic impacts	Local service impacts
retail expenditures	educational services
material and service suppliers	health services
labour market pressures	community services (police, fire)
regional multiplier effects	transportation services and infrastructure
tourism	financial services

Demographic impacts	Socio-cultural impacts
changes in population size	lifestyle changes (family life, seasonality of
changes in population characteristics	employment)
(family size, income, ethnicity)	threats to culture and belief systems
changes in settlement patterns	perceived and actual risks
	social problems (crime rates, substance abuse, divorce)

Health impacts	community stress (conflict, integration, cohesion)
health services availability/access	traditional foodstuffs
disease introduction and transmission	local pride and community perception
physiological impacts (stress, worker satisfaction)	gambling and additions
mental and physical well-being	gender relations

Source: Based on Glasson 2001.

and expert judgment (Table 6.4). Boxes 6.2 and 6.3 provide examples of two different techniques for predicting human environmental impacts. Again, Morris and Therivel (2001) give the subject much more comprehensive treatment.

Table 6.4 Sample Tools Used for Prediction in the Human Environment

Social	demographic models
	participatory mapping
	health-based risk assessment
	intention surveys
Economic	economic multipliers
	total economic productivity models
	input–output analysis
	Monte Carlo analysis
Cultural	traditional knowledge
	participatory mapping
	community dialogues
	analogue techniques

Source: Based on Canter 1996.

Box 6.2 Keynesian Multipliers: An Example of Predicting Economic Impacts

A commonly applied technique for predicting the broader and often indirect economic impact of project development is the **multiplier** technique. A deterministic mathematical model, a multiplier represents a quantitative expression of the extent to which some initial, exogenous change is expected to generate additional effects through linkages within the economic system. The most basic of multipliers and perhaps the most commonly used for predicting local-level impacts is the **Keynesian multiplier**. The basic principle of the Keynesian multiplier is that an injection of money into a local economic system through project development will lead to an increase in the local-level monies by some multiple of the initial injection. For example, $Yr = KrJ$, where "Yr" is the change in the level of income in the project region, "J" is the initial income injection into the local economy as a result of project employment, and "Kr" represents the regional income multiplier.

In principle, the commitment of a proponent to spend locally through local purchase agreements, for example, would provide an initial increase in income to the local labour force. In turn, this would generate extra income for local suppliers, who would spend that extra income in the local region. A multiplier of 1.8, for example, would mean that for each dollar injected directly into the local economy by the project, an extra 80 cents would be generated indirectly (Morris and Therivel 2001).

However, the local project environment is not isolated, and thus economic leakages reduce the multiplier effect. For example, some of the additional income is lost to taxation, individuals may choose to save their extra income rather than spend it, or individuals may choose to spend that extra income in other communities or regions outside the local environment. Thus, a multiplier for predicting local economic impacts of an initial project economic investment might more closely resemble: Yr = [1 / 1 − (1-s)(1-tx-u)(1-m)(1-tl)] x J, where "*s*" is the proportion of additional income generated that is lost to the economy through personal savings, "*tx*" is the proportion of additional income that is lost to taxes on income, "*u*" is attributed to the reduction in government support and transfers as local employment increases, "*m*" is the proportion of additional income spent on goods and services outside the region, and "*tl*" is the proportion of additional income that is lost to taxes on goods and services purchased.

Typically, Keynesian multipliers are limited in EIA to the prediction of indirect economic impacts during peak and trough periods of project construction or operation—that is, best-case and worst-case scenarios.

Box 6.3 Decibel Ratings: An Example of Predicting Cumulative Noise Impacts

Noise is often an impact generated by large-scale projects such as airports, highways, railways, or heavy industries. The intensity of sound, while often expressed as "loudness," is actually measured in decibels (dB). The range of audible sounds is typically from 0 dB to 140 dB. However, because sound level is logarithmic in nature, adding two sounds or doubling the intensity of a sound does not result in a doubling of the decibel rating. For example, adding two identical sounds (e.g., two low-level aircraft passing overhead at once) does not double the decibel rating but increases it by only 3 dB. When two sources of sound that are at different decibel ratings are added together (e.g., sound source 1 = 59 dB; sound source 2 = 53 dB), the graph below can be used to predict the cumulative noise rating (e.g., 59 dB − 53 dB = 6 dB difference; therefore, according to the graph, add 1 dB to the higher level sound = total 60 dB). To add three sounds, each of 50 dB, we add 50 dB + 50 dB = 53 dB, then 50 dB + 53 dB = 54.75 dB; the difference between 50 dB and 53 dB is 3 dB, so according to the graph we add 1.75 dB to the higher rating.

This is a rather simplistic approach to predicting noise impacts. In practice, several additional elements have to be factored into the equation, such

Sound level (dB)	Example	Sound level (dB)	Example
140	pain threshold for human hearing	70	busy street corner
120	pneumatic drill	60	busy office place
110	loud car horn at 1 metre	30	bedroom at night
105	jet flying overhead at 250 metres	10	normal breathing
90	inside a subway car	0	threshold of hearing

continued

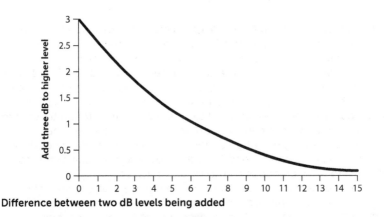

Difference between two dB levels being added

as the level of other sounds in the environment, meteorological effects such as average wind direction and speed, local topography, other physical obstructions between the noise source and the noise receptor, and the sensitivity of the noise receptor. In addition, the impacts of noise are not limited to the human environment or to human assessments of loudness; impacts on wildlife are also important considerations, which may result in disruptions to breeding areas or migration patterns.

Sources: Based on Harrop and Nixon 1999; Morris and Therivel 2001.

Characterizing Predicted Impacts

Impact prediction should provide insight into the specific characteristics of potential impacts. Effects and impacts can be classified in a number of ways. In the Mackenzie Gas Project in Canada's Northwest Territories, for example, predicted environmental effects and impacts in the biophysical environment, specifically fish and fish habitat, were classified based on their direction, magnitude, spatial extent, and duration (Box 6.4). In short, there is no single best set of criteria that can be used to classify all predicted effects and impacts in every project situation, but there are several common classifications of effects and impacts that should be examined for each environmental component affected by a project (Table 6.5). The following is by no means a comprehensive list; rather, it serves to illustrate a variety of classification approaches to predicted effects and impacts.

Order of Environmental Effects

Effects resulting from a development project can be either **direct effects** (first-order) or **secondary effects** (resulting from the direct effects). For example, the flooding of land during construction of a hydroelectric project has a direct effect on terrestrial habitat. Indirectly, human health may be affected as a result of exposure to increased mercury levels in fish caused by mercury release and the flooding of soils and decay of organic matter. The relationship between first-order effects (or impacts)

Box 6.4 Effects of Natural Gas–Gathering Pipelines and Facilities on Fish and Fish Habitat for the Mackenzie Gas Project, Northwest Territories

| Key indicator | Phase when impact occurs | Effect attribute | | | | |
		Direction	Magnitude	Geographic extent	Duration	Significant
Habitat	Construction	Adverse	Low	Local	Short term	No
	Operations	Adverse	Low	Regional	Long term	No
	Decommis-sioning/ abandonment	Adverse	Low	Local	Long term	No
Health	Construction	Adverse	Low	Local	Long term	No
	Operations	Neutral	No effect	N/A	N/A	No
	Decommis-sioning/ abandonment	Neutral	No effect	N/A	N/A	No
Distribution and abundance	Construction	Adverse	Low	Local	Long term	No
	Operations	Neutral	No effect	N/A	N/A	No
	Decommis-sioning/ abandonment	Neutral	No effect	N/A	N/A	No

Effect attributes:

Direction

Adverse: Impact will cause an adverse change in a measurable parameter relative to baseline conditions or trends.

Neutral: Impact will cause no change in a measurable parameter relative to baseline conditions or trends.

Positive: Impact will cause a positive change in a measurable parameter relative to baseline conditions or trends.

Magnitude

No effect: No change is foreseen in the valued component.

Low: An individual or group within a population found in a localized area, such as the local or regional study area, might be affected.

Moderate: Part of a regional population within the local or regional study area might be affected, changing the abundance or distribution of the valued component and affecting opportunities for hunting, trapping, or viewing wildlife as currently practised.

continued

High: An entire population within the local or regional study area might be affected, changing the abundance or distribution to such an extent that the population would not likely return to its previous level, resulting in reduced population viability and unsustainable harvest compared with current practice.

Geographic extent

Local: Terrestrial—the effect on the valued component is measurable within the local study area; Marine—the effect will be limited to within about 10 kilometres of the proposed activity.

Regional: Terrestrial—the effect on the valued component is measurable within the regional study area; Marine—the effect will extend beyond 10 kilometres of the proposed activity to the Canadian Beaufort Sea region.

Beyond regional: Terrestrial—the effect on the valued component is measurable beyond the regional study area.

Duration

Short term: Effect is limited to less than 1 year.

Medium term: Effect lasts for more than 1 year but less than 4 years.

Long term: Effect lasts longer than 4 years, but the valued component will recover not more than 30 years after project decommissioning.

Far future: Effect extends more than 30 years after decommissioning.

Source: Mackenzie Gas Project Environmental Impact Statement, 2004, vol. 5, section 7, 7–183 to 10–27.

and second-order effects (or impacts) is not always straightforward. For example, runoff from an agricultural operation may cause enrichment of a local freshwater body by depositing excess nutrients. The result may be excess plant growth near the shores of the water body, which may create an aesthetically pleasing waterscape. However, the subsequent eutrophication—excess plant decomposition absorbing dissolved oxygen at high rates—generates a negative secondary effect in making water unsuitable for aquatic life. Preston and Bedford (1988) suggest that an "effect" is a scientific assessment of facts and an "impact" denotes a value judgment or the relative importance of the effect (Box 6.5).

Nature of Environmental Effects and Impacts

Incremental effects or impacts are marginal changes in environmental conditions that are directly attributable to the action being assessed. For example, a hydroelectric dam may lead to a slight increase in heavy metal concentrations in the river each year. For incremental effects, it is important to consider the "rate of change" in the affected environmental parameter and to regularly assess total change against determined thresholds, standards, or specified targets.

Table 6.5 Generic Classification System for Environmental Effects and Impacts

Order of effect or impact	Nature of effect or impact	Temporal characteristics	Magnitude	Direction of change in the affected environmental parameter	Spatial extent	Reversibility of change of the affected environmental parameter	Probability of occurrence
a) Direct	a) Incremental	a) Duration	a) Size	a) Increasing	a) On-site	a) Reversible	a) Likelihood
b) Indirect	b) Additive	b) Continuity	b) Degree	b) Decreasing	b) Off-site	b) Irreversible	b) Risk
	c) Synergistic	c) Immediacy	c) Concentration	c) Positive	d) Regional		
	d) Antagonistic	d) Frequency		d) Negative			
		e) Regularity					

Box 6.5 Actions, Causal Factors, Effects, and Impacts

Action	Hydroelectric dam construction on a river system. Assume a simple development project in which a river is modified and dammed for hydroelectric power generation.
Causal Factor	Dredging. The project involves dredging and widening the river channel upstream of the dam site prior to its construction.
Condition Change	Increased sediment flow into the river downstream of the dredging site creates changes in deposition of sediments and increased turbidity.
Effects and Impacts	Direct: Decreased growth rates in fish
	Indirect: Decline in recreational fishery
	Several different impacts may occur as a result of the initial change in condition. For example, deposition of sediments and increased turbidity (project effects) may result in decreased growth rates in river fish (first-order, direct impact), leading to a decline in the region's recreational river fishery (second-order, indirect impact).

Additive effects or impacts are the consequence of separate or related actions that may be minor individually but together can create a significant overall impact (Figure 6.2). For example, chemical "A" may result in 10 per cent mortality in fish over a 30-day exposure period. Chemical "B" may result in the same mortality over the same time period. If chemical "A" and chemical "B" were released together, the additive mortality in fish would be 20 per cent. Additive effects are not only the result of progressive increases in environmental parameters but also include, for example, the actions of several large-scale disturbances, such as multiple logging operations, within a single ecosystem. Thus, while perhaps individually manageable, together such operations present significant adverse additive effects on wildlife because of habitat fragmentation. Additive effects may therefore result from two types of actions:

i)   Where two or more of the same type of actions are affecting the same environmental component. For example, multiple coal-bed methane developments may generate cumulative effects on the landscape as a result of saline water discharge.

ii)   Where a single development action produces multiple condition changes or effects that result in effects on the same environmental component. For example, the development of a mine site may create habitat loss, changes in surface water drainage and quality, and increased noise and emissions from vehicles and operations. In combination, they can have a cumulative negative effect on wildlife.

Synergistic effects or impacts are the result of interactions between effects and occur when the total effect is greater than the sum of the individual effects (Figure 6.3). For example, a single industry located adjacent to a river system may alter water temperatures, change dissolved oxygen levels, and introduce heavy metals. Individually, each effect may be tolerable for fish, but the toxicity of certain heavy metals is multiplied in high water temperatures and low dissolved oxygen content. The effect on fish as a result of the interaction of these effects is thus greater than the sum of the individual project-induced changes.

Antagonistic effects or impacts occur in certain situations in which one adverse effect or impact may partially cancel out, offset, or interrupt another adverse effect or impact. For example, when chemical "A" with a 10 per cent mortality rate in fish is combined with chemical "B," which has the same mortality rate, the antagonistic result is less than a 20 per cent mortality rate. These effects are usually less common in EIA, since additional stressors to the environment more often create further disturbance and degradation. One example, however, is the reduced eutrophication of a water body receiving effluents containing both chlorine and phosphates. While each substance is individually harmful for aquatic life, together and in moderate amounts they may be beneficial for managing eutrophication.

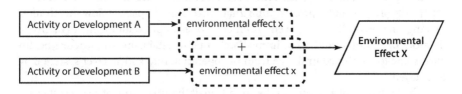

Figure 6.2 ◉ Additive Environmental Effects

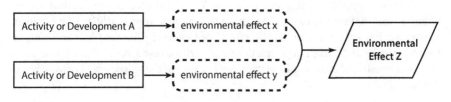

Figure 6.3 ◉ Synergistic Environmental Effects

In terms of the biophysical environment, antagonistic effects can be further defined according to functional, chemical, dispositional, and receptor antagonism. **Functional antagonism** refers to one effect counterbalancing the physiological response of a second effect on the same receptor. **Chemical antagonism**, on the other hand, refers to a reaction between the two effects, such as in a chemical reaction, where the severity of the combined effect is reduced. For example, the total effects of arsenic and mercury are reduced when combined with dimercaprol, which itself is a toxic compound. In the case of **dispositional antagonism**, one effect influences the uptake or transport of the other, such as ethanol enhancing mercury elimination in mammals. **Receptor antagonism** refers to one effect blocking the other, such as a toxicant binding to a receptor and blocking the effects of a second toxicant.

Temporal Characteristics of Effects and Impacts

Since effects and impacts often occur over time, their characteristics may vary. But we can characterize the temporal nature of any effect or impact according to the following factors.

Duration is the length of time that the effect or impact occurs—for example, short-term effects such as the noise associated with a bridge construction site or the long-term effects associated with riverbank erosion and downstream loss of arable land.

Continuity relates to whether or not there are **continuous effects** or impacts, such as energy fields associated with transmission lines. Other effects may be discontinuous and last only for short intervals, such as the noise from blasting at a construction site.

In regard to *immediacy*, different effects and impacts arise at different times during the life of a particular project. Immediate effects and impacts follow shortly after a change in the condition of the environment. For example, odour from an intensive livestock operation would occur immediately after its establishment, whereas health effects due to continued drainage from the operation into a local river system may be delayed.

The *frequency* of a change in the condition of the environment may create very different impacts. For example, local neighbourhood residents may be willing to endure short-term noise from local street maintenance activities, but they may not be willing to tolerate the construction and operation of a new airport near their homes.

Effects or impacts that occur with *regularity* and greater predictability may often be dealt with more easily than those that happen irregularly or that come as a surprise. For example, the "startle effect" of low-level military flight training in Labrador on local Aboriginal populations and caribou resources is of particular concern, not because of the noise level itself but because of the startle effect of low-flying military jets.

Magnitude, Direction

The **magnitude** of an effect refers to its size or degree—for example, the specific concentration of an emitted pollutant. That said, while many impacts can be measured in quantitative terms such as concentration or volume, others, such as the aesthetic impact on a natural area, are subjective matters and not easily indexed or reduced to absolute quantitative terms. It is important to note that the magnitude of an effect is not necessarily related to the significance or importance of the impact. For example, removal of 5 per cent of the forest cover in an area that supports a rare or endangered species may be considered much more significant than removal of 25 per cent of forest cover in an area that supports stable native species populations.

The *direction* of change associated with an effect is closely related to its magnitude. Direction of change refers to whether the affected environmental parameter is increasing or decreasing in magnitude relative to its current or previous state or **baseline condition**. For example, the construction and operation of a large-scale industry may lead to an increase in the local population if it generates employment opportunities and triggers in-migration; however, it may lead to a decrease in local population if smaller industries are displaced or if quality of life in the community decreases as result of environmental change. In the first instance, the impact may be considered positive, whereas in the second, an **adverse effect** is the more likely result. The situation is not always this straightforward. In some cases, effects are multi-directional. For example, some individuals in the affected community may see an increase in local population as a positive impact and an improvement in their quality of life, whereas others may see population increase due to industrial growth as having an adverse impact on their quality of life. When characterizing environmental change, it is important to give at least some indication of direction, even if it cannot be quantified, in order to better understand the magnitude of the effect and impact.

Spatial Extent

The **spatial extent** of an environmental effect or impact will vary depending on the specific parameters involved and the project's environmental setting. For example, soil erosion may be highly localized and agricultural drought may be regional, while economic impacts associated with continuous drought may be national or international in extent. In the case of the Jack Pine mine project, a proposal by Shell Canada to mine bitumen deposits in the Athabasca oil sands of Alberta, the spatial extent of the assessment was defined by "local study areas" and "regional study areas." Local study areas were identified as those directly affected by project development. Regional study areas were identified as those seen from a larger geographic and ecological perspective that would experience direct and indirect impacts.

Sometimes spatial extent is classified according to **on-site** and **off-site impacts**, a recognition that the impacts of any particular development or action may extend well beyond the specific project site. On-site actions may trigger environmental change, which in turn affect off-site environmental components. For example, chemical discharges at an industrial complex generate on-site environmental impacts due to soil contamination, which in turn may create off-site impacts through groundwater contamination. Such off-site impacts may go beyond biophysical impacts alone. **Fly-in fly-out** employment arrangements, often associated with remote mining projects, have impacts that extend well beyond the mine project site to the communities the workers come from and affect personal matters such as family life and relationships. Determining the spatial extent of EIA is a key element of the scoping process, discussed in Chapter 5, and is critical to ensuring that all necessary impacts are considered within the scope of the assessment.

Box 6.6 Impacts of Fly-in Fly-out Mining Operations

Fly-in fly-out is now a common practice in mining operations globally, particularly in Australia and Canada. Storey (2010) defines fly-in fly-out as long-distance commuting work practices whereby workers travel by air or other mode of transport to and from worksites that are typically in remote areas and often at a distance from existing communities. Storey explains that fly-in fly-out was encouraged by the expansion of mining into increasingly remote areas at a time when corporate interests were focusing on leaner and more flexible modes of production and when governments were unwilling to support the development of new single-industry communities in remote areas.

There are several fly-in fly-out mining operations in northern Canada including:

- The Diavik diamond mine in the Northwest Territories, located 300 km north of Yellowknife
- Cameco's Key Lake Uranium Mine in Saskatchewan, located 570 km north of Saskatoon by air and 220 km by road from the nearest village
- CanTung, owned by the North American Tungsten Corporation, located about 300 km northeast of Watson Lake in Yukon
- The Meadowbank gold mine in Nunavut, located 300 km west of Hudson Bay and 70 km north of Baker Lake, the nearest town

Some of the characteristics of fly-in fly-out work practices in the mining industry include working in relatively remote locations where the mining company typically provides accommodations for the workers at or near the mine site and a work roster based on a fixed number of days at the worksite and a fixed number of days at home, such as a two-week work rotation. Although fly-in fly-out work arrangements ensure affordable accommodation and the necessary social and infrastructure services for workforces in remote areas, there are also a number of potentially adverse impacts, both on-site (in the host community or at the work camp) and off-site (in the worker's home community and household). For example, resident communities housing a fly-in fly-out workforce can experience additional demand on local social infrastructure and services. There is the potential for adverse impacts on the workers themselves, including fatigue and other health and safety risks associated with long-distance commuting to and from often remote areas. The worker's household may also experience potentially adverse impacts, including family disruption and reduced socialization with family and friends in the home community. If workers are Aboriginal and still engaged in traditional hunting and fishing activities, there is the potential for cultural disruption.

Such impacts are difficult to predict, and methods for doing so often tend to draw on experiences from other, similar projects and work arrangements. But these off-site and on-site impacts associated with fly-in fly-out work arrangements need to be identified and effectively managed as part of the impact prediction and mitigation strategies for remote mining operations.

Source: Storey 2010; Morris 2012

Reversibility and Irreversibility

An additional characteristic to consider when examining environmental effects and impacts is the degree of reversibility or irreversibility. This is particularly important for determining the nature and effectiveness of impact management strategies. An effect or impact is considered a **reversible impact** when it is possible to approximate the pre-disturbed condition. For example, in the case of a sanitary landfill operation, the surface can be remediated to resemble its pre-project condition, or traffic congestion may return to its original level when a bridge reconstruction project ends. Reversing an environmental effect, however, does not necessarily mean restoring a biophysical or socio-economic environment to its *exact initial* condition, since that is neither always possible nor always desirable. A mine site, for example, may be remediated and reforested based on current forest stand composition in the project area because it may neither be desirable nor feasible to return the site to its pre-project forest stand composition. The environment is a moving target; it changes and adapts irrespective of project actions. Restoring disturbed environments to their pre-disturbed condition is at best an approximation of what the affected environment might look like had the development action not taken place.

Some effects and impacts are completely **irreversible impacts** and are thus of considerable importance when determining significance. For example, the extinction of a rare plant species during site-clearing is irreversible. Other types of impact may be reversible from a technical standpoint, but it may be impractical or economically unfeasible to attempt to do so. For example, it is common practice in large-scale mining initiatives to drain a local pond or lake for tailings disposal. While it is *possible* to restore the pond or lake following mine decommissioning, restoration is often neither practical nor economically feasible. Consequently, these effects are often considered irreversible from any practical impact management perspective. Instead, efforts are made elsewhere to compensate for the loss of aquatic habitat, either through the creation of new habitat or the restoration of damaged habitat. Although these effects are considered irreversible, that does not mean that there is no need for managing such impacts or for site cleanup. The failure to perform site restoration and cleanup following project operations has been particularly problematic in Canada's mining sector (Box 6.7).

Likelihood of Effects and Impacts

Whether an environmental effect or impact is considered important during the course of an EIA rests considerably on the likelihood of it occurring. If, for example, a particular impact is almost certain to occur, then it is more likely to be considered in the decision-making process than one that is determined to be highly improbable. The likelihood or probability of an effect or impact occurring is one component used to measure **risk**—an uncertain situation involving the possibility of an undesired outcome. Risk combines the likelihood of an adverse event occurring with an analysis of the severity of the consequences associated with that event. In other words, risk can be characterized as a function of damage potential, exposure to dangerous

Box 6.7 Reclaiming Abandoned Mine Sites in Northern Canada

The Lorado uranium mine/mill site, located approximately eight kilometres south of Uranium City in northern Saskatchewan, ceased operation in 1960. The mine site was owned by what is now EnCana West Ltd, a Calgary-based company and formerly AEC West Ltd, renamed after the 2002 merger of PanCanadian Energy Corporation and Alberta Energy Company Limited. The mine pre-dated any formal requirements for EIA or post-closure site reclamation, and, as a result, left behind are exposed tailings—waste rock left over after the ore is extracted—containing metals, radioactive elements, and generate acid. The Lorado mine is just one of 65 abandoned mine sites in northern Saskatchewan for which no corporation or government agency is clearly responsible for site restoration and cleanup.

Saskatchewan's Lorado mine site is not a unique case; in fact, thousands of abandoned mine sites in northern Canada have not been reclaimed and continue to pose serious threats to the environment and to human health and safety. In the past, little could be done to ensure that mine sites were reclaimed or to prevent bankrupt companies from "walking away" from their operations. As a result, the federal government currently spends millions of dollars each year to contain pollutants left behind at abandoned mine sites. In 2002, for example, the prevention of water contamination from abandoned mine sites cost Canadian taxpayers an estimated $25 million.

While the burden for cleanup and restoration of currently abandoned mine sites rests with the various federal and provincial governments, longer-term and viable solutions for environmental protection in Canada's mining industry now exist. Under the Canadian EIA system and most provincial systems, mining companies are responsible for preparing mine closure plans for approval for both new and active mines and, in addition, must provide post-closure financial assurance for environmental rehabilitation or stabilization. For mining companies operating in Canada's North, a financial security deposit is required at the time of project start-up and during operation. Should a mining company declare bankruptcy, Aboriginal Affairs and Northern Development Canada, the department responsible for land administration in northern Canada on behalf of the federal government, will use the security deposit to cover the eventual costs of repair, maintenance, cleanup, and closure of the mine site. If a mining company conducts the proper cleanup and site reclamation, the financial security deposit is returned.

substances such as toxic chemicals, opportunity for exposure, and the characteristics of the population at risk. For example, the elderly and young children may be much more vulnerable to exposure than other segments of the population. **Risk assessment**, then, refers to the process of accumulating information, identifying possible risks, risk outcomes, and the significance of those outcomes, and assessing the likelihood and timing of their occurrence. The acceptability of a specified level of risk depends on several factors, including the catastrophic potential associated with the risk event, scientific uncertainty, distribution of risk outcomes, and understanding of and familiarity with the risk. In terms of the latter, for example, an individual living on a flood plain may have perceptions of flood risk significantly different from those of individuals living at a distance from the same flood plain.

Good Practice Requirements for Impact Prediction

What makes a "good" impact prediction? There is no set of laws or regulations that prescribe *how* impact predictions must be made or that can guarantee *quality* impact predictions. In many cases, the quality of an impact prediction depends on the skill and experience of the practitioner, the nature and availability of baseline data, quality of the predictive tool or model used, and the degree of complexity or uncertainty in the environmental or socio-economic system of concern. There are, however, a number of "good practice" requirements for impact prediction that apply to all projects and contexts.

Accuracy and Precision

Although prediction is the cornerstone of EIA, relatively little attention is given to the manner in which predictions about anticipated effects are phrased. Impact predictions such as "slight reduction" or "minor effect" are of little value for under-standing, monitoring, and managing the actual effects of project development if a "slight reduction" or "minor effect" is neither quantified nor qualified. In other words, it is possible to generate quite accurate predictions when such predictions are couched in vague language. For example, a prediction stating "there will be a slight increase in methyl mercury concentrations in the reservoir due to flooding" is difficult to prove "wrong"—"slight increase" is a vague concept, the increase is not quantified relative to a baseline condition, and no time frame for the impact is identified. Such predictions are of limited value in terms of avoiding or managing actual project effects.

    Accuracy in impact predictions refers to the extent of system-wide bias in impact predictions, or the closeness of a predicted value to its "true" value. **Precision** refers to the level of preciseness or exactness associated with an impact prediction (Figure 6.4). Accuracy, then, depends on the level of precision required. One can be con-sistently accurate when predictions are imprecise—for example, a "slight increase in methyl mercury concentrations." As the desired level of precision increases, then the possibility of being wrong in impact predictions also increases—for example, "methyl mercury concentrations will increase by 7.6 per cent above the current baseline level 14 days after reservoir flooding." The fact that an impact prediction is accurate does not necessarily constitute a useful impact prediction if the level of precision is relaxed. However, if more precise impact predictions are desired, then accuracy may be forfeited (De Jongh 1988).

    There are two dimensions to predictive accuracy: (1) the logical validity of the impact prediction; and (2) the relative severity of the actual versus predicted impact.

Buckley (1991) explains that in cases where actual impacts prove less severe than anticipated, predictions may still be logically incorrect if they were expressed as "impact parameter will be equal to predicted numerical value" rather than "impact parameter will be less than or equal to predicted numerical value." However, whether an impact prediction is logically correct is perhaps less meaningful than the accuracy of the relative severity of actual versus predicted impacts—that is, impacts less or more severe than initially predicted.

Good impact predictions are ones that can be validated. But in an international study of the effectiveness of environmental assessment, Sadler (1996) found that 60 per cent of the time, assessments are unsuccessful to only marginally successful in making precise, verifiable impact predictions and that 75 per cent of the time, assessments fail to indicate confidence levels for the data used for impact predictions. Little seems to have changed since Bisset and Tomlinson (1988) found that 697 of a total 791 predicted impacts across a survey of four UK impact assessments could not be confirmed, partly because of the vagueness of impact predictions. Similarly, Bailey, Hobbs, and Saunders (1992), in an environmental audit of waterway developments in Western Australia, found that 90 per cent of impact predictions did not indicate a specific time scale in which the predicted impact was expected to occur. This was

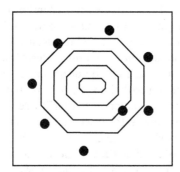

Inaccurate and imprecise

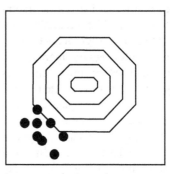

Inaccurate but precise

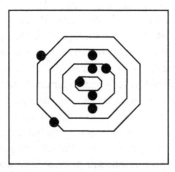

Accurate but imprecise

Accurate and precise

Figure 6.4 ◉ Accuracy and Precision

confirmed by Morrison-Saunders and Bailey (1999), who analyzed six Australian cases and found little evidence of impact quantification or precision in prediction, with most predictions being vague and qualitative in nature.

Verifiable Impact Predictions

Closely related to the above, impact predictions must also be stated in such a way that they can be followed up and verified. This can be accomplished by one of two methods depending on the nature and availability of baseline data.

Hypotheses-Based Approaches
When impacts are predicted in environmental assessment, they should, where at all possible, be based on rigorous and falsifiable null-impact hypotheses stating the relevant affected variables, impact magnitude, spatial and temporal extent, probability of occurrence, significance, and associated confidence intervals. Ringold et al. (1996) suggest that impact predictions should be couched in quantitative terms, should identify the spatial and temporal characteristics of interest, and should state the Type I and Type II probabilities. This is consistent with recommendations of the panel reviewing the Voisey's Bay nickel mine project proposal for northern Labrador, which suggested that a hypothesis-driven approach is the preferred formula for impact prediction (Voisey's Bay Panel 1999). Such an approach to impact prediction allows practitioners to measure variables and either accept or reject impact predictions and environmental change within specified confidence limits. However, it is important to note that a hypothesis-driven approach to impact prediction does not supplant the role of professional judgment when considering what constitutes a significant or acceptable deviation from the null and that it inevitably relies on the availability and quality of data.

Threshold-Based Predictions
Impact predictions often turn out to be inaccurate because of the mix of assumptions that normally have to be made and the multiplicity of exogenous factors involved (Mitchell 2002, 56). Projects are almost always modified during the development phase, thereby making initial impact predictions less valuable from a follow-up and learning perspective, and given the nature of constantly changing and often unpredictable human environmental systems, the environmental impacts of development can rarely be predicted with any degree of certainty.

However, not all environmental effects need to be predicted precisely; in many cases, particularly when baseline data are inadequate, it is recognized in the assessment that outcomes should not exceed specified threshold levels, in which case management practices, the setting of targets, and the determination of threshold levels become the focus of attention (Storey and Noble 2004). It is often the case, for example, that a threshold or particular target is set and a monitoring framework is established to ensure that negative impacts do not exceed certain thresholds and that positive impacts meet specified expectations. Where hypothesis-based approaches are not suitable, **threshold-based predictions** may rely on previous experience with similar projects, similar types of impacts in different environments, or regulatory standards. For example, it is known that certain levels of noise can have a negative effect on human hearing and that certain project construction activities (such as heavy machinery or blasting) generate particular levels of noise. One approach would be to compare the level of cumulative noise to specified thresholds for human hearing in order to predict the impact of project construction activities.

When comparative examples or regulatory standards are not readily available, an alternative approach is to base impact predictions on a desirable level or maximum level of change. **Maximum allowable effects levels** (MAEL), for example, reflect an approach to impact prediction in which the impact is stated in the form of a hypothesis such as: "project impact *i* will not exceed a particular threshold or desired effects level for a particular impact indicator *j*." This approach was used in the Hibernia offshore oil project for predicting potential project impacts on the biophysical environment, and it is particularly useful for managing project outcomes and meeting desired sustainability objectives. It is also a useful approach for socio-economic effects, particularly for those phenomena that depend on individual and community behaviours and are almost impossible to predict. For example, in the case of a small community subject to temporary worker influx due to the construction of a hydroelectric project, predicting additional demand on local health care services is a highly uncertain task. It depends, in part, on worker health and safety policies and practices, the health of the incoming workforce itself, and the social behaviours of the workforce in the community—such as alcohol or drug consumption. One approach to impact prediction could involve setting an acceptable population-to-physician ratio based on the policy of the regional health authority or on what the affected community deems an acceptable ratio in comparison to current baseline conditions. An MAEL prediction is then made that the population-to-physician ratio will not be exceeded as a result of worker influx to the local community. As a result, the focus shifts from prediction per se to monitoring to ensure that adverse effects beyond the MAEL do not occur (see Environmental Assessment in Action: Health Impact Assessment and the Lower Churchill Hydroelectric Generating Project).

Environmental Assessment in Action

Health Impact Assessment and the Lower Churchill Hydroelectric Generating Project

The inclusion of health impacts in project assessment is receiving increased attention from EIA and health practitioners alike, and many health authorities, including the World Health Organization and Health Canada, have recognized the need for and benefits of addressing health in EIA. At the same time, health and other human impacts on community well-being have not been given adequate treatment in project EIA. Part of the challenge is the complexity of pathways that link project development, environmental change, and human health and well-being. Recent literature on health impact assessment points to the need to focus attention not on direct cause–effect relationships for human health impact prediction but instead on the linkages between project actions and the "determinants" of health and well-being. Determinants of health are not themselves "health impacts"; rather, they are factors that influence or provide an indication of health and well-being—factors such as income, physical environments, health services, and social support networks. Shifting attention away from predicting impacts on human health and focusing on health determinants should simplify the complexity of pathways that connect project actions to human health impacts. Several countries, including Canada (Health Canada), have recently adopted health determinant frameworks for

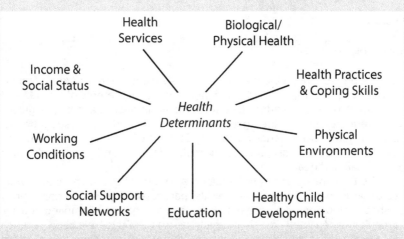

EIA to provide practitioners with advice on addressing the health implications of project development.

In 2006, Nalcor Energy, a provincial Crown corporation of Newfoundland and Labrador, filed an application for the development of the Lower Churchill Hydroelectric Generating Project on the lower Churchill River, central Labrador. The project would consist of the development of two hydroelectric facilities totalling 3074 megawatts and associated dams and reservoirs; two interconnecting transmission lines between the two facilities and the existing grid totalling 263 kilometres; and the construction of work camps to accommodate the anticipated 2000-member workforce during the peak of construction activity. The project location in the Upper Lake Melville region of central Labrador is relatively remote. There is one larger centre in the project's primary impact area, Happy Valley–Goose Bay, with a population of approximately 7500, and three smaller centres: Mud Lake, North West River, and Sheshatshiu—the largest Innu community in Labrador, with an estimated population of about 1200 people.

continued

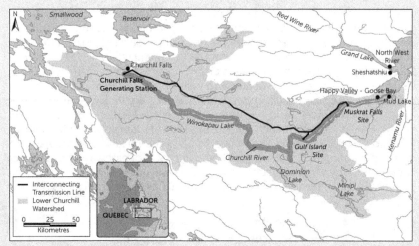

Source: Nalcor Energy 2009.

The project was subject to a joint review panel under the environmental assessment legislation and guidelines of the governments of Canada and Newfoundland and Labrador. The environmental impact statement (EIS) guidelines were prepared by the governments in 2008, and Nalcor submitted its EIS to the joint review panel in February 2009. A community health baseline study and impact assessment were completed for the project EIS, based on a determinants of health framework.

The baseline assessment indicated a history of social and community health problems in Upper Lake Melville, particularly in the Innu community of Sheshatshiu, including solvent abuse and alcoholism. Gambling activity, particularly video lottery terminal addictions, was found to have become much more prevalent in recent years, with the number of gambling addictions in the region reported to be significantly higher than the provincial average. The baseline study also identified women in the Upper Lake Melville area as having children at considerably younger ages than women in the province as a whole, with births to teenage mothers most pronounced in Sheshatshiu. The baseline study also reported an increased demand for social and mental health services and treatment facilities in the region, with current services already at or beyond capacity. Primary health care delivery in the Upper Lake Melville area, for example, as measured by the family-physician-to-population ratio, was found to be substantially lower than for the province as a whole.

Some of the primary issues of concern in the community health impact assessment were thus issues related to the potential for increased alcohol and drug use and the implications for criminal activity, the health of the individual and the family, and the ability to provide social and health services. The challenge for the proponent was that many of these were ongoing issues of concern with or without the proposed project and many of the health and social services in the region were already either at or beyond capacity. Most impacts to community health as a result of the project would be experienced indirectly through demographic change, increased disposable income, and any in-migration to and worker–community interactions within the Upper Lake Melville area.

As with most assessments, the first source of information for predicting impacts was to draw on experiences from elsewhere, in this case similar hydroelectric projects in Quebec, Manitoba, and British Columbia. However, results from these projects showed a complex mix of both positive and potentially adverse impacts. Further, lessons from the socio-economic environments of those projects were not easily transferable to the Innu health, cultural, and historical context of the Lower Churchill project. The proponent recognized early on that not all things can be predicted. For example, most adverse effects on mental health, addictions, and counselling services would likely be associated with an increase in disposable income as a result of project employment and associated local spending patterns. An additional concern raised by the Innu community was the potential for an increased incidence of teenage pregnancies as a result of worker interaction with the community. Many such impacts depend on individual and community behaviour as well as on personal and community coping skills.

The proponent stated in its EIS that many such impacts cannot be predicted in any quantifiable way and any predictions that could be made would be subject to considerable uncertainty. The view was that it was better to plan to adapt than to make inaccurate impact predictions or ones that simply could not be verified. Given a baseline assessment that indicated that any increase in demand for community health services associated with the project was likely to be beyond the current system capacity, the proponent focused instead on what might be deemed an acceptable level of impact, resembling threshold or maximum allowable effects levels (discussed earlier in this chapter). Emphasis was placed on co-ordinated monitoring programs between the proponent and the regional health authority and impact mitigation to identify and avoid potentially adverse impacts on community health in the event they should occur.

The project was approved in 2012 by the governments of Canada and Newfoundland and Labrador.

Source: Aura Environmental Research and Consulting Ltd 2008; Nalcor Energy 2009.

Addressing Uncertainty

Uncertainty is simply the difference between the "known" and the complete "unknown." How certain are we of a predicted impact? Uncertainty is inevitable in impact predictions, since no model or predictive technique can accurately predict what might happen in a dynamic environment. Practitioners of EIA are expected to predict the impacts of proposed projects and communicate their findings in a way that can be used by decision-makers to make sound environmental decisions. Notwithstanding consistent messages that impact predictions often prove to be incorrect and mitigation less successful than anticipated, a degree of confidence is presented in EIA and decision-making that simply is not warranted. Since EIA *appears* to be founded on certainties, the challenge is where and how to accommodate uncertainties.

A survey of EIA reports in the UK by Wood, Dipper, and Jones (2000) found impact predictions on ecosystems to be accurate in only 50 per cent of the cases examined. Little has changed. Tennoy, Kvaerner, and Gjerstad (2006) reviewed 22 EIAs in Norway and found that uncertainty disclosure diminished over the course of the EIA process—from scoping to final approval. Wood (2008) reviewed 30 EIAs in the UK and found that only 23 per cent considered the uncertainty associated with prediction methods; precautionary-based approaches were used in less than one-third of all cases, and limited consideration was given to uncertainties about the effectiveness of mitigation measures. Wood concluded that EIA can impart a greater sense of certainty than is genuinely warranted, including uncertainty related to the confidence in predictions.

The EIA process demands that impacts are predicted, even though it is understood that the future is uncertain. Omitting or underestimating uncertainties can result in a systematic bias in the decisions taken and overconfidence in the assessment process, including the appropriateness of planned mitigation, almost certainly resulting in compromised protection of the environment. There are, however, four principal measures that can be taken to address uncertainty in impact prediction. The first three measures assume that the level of uncertainty can be assessed and quantified. The fourth measure accounts for the reality that more often than not, uncertainty cannot not be expressed as a probability or explained by the tools and techniques used to predict impacts. In many cases, there is inherent uncertainty in environmental and socio-economic systems.

1. **Probability analysis** involves quantifying the probability that an impact or range of impacts is likely to occur and under what conditions, as well as the confidence in the prediction. Munro, Bryant, and Matte-Baker (1986), for example, identify a range of probability and confidence intervals associated with impact predictions (Table 6.6). The lower the associated confidence level, the greater the likelihood that effects monitoring will become important as the project proceeds.

Table 6.6 Impact Predictions and Confidence Limits

Confidence levels	Data characteristics	Knowledge	Permitted approach
high/factual (95%)	reliable	proven cause–effect relationship	statistical prediction
fairly high	sufficient	evidence for hypotheses	quantitative simulation
fairly low	insufficient	postulated linkages	conceptual modelling
low/intuitive (50%)	absent or unreliable	speculation	professional opinion

Source: Munro, Bryant, and Matte-Baker 1986.

2. **Sensitivity analysis** examines the sensitivity of the prediction to minor variations in input data, environmental parameters, and assumptions.

For example, one might examine how sensitive a prediction concerning the impacts of worker influx might be on a local population during the construction phase of a project by testing a variety of different scenarios concerning labour demands and the composition of the local workforce. Sophisticated techniques are available for evaluating the sensitivity of impact predictions, such as **Monte Carlo analysis**, a mathematical technique using random samples, while others simply involve providing a range of impact predictions for a particular impact indicator based on minor variations in the data and parameters on which those predictions are based.

3. **Confirmatory analysis** is used to test for uncertainty regarding the predictive technique itself and to ensure that the predicted impact is not solely a product of the particular technique used. In other words, the objective is to determine the extent to which different techniques provide similar predicted outcomes.

4. **Uncertainty disclosure** requires that practitioners disclose their assumptions and uncertainties about impact predictions and mitigation. Regardless of whether uncertainty can be quantified or not, the disclosure of uncertainty is important to ensuring the usefulness of impact predictions when making determinations about suitable impact management measures and the need for precautionary approaches. Unfortunately, uncertainty disclosure in EIA has not been common practice. Tennoy, Kvaerner, and Gjerstad (2006), for example, found that decision-makers are often not made aware of uncertainty in EIAs and recommend that practitioners be more explicit about their assumptions and knowledge gaps, disclosing uncertainties such that decision-makers can make informed decisions. Duncan (2008) adds that proponents, consultants, and decision-makers have a vested interest in making EIAs and their decisions appear defensible and politically palatable, resulting in practices that systematically seek to minimize uncertainty disclosure. What has become standard practice, but arguably poor practice, in EIA is to systematically diminish uncertainty disclosure, resulting in overconfidence in impact predictions.

Key Terms

accuracy	direct effects
additive effects	dispositional antagonism
adverse effects	fly-in fly-out
antagonistic effects	functional antagonism
baseline condition	incremental effects
chemical antagonism	irreversible impacts
continuous effects	Keynesian multiplier
confirmatory analysis	magnitude

maximum allowable effects level
Monte Carlo analysis
multiplier
off-site impacts
on-site impacts
precision
probability analysis
receptor antagonism

reversible impacts
risk
risk assessment
secondary effects
sensitivity analysis
synergistic effects
threshold-based prediction

Review Questions and Exercises

1. Discuss the relationship between accuracy and precision in impact prediction.
2. What are some of the challenges of making impact predictions concerning bio-physical and social change?
3. Consider a condition with the following baseline noise environment:

    i)   average daily road traffic   55 dB
    ii) average daily rail traffic   48 dB
    iii) average daily air traffic   58 dB
    iv) average daily noise from factories 62 dB

    What would be the total noise impact of a proposal to develop a heavy manufacturing plant in the region with equipment operating at average daily noise levels of 68 dB? Would you consider this a significant impact? Explain. How might you manage such an impact?
4. Obtain a completed project EIS from your local library or government registry, or access one online. Scan the document for stated impact predictions, and examine how these predictions are stated. Are the statements based on verifiable hypotheses? Are thresholds or maximum allowable effects levels stated? Given the nature of the predictions and the way in which they are stated, do you think that they can be followed up and verified? Compare your findings with those of others.
5. Obtain a completed project EIS from your local library or government registry, or access one online. Scan the document, and generate a list of the techniques used to predict, describe, or assess impacts on the biophysical and human environments. Compare your results with those of others. Is there a common set of techniques that emerge for various environmental components, such as water quality, air quality, or employment?
6. Consider a proposal for the construction of an industrial complex that requires forest clearing of the proposed site adjacent to a river system. Following the example presented in Box 6.5, identify potential causal factors, condition changes, and direct and indirect impacts of the proposed action.
7. Provide an example for each of incremental, additive, and synergistic impacts.
8. Suppose a proponent submits an application for the development of a large-scale open-pit gold mine operation in your region. In small groups, brainstorm the potential "on-site" and "off-site" impacts. Use the "impact characteristics" discussed in

this chapter to identify the impacts you believe are most important to consider among both on-site and off-site impacts. Are there any additional criteria or characteristics that should be considered? Compare the results among groups.

References

Aura Environmental Research and Consulting Ltd. 2008. *Community Health Baseline Study: Lower Churchill Hydroelectric Generating Project.* Report prepared for Minaskut Limited Partnership. St John's, NL.

Bailey, J., V. Hobbs, and A. Saunders. 1992. "Environmental auditing: Artificial waterway developments in Western Australia." *Journal of Environmental Management* 34: 1–13.

Bisset, R., and P. Tomlinson. 1988. "Monitoring and auditing of impacts." In P. Wathern, ed., *Environmental Impact Assessment: Theory and Practice,* 117–26. London: Unwin Hyman.

Buckley, R. 1991. "Auditing the precision and accuracy of environmental impact predictions in Australia." *Environmental Monitoring and Assessment* 18: 1–23.

Canter, L.W. 1996. *Environmental Impact Assessment.* 2nd edn. New York: McGraw Hill.

De Jongh, P. 1988. "Uncertainty in EIA." In P. Wathern, ed., *Environmental Impact Assessment: Theory and Practice.* London: Unwin Hyman.

Duinker, P.N., and G.E. Baskerville. 1986. "The significance of environmental impacts: An exploration of the concept." *Environmental Management* 10 (1): 1–10.

Duncan, R. 2008. "Problematic practice in integrated impact assessment: The role of consultants and predictive computer models in burying uncertainty." *Impact Assessment and Project Appraisal* 26 (1): 53–66.

Elsom, D.E. 2001. "Air quality and climate." In P. Morris and R. Therivel, eds, *Methods of Environmental Impact Assessment,* 2nd edn. London: Taylor and Francis Group.

Glasson, J. 2001. "Socio-economic impacts: overview and economic impacts." In *Methods of Environmental Impact Assessment,* 2nd ed. Edited by P. Morris and R. Therivel, 20–41. London: UCL Press.

Harrop, D.O., and A.J. Nixon. 1999. *Environmental Assessment in Practice.* Routledge Environmental Management Series. London: Routledge.

Mackenzie Gas Project. *Environmental Impact Statement for the Mackenzie Gas Project,* 2004, vol. 5, section 7, 7–183 to 10–27.

Mitchell, B. 2002. *Resource and Environmental Management.* 2nd edn. New York: Prentice-Hall.

Morris, P., and R. Therivel, eds. 2001. *Methods of Environmental Impact Assessment.* 2nd edn. London: Taylor and Francis Group.

Morris, R. 2012. *Scoping Study: Impact of Fly-in Fly-out/Drive-in Drive-out Work Practices on Local Government.* Sydney: Australian Centre of Excellence for Local Government, University of Technology.

Morrison-Saunders, A., and J. Bailey. 1999. "Exploring the EIA/environmental management relationship." *Environmental Management* 24 (3): 281–95.

Munro, D., T. Bryant, and A. Matte-Baker. 1986. *Learning from Experience: A State of the Art Review of Environmental Impact Assessment Audits.* Ottawa: Canadian Environmental Assessment Research Council.

Nalcor Energy. 2009. *Environmental Impact Statement. Lower Churchill Hydroelectric Generating Project.* St John's, NL: Nalcor Energy.

Preston, E., and B. Bedford. 1988. "Evaluating cumulative effects on wetland functions: A conceptual overview and generic framework." *Environmental Management* 12 (5): 565–83.

Ringold, P.R., et al. 1996. "Adaptive monitoring design for ecosystem management." *Ecological Applications* 6 (3): 745–7.

Sadler, B. 1996. *Environmental Assessment in a Changing World: Evaluating Practice to Improve Performance.* Final report of the International Study of the Effectiveness of Environmental Assessment. Fargo, ND: IAIA.

Storey, K. 2010. "Fly-in/Fly-out: Implications for community sustainability." *Sustainability* 2: 1161–81.

———, and B. Noble. 2004. *Toward Increasing the Utility of Follow-up in Canadian EA: A Review of Concepts, Requirements and Experience.* Report prepared for the Canadian Environmental Assessment Agency. Gatineau, QC: CEAA.

Tennoy, A., J. Kvaerner, and K.I. Gjerstad. 2006. "Uncertainty in environmental impact assessment predictions: The need for better communication and more transparency." *Impact Assessment and Project Appraisal* 24 (1): 45–56.

Voisey's Bay Mine and Mill Environmental Assessment Panel. 1999. *Report on the Proposed Voisey's Bay Mine and Mill Project / Environmental Assessment Panel.* Hull, QC: Canadian Environmental Assessment Agency.

Wood, C., B. Dipper, and C. Jones. 2000. "Auditing the assessments of the environmental impacts of planning projects." *Journal of Environmental Planning and Management* 43 (1): 23–47.

Wood, G. 2008. "Thresholds and criteria for evaluating and communicating impact significance in environmental impact statements: See no evil, hear no evil, speak no evil?" *Environmental Impact Assessment Review* 28: 22–38.

Impact Management

The utility of the EIA process lies not so much in the accuracy of the predicted environmental impacts but in the scope and effectiveness of impact management (Bailey, Hobbs, and Saunders 1997). Once potential environmental effects and impacts have been identified and characterized, the next phase of the EIA process is to design management strategies to address those effects and impacts. Impact management is foundational to the entire EIA process in that it requires the identification of impact management measures that translates the findings from an EIS into recommendations to enhance positive outcomes and avoid, minimize, or offset potentially adverse impacts (Tinker et al. 2005). In other words, the focus of impact management is on plans or strategies designed to avoid or alleviate anticipated impacts generally perceived as undesirable and to generate or enhance effects seen as beneficial.

Managing Adverse Impacts

Many, arguably most, impacts that are the focus of EIA are adverse in nature. There is a hierarchy of approaches that should be adopted for managing potentially adverse impacts, starting with impact avoidance early in the project design stage to compensating for impacts that simply cannot be avoided, mitigated, or rectified (see Mitchell 1997).

Avoidance

Avoiding potentially adverse environmental effects and impacts at the outset is the most desirable approach—if an impact can be avoided, then the time and financial resources required by impact mitigation or compensation can also be avoided. Methods of avoiding potentially adverse impacts can include such measures as setting regulatory standards concerning the use of toxic substances; scheduling project construction activities so that they do not conflict with daily patterns of local socio-economic activity; routing pipelines, roads, and other linear features to avoid sensitive habitat or cultural features; and the construction of self-contained work camps to avoid potentially negative socio-economic effects that might be caused by site worker–community interaction. To be effective, **impact avoidance** must enter the EIA equation early, since most impact avoidance opportunities are presented in the scoping process through identifying alternative locations, project designs, or implementation strategies. For example:

Activity: road construction
Causal factor: surface disturbance, water crossings
Effect/impact: loss of wetland habitat and function, stream sedimentation
Avoidance: re-routing to avoid wetland habitat and stream crossings

Mitigation

Not all potentially adverse impacts are recognized in advance. While mitigation is often used in a generic way to refer to impact management in general, strictly speaking, **impact mitigation** is defined as the application of project design (Box 7.1), construction, operation, scheduling, and management principles and practices to *minimize* potentially adverse environmental impacts. For example, forest-harvesting operations can lead to soil erosion and excessive runoff, which in turn may affect the quality of aquatic environments. The maintenance of **buffer zones**, or areas of undisturbed vegetation, is thus a desired impact mitigation practice in forestry. While buffer zones do not *prevent* soil erosion or surface runoff, they do reduce the severity of the impacts of erosion and runoff on aquatic environments. For example:

Activity: forest clearing
Causal factor: erosion and runoff
Effect/impact: increased turbidity in local stream
Mitigation: stream buffer zones

Remediation

Not all adverse environmental effects can be avoided or mitigated. In certain cases, environmental components will inevitably be temporarily damaged. **Remediation** refers to restoring environmental quality, rehabilitating certain environmental features, or restoring environmental components to varying degrees. For example,

---

Box 7.1 Setting a Higher Mitigation Standard through Technological Design: The Al-Pac Pulp Mill, Alberta

The Alberta-Pacific (Al-Pac) Pulp Mill is a bleached kraft pulp mill that was proposed for development in 1989 in north-central Alberta, on the Athabasca River. At the time, Al-Pac would be the largest pulp mill in the world when completed, with a capacity to produce 1500 tons per day of bleached hardwood pulp. The proposed project would also be the cleanest bleached kraft pulp mill, based on the lowest emission of chlorinated organic compounds per unit of pulp produced.

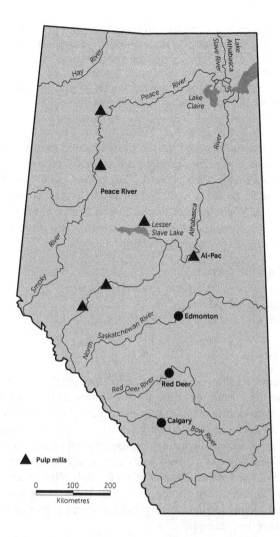

The project was subject to review by a joint federal–provincial assess-ment panel. A major concern noted by the panel was the potential for adverse impacts of mill effluent on the Peace-Athabasca Rivers system. Numerous indigenous communities live in the vicinity of the Peace-Athabasca Delta, which would receive effluent from Al-Pac as well as from other existing kraft mills. Of particular concern was the potential adverse and cumulative effects of effluent from the Al-Pac mill, combined with that of other mills on the Peace and Athabasca Rivers (e.g., Proctor & Gamble, Diashowa, Alberta Energy, Alberta Newsprint, Weldwood).

In 1990, the review panel issued a report recommending that Al-Pac not be approved pending further assessment of the potential adverse effects of pulp mill operations. In response, Al-Pac submitted a revised proposal for

continued

its bleaching process such that the amount of chlorinated organic compounds discharged to the Athabasca River would be reduced from 1.3 to 0.35 kilograms of pulp, or 26 per cent of the originally proposed effluent level (Alberta-Pacific EIA Review Board, 1990). A scientific review panel was convened by the Alberta government, which recommended in favour of the revised proposal, and the project was approved.

in cases where the construction phase of project development requires clearing of forested landscape, impact management efforts can focus on restoring the landscape post-construction to resemble the pre-disturbed state. In the case of a **sanitary landfill** site, for example, the site is typically returned to a vegetative state once the landfill is full and covered. While the remediated site condition may not exactly resemble the pre-disturbed condition, the objective is to return it to a more desirable condition than what was created by the project actions. For example:

Activity: landfill operation
Causal factor: site preparation and waste disposal
Effect/impact: habitat loss
Reclamation: vegetating following decommissioning

Compensation

Some environmental effects cannot be avoided, mitigated, or rectified. In such cases, the typical action is **compensation** for the unavoidable, residual, or irreparable impacts that remain after other impact management options have been exhausted or for which no management alternative exists. Compensation sometimes involves monetary or other benefit payments to those affected by the damage caused by the project, while in other cases it involves measures to recreate environmental habitats at an alternative site. In Canada, for example, a federal policy on aquatic-based habitat declares that there should be "no net loss." This is not to say that projects posing a threat to aquatic habitat will not be approved; rather, any habitat that is lost must be compensated for. For example:

Activity: open-pit mine development
Causal factor: drainage of a nearby lake for tailings disposal
Effect/impact: loss of fish habitat and aquatic life
Compensation: creation of new habitat or restoration of degraded habitat

Managing Positive Impacts

Simply making impacts less severe is not good enough. As Gibson (2011) argues, "Ultimately, the enhancement we need to deliver through environmental assessment is confidence that every approved undertaking will move us positively towards a desirable and durable future." In order to accomplish this, there is a need to focus as well on delivering positive benefits from project development through the EIA process.

Creating Benefits

Impact management provides the necessary means not only to reduce or avoid potentially negative impacts but also to create positive benefits from project development. The most desirable approach to managing positive impacts is to ensure that the project makes a positive contribution to the environment and society through the creation of new benefits, such as community economic growth or environmental improvement. For example:

> *Activity:* hydroelectric construction project
> *Causal factor:* materials and services purchasing
> *Effect/impact:* increased local expenditures
> *Benefit creation:* small business development support centre

Enhancing Benefits

Development projects often create as many positive impacts, particularly economic ones, as they do negative impacts. Indeed, if this were not the case, democratic societies and their governments would be hard-pressed to justify the continued approval of such projects. Thus, an important management strategy is to enhance the benefits of potentially positive impacts and to maximize the duration of those impacts. This might involve ensuring the greatest possible distribution of financial benefits among affected communities over the longest period of time. In the case of the Mackenzie Gas Project, for example, a proposed 1200-kilometre gas pipeline from the Mackenzie Delta to northern Alberta, the federal, territorial, and Aboriginal government joint review panel charged with reviewing the project was directed to consider, among other things: (i) the extent to which the project makes a positive overall contribution toward environmental, social, cultural, and economic sustainability; (ii) the attainment and distribution of lasting and equitable social and economic benefits from the project; (iii) the rights of future generations to the sustainable use of renewable resources; and (iv) the capacity of social and economic systems to achieve, maintain, or enhance conditions of self-reliance and diversity.

Other, previous projects have also attempted to emphasize enhancing project impacts, including the Voisey's Bay nickel mine project, discussed in Chapter 1, which for the first time in Canadian EIA adopted an explicit sustainability mandate requiring the project proponent to enhance potentially positive impacts by committing to community infrastructure investment and worker training programs beyond the life of the mine project. Other common approaches include adopting "buy local" and "hire local" strategies. This was an important part of enhancing the benefits of the Hibernia offshore oil platform construction project at Bull Arm, Newfoundland, in which Mobil Oil committed to ensuring that a minimum number of construction contracts would be let to local businesses and that a certain percentage of workers employed at the site would be Newfoundlanders. For the Ekati diamond mine in the Northwest Territories, similar impact enhancement strategies were adopted to maximize the benefits of project employment. For example:

Activity: diamond mine development and operation
Causal factor: job creation
Effect/impact: increased local employment and income
Benefit enhancement: provision of free financial counselling to employees

Measures That Support Impact Management

In addition to project-based impact management actions, a number of broader pro-grams, measures, and approaches can facilitate management of the environmental impacts of industry operations. Perhaps the most internationally recognized measure is the environmental management system.

Environmental Management Systems

The International Organization for Standardization, commonly referred to as ISO, is a global federation of more than 100 countries and is headquartered in Geneva, Switzerland. The organization was formed in 1947 to promote the development of international standards, primarily in product manufacturing. The first system for management standardization (ISO 9000 series) was introduced in the early 1990s. By the mid-1990s, the ISO 14000 series was introduced to promote international industry standards in environmental management, and in 1996 the **ISO 14001** standardization for **environmental management system** (EMS) certification was approved (Box 7.2). Currently, ISO 14001 specifies the requirements and standards for an EMS, and ISO 14004 provides the guidelines for EMS implementation. To date, more than 130,000 ISO 14001 certificates have been issued in 140 countries, with more than 23,000 certificates issued in Japan, followed by China with nearly 20,000. As of 2006, the United States ranked seventh globally, with more than 5500 certificates, and 1700 certificates had been issued in Canada (International Organization for Standardization 2007). EMSs are currently among the most widely recognized tools for managing the environmental affairs of industry. An EMS is a voluntary industry-based manage-ment system by which an organization can:

- identify and control the environmental impact of its operations and activities;
- continually improve its environmental performance;
- implement a systematic plan to set environmental targets and objectives and to clearly demonstrate to the public that those targets and objectives have been achieved.

By putting an EMS into effect, a company is seeking to minimize the environmental impacts of its operations. EMSs emerged in response to industry's realization of the need for an integrated and proactive approach to managing industry environmental issues to assist in ensuring that it is complying with environmental regulations, adopting and following environmental objectives, and receiving economic benefits from improved environmental performance (Strachan, McKay, and Lal 2003). A well-designed EMS should help an organization to develop a proactive approach to

Box 7.2 Environmental Management Systems

An EMS is ideally a cyclical process of continual improvement in which an organization is constantly reviewing and revising its management system. This consists of a "plan-do-check-act" cycle in which the first step is developing a management policy. Through policy development, the organization's environmental goals and objectives are formulated, reflecting the important environmental aspects and targets as well as the legal requirements of the organization. This is followed by planning, implementation, monitoring, and program review. In this sense, not only can an EMS act as a regulatory system by which an organization seeks to meet industry standards, it can also be a potentially valuable tool for environmental management and improvement of industry operations.

environmental management, achieve a balanced view across all aspects of its operations, enable effective and directed environmental goal-setting, and contribute to a more effective environmental auditing process. That said, the link between meeting ISO standards and genuine improvement in environmental performance has not been clearly established. A study of toxic emissions from US automobile assembly facilities revealed that facilities with ISO 14001 EMS certification often fared worse than those without certification (Matthews, Christini, and Hendrickson 2004).

Environmental Protection Plans

A second measure to ensure impact management effectiveness is an **environmental protection plan** (EPP). While an EMS is a voluntary industry initiative, an EPP is often a mandatory requirement in project-based EIA. As part of the EIA process, key impacts, issues, and management measures to address those impacts will have been identified. In certain Canadian jurisdictions, if a project is approved, then impact management measures may be further articulated in the form of an overall EPP designed to detail the specific impact management actions and the ways in which they are to be implemented. The particular components that make up an EPP are project-specific and custom designed to fit the project context. EPPs may provide general information on the management, construction, operation, and decommissioning of certain types of project, may be developed for different stages of the project, or can provide specific information concerning management techniques relevant to an individual project.

EPPs vary in their nature and requirements from one jurisdiction to the next and by industrial sector. In Saskatchewan, for example, the Ministry of Environment has implemented guidelines under the Environmental Assessment Act for the preparation of EPPs for oil and gas projects. Oil and gas projects that have a potential to trigger a full EIA require a detailed EPP, which undergoes interagency review co-ordinated by the Environmental Assessment Branch. The EPP is prepared by the proponent and describes how the project will be undertaken, including such aspects as a description of the project environment, potential conflicts, and environmental protection measures that the proponent will take to avoid or minimize those conflicts. Saskatchewan's "Guidelines for preparation of an environmental protection plan for oil and gas activities" can be accessed on the province's publications website at http://www.publications.gov.sk.ca/index.cfm.

Adaptive Management Strategies

At a 2002 US Department of Defense news briefing, Secretary of Defense Donald Rumsfeld made the following statement:

> There are known knowns. These are the things we know that we know. There are known unknowns. That is to say, these are the things that we know we don't know. But there are also unknown unknowns. These are the things we don't know we don't know.

Identifying programs and actions to manage the predicted impacts of a project yet to be developed is full of "known unknowns" and "unknown unknowns." The determination of potential effects and impacts identified through the EIA process is often imprecise because of uncertainties surrounding such factors as the project's design and timetable, inherent uncertainties due to the complexity of environmental and socio-economic systems, or exogenous factors such as the emergence of new technology or changing political or market conditions. Thus, a management system designed to address central impact issues in EIA should be flexible enough to respond to unanticipated impacts as well as the differences between the actual and the predicted nature, level, or severity of the impacts.

The success of a prescribed impact management measure is never truly known until it is implemented and tested in practice. As such, it should be expected that any management measure prescribed in an EIA could be wrong or less than effective. Important to sound environmental management practices is the adoption of **adaptive environmental management** (AEM) strategies. Although often described as "learning by doing," AEM is not a haphazard approach. AEM is a structured, well-planned approach to environmental management that treats management prescriptions as experiments to test hypotheses, monitor the outcomes, and subsequently adapt management actions as new knowledge and understanding are gained (Taylor, Kremsater, and Ellis 1997). Adaptive management is a deliberative process that explores alternative management actions and makes explicit forecasts about their outcomes, carefully designs monitoring programs to provide feedback and understanding of the reasons underlying outcomes, and then adjusts objectives or management actions based on this new understanding (Canter 2008). In other words, and specifically within the context of EIA, AEM is a structured approach for proceeding with management actions despite uncertainties about the best course of action (Schreiber et al. 2004). Even the most tested and thought-out management prescriptions can fail; AEM encourages practitioners to approach management with an expectation that they may be wrong but that new knowledge can be gained and management actions improved through careful, planned monitoring and evaluation of management programs. AEM should not be used as a substitute for committing to specific impact management measures, or to attempt to cover a situation where a proponent is not sure how to mitigate a negative environmental impact, but is committed to finding the technology or science in the future if a problem arises (Kwasniak 2010)

Impact Benefit Agreements

A fourth approach to facilitating impact management is an **impact benefit agreement (IBA)**, sometimes referred to as a socio-economic agreement, environmental agreement, impact management agreement, or simply a negotiated agreement. IBAs are particularly useful for addressing the impacts generated by development projects on local communities and are usually associated with compensation measures. IBAs are legally binding agreements between a proponent and a community that serve to ensure that communities have the capacity and resources required to maximize the potential positive benefits stemming from project development. Typically negotiated to establish benefit streams and forms of compensation and impact management in relation to a resource development project, many recent agreements also seek to engage indigenous communities directly in the environmental monitoring and management of the impacts of development (Noble and Fidler 2011) (see Environmental Assessment in Action: Negotiated Agreements and EIA Governance in the British Columbia and Saskatchewan Mining Sectors).

Negotiated IBAs and environmental agreements are now almost commonplace in the mining sectors of Canada, Australia, and the United States and are beginning to emerge in developing nations and in other industry sectors (O'Faircheallaigh and Corbett 2005). In the Canadian context, five agreements for major projects have been implemented during the past decade—for the Ekati, Diavik, and Snap Lake diamond mines in the Northwest Territories; the Voisey's Bay nickel mine in Newfoundland

and Labrador; and the Horizon Oil Sands Project in Alberta. All five agreements establish goals and mandates that relate to Aboriginal participation in environmental management, follow-up, and adaptive management; only two (Diavik and Snap Lake) included government as a signatory; none emerged because of a legal requirement for the negotiation of an agreement. Arguably, the rise of such agreements is a reflection of the privatization of environmental governance.

Some of the first IBAs in Canada, such as the one between Ontario Hydro and the Township of Atikokan, Ontario, in 1978, focused primarily on local employment opportunities and community investment. More recent agreements, however, also include environmental restrictions, social and cultural programs, dispute resolution mechanisms, and provisions for revenue-sharing with others (Sosa and Keenan 2001). In Canada, IBAs are especially important with regard to industry obtaining the cooperation of Aboriginal communities and for these communities to achieve certain guaranteed benefits. As Veiga, Scoble, and McAllister (2001, 191) explain, "mining companies must now pursue their interests in a way that also promotes those of the local communities in regions where they are operating." They go on to suggest that "the challenge for any mining company is to engage in an equitable partnership with the associated community and thus leave a lasting legacy of sustainability and well-being to the community, avoiding environmental degradation and social dislocation."

The rise in IBAs in recent years may be attributed in part to the deficiencies of EIA in negotiating community issues, impacts, and benefits at the time of the development proposal and impact assessment. In a review of impact assessment and benefit agreements in the Mackenzie Valley of the Northwest Territories, for example, Galbraith, Bradshaw, and Rutherford (2007) found that IBAs addressed many issues of concern to the local community, including community–industry relationships and benefits-sharing, that project EIAs did not. In many respects, the emergence of such agreements is simply the reality of doing business, and companies are expected to pursue their interests in ways that promote the interests of the communities and regions in which they operate (Jenkins 2004). In so doing, companies can avoid, mitigate, or compensate for land and resource conflicts so that business can proceed with much greater certainty (Noble and Fidler 2011). Negotiated agreements have the potential to play an important role in fostering a collaborative vision for resource development that, in many cases, goes beyond conventional EIA and impact management.

IBAs are not meant to replace project EIA but to complement the existing regulatory process. These agreements are negotiated for different reasons depending on the particular community's interests or First Nations' land and resource rights, the regulatory framework in place, and the relationship that exists between the community and the company (Sosa and Keenan 2001). A major challenge to the impact assessment community in understanding the value added by these agreements to the EIA process is that IBAs and similar negotiated agreements occur outside the public realm and little is known about their content, benefits-sharing and impact management details, and overall efficacy.

Environmental Assessment in Action

Negotiated Agreements and EIA Governance in the British Columbia and Saskatchewan Mining Sectors

Galore Creek Project, British Columbia

The Galore Creek Project, a proposed open pit porphyry-copper-gold-silver mine, is located in northwest British Columbia—an area referred to as the "Golden Triangle" because of its rich mineral deposits. The Galore Creek property was acquired by NovaGold Resources Inc. in 2007 and is currently owned jointly by NovaGold and Teck Resources. The project is situated within the traditional territory of the Tahltan First Nations, a group with a long history of involvement with the mining sector. Before commencement of an EIA for the project, an agreement was signed between the Tahltan and NovaGold, defining how the Tahltan and NovaGold would collaborate to achieve EIA approval—specifically, how the Tahltan and NovaGold would work together in identifying the potential environmental effects of the project on the Tahltan Nation and determining how best to avoid or minimize such impacts.

The Galore Creek Project agreement, although voluntarily negotiated, was a legal framework and enforceable contract between the mining company and the community to which the government was not privy. Its purpose was to create certainty, for both parties, for investment, access, extraction, and ownership of mineral rights. The agreement was pivotal in defining how the Tahltan and NovaGold would collaborate during the mine's lifecycle, specifically through the EIA approval phase. The agreement sought to create a framework for communication and partnership to ensure benefits to the Tahltan from the project and to guarantee the Tahltan's support of the project. The agreement captured many of the same issues protected under environmental legislation but went further to provide additional investment security for the signatories and benefits to the Tahltan, including environmental monitoring, heritage resources, an ongoing review of the closure plan, scholarships, employment training, and contracting opportunities.

Athabasca Basin Uranium Mining Operations, Saskatchewan

The Athabasca Basin of northern Saskatchewan is one of the world's most productive uranium mining regions, contributing approximately 20 per cent of global uranium supply. The region is also home to seven indigenous communities. Uranium mining operations are "fly-in fly-out," and local economies are based primarily on traditional hunting, trapping, fishing, and guiding activities. Cameco Corporation (www.cameco.com) and AREVA Resources Canada (www.arevaresources.com) are the two uranium producers in Saskatchewan and the main producers in Canada. In 1991, the Canadian federal and Saskatchewan provincial governments established a Joint Federal–Provincial Panel on Uranium Mining Developments in Northern Saskatchewan to review industry proposals for uranium mine development and existing mine

continued

expansion in the Athabasca Basin. The joint panel concluded that, among other things: (i) many of the communities in the Athabasca Basin wanted to participate in, and receive benefits from, uranium mining activities; (ii) although monitoring systems were in place that met regulatory requirements, there were enduring concerns about their effectiveness in determining the impacts of mining activities; and (iii) because of the proximity of mining operations to communities, community involvement should extend beyond consultation.

In response to the recommendations of the joint panel, the uranium industry established the Athabasca Working Group (AWG) in 1993—a private partnership between the two uranium mining companies and the seven communities of the Athabasca Basin. The rationale for forming the AWG was largely the recognized need to build relationships with the local communities and to build an environment of corporate policy and responsibility to ensure that stakeholders are involved in and familiar with mining operations. The AWG itself has no legal authority. It was created by the industry to address the concerns of the Athabasca communities about the impacts of uranium mining; to ensure the engagement of communities in mining related activities, including impact management; and to facilitate discussions about the sharing of benefits of mining activity. An Impact Management Agreement (IMA) was later signed between Cameco, Areva, and the local communities. The IMA is focused on local employment, training and business development, benefit-sharing, and environmental protection. As part of the agreement's commitment to environmental protection, and in part because of the joint panel's recommendation for more direct community involvement, a community-based environmental monitoring program was established to monitor the "off-site" impacts of uranium mining operations. Community members, appointed by the community, are responsible for identifying monitoring locations and for monitoring data collection.

Lessons and Implications for EIA Governance

In the Galore Creek case, the agreement was established in large part to formalize a process of working together through the EIA process. In the Saskatchewan case, the agreement was established post-EIA to engage communities in environmental monitoring and impact management programs. Both cases illustrate going beyond the prescribed requirements of EIA and, in doing so, reveal several lessons for improving the governance of EIA as a tool for environmental management:

- Successful agreements hinge on community trust, the result of which is enhanced corporate image and a social licence to operate. Enhanced community trust can lead to increased community confidence in the mining company and in the EIA process.
- Building this level of trust requires genuine community participation pre- and post-EIA in the design, implementation, and evaluation of impact management strategies. There is an opportunity through negotiated agreements to enhance the capacity of indigenous communities to become engaged in resource development by, for example, providing funding for participation in the EIA process, ensuring employment

and ongoing skills development and training opportunities, and involving communities in post-EIA monitoring and management programs.
- There is a danger with negotiated agreements of blurring the lines of responsibility and accountability between the mining company and government with respect to indigenous communities and commitments to impact management. Agreements negotiated prior to or parallel with the EIA process provide an opportunity to chart out how regulatory approval can be achieved but can be criticized for undermining the public EIA process. When agreements are negotiated post-EIA, however, communities may have much less control over impact management and benefit enhancement strategies as conditions for granting their approval of the project.

Source: Based on Noble and Fidler 2011; Noble and Birk 2011

EIA and Impact Management in Canada

EIA is more than just a tool to identify and predict potential environmental impacts; it should also play an important environmental management role. Impact management must be planned in an integrated and coherent fashion to ensure that such measures are effective and non-contradictory and that they do not shift the problem from one environmental component or sector of society to another (Glasson, Therivel, and Chadwick 1999). Thus, impact management is not limited to any one point in the EIA process. From the outset of project design and scoping, measures should be considered for avoiding potentially adverse impacts and for creating or enhancing positive ones.

In practice, the focus of impact management tends to be on reducing or making less severe the adverse environmental effects of development. Storey and Noble (2004) note that while most biophysical effects arising from human activities are adverse, many social, economic, and other human effects are not. Increased employment, training, and business development, for example, are usually regarded as positive outcomes. The focus of impact management in Canada, under the Canadian Environmental Assessment Act, 2012, is "mitigation," defined under section 2(1)(h) of the act as "measures for the elimination, reduction or control of the adverse environmental effects of a designated project, and includes restitution for any damage to the environment caused by those effects through replacement, restoration, compensation or any other means." Consistent with the former Canadian Environmental Assessment Act, the focus of attention is on making adverse impacts less severe. While this is important, it is only one of several types of action that those responsible for managing project effects need to take. The emphasis on simply reducing the severity of effects reflects a "damage-control" perspective rather than a proactive one in which actions taken might conceivably have a positive effect. If sustainable development is truly a goal of EIA, then there is a need to adopt a broader perspective on management strategies for all types of potential effects.

Rather than simply aiming to mitigate adverse effects, practitioners and regulators should consider a full range of management options, from avoidance to creating positive impacts, and the language of EIA should reflect this. From a broader sustainability perspective, the priority focus of impact management should be on the creation and enhancement of overall, long-term positive benefits to the environment and society and recognition of the trade-offs involved in doing so. The positive news is that although the scope of impact management under the Canadian Environmental Assessment Act, 2012 is to make adverse impact less severe, there is still an opportunity for impact management to focus on ensuring an overall positive contribution of a project development to environment and society. Included among the purposes of the act, for example, section 4(1)(h), is to encourage federal authorities to take actions that promote sustainable development in order to achieve or maintain a healthy environment and a healthy economy. Gibson (2012) reports that five joint review panels established under the former Canadian Environmental Assessment Act and other provincial, territorial, or Aboriginal authorities have applied the "positive contribution" test, namely, the Voisey's Bay Nickel Mine, White Points Quarry and Marine terminal, Mackenzie Gas Project, Kemess North Copper–Gold Mine, and the Lower Churchill Hydroelectric Generation Project.

Key Terms

adaptive environmental management
buffer zone
compensation
environmental management system
environmental protection plan
impact avoidance

impact benefit agreement
impact mitigation
ISO 14001
remediation
sanitary landfill

Review Questions and Exercises

1. Suppose you are responsible for negotiating an IBA for a small community (population less than 5000) about to be the recipient of a large mining operation. What items might you want to negotiate with the proponent and the government for inclusion in the IBA?

2. Assume a simple highway construction project through a small community. Develop a list of project activities and potential impacts, and propose as many different types of impact management measures as possible, from avoidance to enhancement.

3. Obtain a completed project EIS from your local library or government registry, or access one online. Document the nature of impact management measures. For example, are most impact management measures based on avoidance, mitigation, rectification, or compensation? Are there any impact management measures that emphasize creating or enhancing positive project impacts? Generate a list, and compare your results with those of others.

4. Visit the International Organization for Standardization website at www.iso.org.
   a) How many countries were ISO members this past year?

b)   Is your country an ISO member?
c)   What is the trend in ISO 14001 certification over the past five years?
d)   What does ISO identify as the principal benefits of ISO 14001 certification to businesses?

References

Alberta-Pacific EIA Review Board. 1990. *The Proposed Alberta-Pacific Pulp Mill*. Report of the EIA Board. Edmonton: Alberta Environment.

Bailey, J., V. Hobbs, and A. Saunders. 1997. "Environmental auditing: Artificial waterway developments in Western Australia." *Journal of Environmental Management* 34: 1–13.

Canter, L. 2008. "Adaptive management for integrated decision-making: An emerging tool for cumulative effects management." Paper presented at Assessing and Managing Cumulative Environmental Effects, Special Topic Meeting of the IAIA, 6–9 November, Calgary.

Galbraith, L., B. Bradshaw, and M. Rutherford. 2007. "Towards a supraregulatory approach to environmental assessment in northern Canada." *Impact Assessment and Project Appraisal* 25 (1): 27–41.

Gibson, R.B. 2011. "Application of a contribution to sustainability test by the Joint Panel for the Canadian Mackenzie Gas Project." *Impact Assessment and Project Appraisal* 29 (3): 231–44.

———. 2012. "In full retreat: The Canadian government's new environmental assessment law undoes decades of progress." *Impact Assessment and Project Appraisal* 30 (3): 179–88.

Glasson, J., R. Therivel, and A. Chadwick. 1999. *Introduction to Environmental Impact Assessment: Principles and Procedures, Process, Practice and Prospects*. 2nd edn. London: University College London Press.

International Organization for Standardization. 2007. "The ISO survey of certification: 2006." http://www.iso.org/iso/store.htm, under "Products and services/additional publications/management standards."

Jenkins, H. 2004. "Corporate social responsibility and the mining industry: Conflicts and constructs." *Corporate Social Responsibility and Environmental Management* 11: 23–34.

Kwasniak, A.J. 2010. "Use and abuse of adaptive management in environmental assessment law and practice: A Canadian example and general lessons." *Journal of Environmental Assessment Policy and Management* 12 (4): 425–68.

Matthews, D.H., G.C. Christini, and C.T. Hendrickson. 2004. "Five elements for organizational decision-making with an environmental management system." *Environmental Science and Technology* 38 (7): 1927–32.

Mitchell, B. 1997. *Resource and Environmental Management*. Harlow, UK: Addison Wesley Longman.

Noble, B.F., and J. Birk. 2011. "Comfort monitoring? Environmental assessment follow-up under community–industry negotiated environmental agreements." *Environmental Impact Assessment Review* 31: 17–24.

Noble, B.F., and C. Fidler. 2011. "Advancing indigenous community–corporate agreements: Lessons from practice in the Canadian mining sector." *Oil, Gas and Energy Law Intelligence* 9 (4): 1–30.

O'Faircheallaigh, C., and T. Corbett. 2005. "Indigenous participation in environmental management of mining projects: The role of negotiated agreements." *Environmental Politics* 14 (5): 629–47.

Schreiber, E.S.G., et al. 2004. "Adaptive management: A synthesis of current understanding and effective application." *Ecological Management and Restoration* 5 3: 177–82.

Sosa, I., and K. Keenan. 2001. "Impact benefit agreements between Aboriginal communities and mining companies: Their use in Canada." http://www.cela.ca/publications/impact-benefit-agreements-between-aboriginal-communities-and-mining-companies-their-use.

Storey, K., and B. Noble. 2004. *Toward Increasing the Utility of Follow-up in Canadian EA: A Review of Concepts, Requirements and Experience.* Report prepared for the Canadian Environmental Assessment Agency. Gatineau, QC: CEAA.

Strachan, P.A., I. McKay, and D. Lal. 2003. "Managing ISO 14001 implementation in the United Kingdom Continental Shelf (UKCS)." *Corporate Social Responsibility and Environmental Management* 10: 50–63.

Taylor, B., L. Kremsater, and R. Ellis. 1997. *Adaptive Management of Forests in British Columbia.* Victoria: British Columbia Ministry of Forests.

Tinker, L., et al. 2005. "Impact mitigation in environmental assessment: Paper promises or the basis of consent conditions?" *Impact Assessment and Project Appraisal* 23 (4): 265–80.

Veiga, M., M. Scoble, and M. McAllister. 2001. "Mining with communities." *Natural Resources Forum* 25: 191–202.

Impact Significance

Determining the significance of environmental effects is one of the most critical components of the EIA process but also the most complex. The determination of whether a project is likely to cause a significant adverse environmental effect is dependent on several factors, including the appropriate baseline information being available, adopting the right spatial and temporal boundaries for the assessment, and technically and economically feasible mitigation measures being identified. Several attempts have been made to define significance in relation to environmental effects; however, there is still no accepted interpretation of significance or standard methods for significance determination (Table 8.1).

Significance is dynamic, contextual, political, and uncertain (Lyhne and Kørnøv 2013). Part of the complexity surrounding significance is that there is no generic standard that defines a "significant impact," there will always be site-specific issues to take into consideration, and there is also considerable variation in the types of activities and impacts being assessed as significant in any given EIA. Under the Canadian Environmental Assessment Act, 2012, for example, paragraph 52(1) requires a decision-maker to determine whether, taking into account the implementation of any mitigation measures considered appropriate, the proposed project is likely to cause significant environmental effects. Only "environmental effects" as defined under section 5 of the act (see Chapter 4) are to be considered; however, consideration is also given to various other factors under section 19 of the act, such as accidents or malfunctions associated with the project, potential cumulative effects, and Aboriginal traditional knowledge.

Despite the lack of a single definition of "significance," Sippe (1999) recognizes that some degree of commonality exists across the various interpretations of the concept—namely, significance:

1. is an informed judgment;
2. depends, in part, on the characteristics of anticipated environmental impacts and the nature of the receiving environmental components;
3. is both biophysical and socio-economic in context;
4. involves some level of change, including cumulative change, which is perceived to be acceptable to the interested parties.

Table 8.1 Interpretations of "Significance"

Canter and Canty (1993)	Significance can be considered on three levels: (1) significant and not mitigatible; (2) significant but mitigatible; and (3) insignificant. Significance is sometimes based on professional judgment, executive authority, the importance of the project/issue, sensitivity of the project/issue, and context, or by the controversy raised.
US Council on Environmental Quality (1987)	The US NEPA requires significance to be determined within the framework of context and intensity. Context—the significance of an action must be analyzed in several contexts such as society as a whole, the affected region, the affected interests, and the locality. Intensity—refers to the severity of impact.
Duinker and Beanlands (1986)	Significance of environmental impacts is centred on the effects of human activities and involves a value judgment by society of the significance or importance of these effects. Such judgments, often based on social and economic criteria, reflect the political reality of impact assessment in which significance is translated into public acceptability and desirability.
Haug et al. (1984)	Determining significance is ultimately a judgment call. The significance of a particular issue is determined by a threshold of concern, a priority of that concern, and a probability that a potential environmental impact may cross the threshold of concern.
Sadler (1996)	The evaluation of significance is subjective, contingent upon values, and dependent upon the environmental and community context. Scientific disciplinary and professional perspectives frame evaluations of significance. Scientists therefore evaluate significance differently from one another and from local communities.

Determination of impact significance then essentially involves making judgments about the importance of environmental effects. Significance reflects the degree of importance placed on the effects in question and consists of two components:

1. the significance of predicted project effects;
2. the significance of predicted effects following impact management or mitigative measures (see Chapter 7).

Effects associated with the latter are typically referred to as **residual effects**, or impacts that remain after all management and mitigation measures have been implemented. Residual effects are typically the focus of significance determinations in EIA practice.

The determination of significance begins at the outset of the EIA process when a decision is being made as to whether the proposal requires a formal assessment and extends throughout the scoping, prediction, mitigation, and follow-up stages (Table 8.2). The information and characteristics used to define and support significance determination during the early stages of screening and scoping differ from the conceptualization and determination of significance during impact prediction, evaluation, and verification of impact mitigation effectiveness (e.g., see Wood and Becker 2004; Sadler

Table 8.2 Interpretations of Significance in the EIA Process

EA Activity	Significance Interpretation
Screening	If (and what) EIA requirements are to be applied Criteria and procedures for making screening decisions
Scoping	Alternatives that are reasonable and criteria for comparing them Analysis of boundaries and components to focus on Public and agency issues Proposal characteristics most likely to induce significant effects Proposal characteristics that warrant mitigation and monitoring
Baseline analysis	Criteria for determining environmental significance and sensitivity Choice of valued ecosystem components
Impact analysis	Potential impacts to analyze and at what level of detail and interpretation Impact magnitude and impact significance criteria and thresholds Impact acceptability and significance interpretations
Cumulative effects analysis	Cumulative effects criteria and thresholds effects analysis Acceptability of cumulative effects Cumulative effects significance interpretations
Decision-making	Proposal acceptability and compliance with standards, policies, etc. Basis for decision-making
Mitigation	Impacts that warrant mitigation, compensation, or benefits Choice of measures and significance of residual impacts
Monitoring and auditing	Impacts that should be monitored and choice of monitoring methods Effectiveness of mitigation and monitoring

Source: Based on Lawrence 2000.

1996; Westman 1985). It is possible, explains Lawrence (2005), to apply significance thresholds and criteria to determine whether an EIA is required and to focus or scope the EIA process. Following screening and scoping, significance determination can involve applying different types of quantitative thresholds and criteria to individual and cumulative effects, taking into account context, stressor characteristics, public interest, uncertainty, and the characteristics of the receiving environment.

Determining Significance

There are three basic considerations when determining whether a project is likely to cause significant adverse environmental effects under the Canadian Environmental Assessment Act, 2012, namely: (i) determining whether the environmental effects are adverse; (ii) determining whether the adverse environmental effects are significant; and (iii) determining whether significant adverse environmental effects are likely. However, notwithstanding the importance of determining significance, there is little consistent guidance available for practitioners. Generally speaking, effects are more likely to be considered significant if they are:

- adverse;
- intensive in concentration or associated with significant levels of change;
- associated with a high degree of probability;
- frequent and long-lasting;
- likely to occur at a broad spatial scale;
- irreversible;
- associated with cumulative change;
- going to detract from the sustainability of environmental and socio-economic systems;
- likely to affect ecological functions or exceed **assimilative capacity** of the environment;
- associated with variables of societal importance and public concern;
- not in compliance with existing standards or regulations;
- likely to exceed desired levels of change.

Impact significance, then, is a function of the characteristics of the environmental effect or impact and the importance or value attached to the affected component (Figure 8.1).

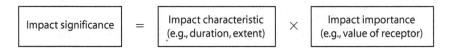

| Impact significance | = | Impact characteristic (e.g., duration, extent) | × | Impact importance (e.g., value of receptor) |

Figure 8.1 ⊚ Definition of Impact Significance

Effect and Impact Characteristics

For each prediction, several questions concerning the characteristics of environmental effects and impacts are typically asked (see Lawrence 2004; Canter 1996; Wood 1995). The specific nature of these questions varies from project to project but usually involves the identification of "major," "moderate," "minor," and "negligible" effects based on several characteristics (Table 8.3). Many of these characteristics were discussed in Chapter 6 and are reviewed only briefly here, with particular attention to other characteristics not previously introduced.

Adverse Effect
A first step in determining significance is to determine whether effects are likely to be adverse following effects mitigation. This is typically done by comparing the predicted future status or condition of a VEC, or indicator of that VEC, with and without the proposed project. Some major factors that should be used to determine whether environmental effects are adverse are described by FEARO (1994) and summarized in Table 8.4. The most common way of determining whether effects are adverse is to compare the quality of the environment with and without the project, using as variables the relevant environmental components identified in Table 8.4. This does, however, imply that baseline data are available or can be established for each of the variables concerned.

Table 8.3 Selected Impact Significance Criteria for the Ekati Diamond Mine EIA

Impact significance	Type of environmental component		
	Physical	Biological	Socio-economic
Major	Parameter affected within most of ecozone for several decades	Whole stock or population of ecozone affected over several generations	Whole population of region affected over several generations
Moderate	Parameter affected within most of eco-region for one or more decades	Portion of population of eco-region affected over one or more generations	Community affected over one or more generations
Minor	Parameter affected within most of eco-region during less than one decade	A specific group of individuals within an ecosystem affected during less than one generation	A specific group of individuals within a community affected during less than one generation
Negligible	Parameter affected within some part of eco-region for a short time period	A specific group of individuals within an eco-region affected for a short time period	A specific group of individuals within a community affected for a short time period

Source: Based on BHPB 1998.

Table 8.4 Selected Factors in Determining Adverse Environmental Effects

Changes in the environment	Human impacts resulting from changes
Negative effects on the health of biota, including plants, animals, and fish	Negative effects on human health, well-being, or quality of life
Threat to rare or endangered species	Increase in unemployment or shrinkage in the economy
Reduction in species diversity or disruption of food webs	Reduction of the quality or quantity of recreational opportunities or amenities
Loss or damage to habitats, including habitat fragmentation	Detrimental change in the current use of lands and resources for traditional purposes by Aboriginal persons
Discharges or release of persistent and/or toxic chemicals, microbiological agents, nutrients (e.g., nitrogen, phosphorus), radiation, or thermal energy (e.g., cooling waste water)	Negative effects on historical, archaeological, paleontological, or architectural resources
Population declines, particularly in top-predator, large, or long-lived species	Decreased aesthetic appeal or changes in visual amenities (e.g., views)
The removal of resource materials (e.g., peat coal) from the environment	Loss of or damage to commercial species or resources
Transformation of natural landscapes	Foreclosure of future resource use or production
Obstruction of migration or passage of wildlife	
Negative effects on the quality and/or quantity of the biophysical environment (e.g., surface water, groundwater, soil, land, and air)	

Source: Based on FEARO 1994.

Magnitude

This refers to the amount of change in a measurable parameter relative to baseline conditions (e.g., percentage habitat change, amount of change in concentration of a contaminant, percentage change in land use) or other target. Magnitude itself does not always equate directly with significance. For example, a large increase in mercury levels in a local water supply reservoir may not be considered significant if the mercury levels are still within the quality standards for drinking water. Further, magnitude can be affected by several factors, including natural variability or normal fluctuations in baseline conditions, the scale at which the effect is being considered, and the resiliency of the affected environmental component to change. The significance of an effect, in terms of its magnitude, is typically assessed in comparison to deviation from a pre-project baseline condition or from an alternative predetermined measurement point. For example, in the Jack Pine mine EIS, impact magnitude measures for noise, groundwater hydrology, surface water hydrology, and wildlife health were determined on the basis of specified measured baseline conditions (Table 8.5). Magnitude should be expressed in measureable or quantifiable terms when possible. When using qualitative descriptions of magnitude, such as low, medium, or high, clear definitions should be provided.

Table 8.5 Selected Magnitude Criteria for the Jack Pine Mine EIA

Factor	Impact magnitude criteria and indices
Noise	Negligible (0): no projected increase in ambient noise levels
	Low (+5): increased noise levels do exist but do not exceed the nighttime noise criteria
	Moderate (+10): increased noise levels exceed the nighttime criteria
	High (+15): increased noise levels exceed the daytime noise criteria
Groundwater hydrology	Negligible (0): no change from the baseline case
	Low (+5): near (slightly above) the baseline case
	Moderate (+10): above the baseline case
	High (+15): substantially above the baseline case
Surface water hydrology	Negligible (0): < 5% change
	Low (+5): 5–10% change
	Moderate (+10): 10–30% change
	High (+15): > 30% change
Wildlife health	Negligible (0): no appreciable increase in hazard compared to baseline
	Low (+5): possible small increase in hazard compared to baseline
	Moderate (+10): possible moderate increase in hazard compared to baseline
	High (+15): substantial increase in hazard compared to baseline

Source: Shell Canada Ltd 2002.

Probability

If an environmental effect or impact is not likely to happen, then it may not be significant. The determination of likelihood is based on two criteria: (1) the probability of occurrence and (2) scientific certainty. In practice, the likelihood of an attribute of significance is often rated on a scale of "none" (0 per cent chance), "low" (< 25 per cent chance), "moderate" (25–75 per cent chance), and "high" (> 75 per cent chance). While sometimes based on statistical significance, this is only one means of determining probability (Table 8.6).

Table 8.6 Impact Probability Classifications

High	Previous research, knowledge, or experience indicates that the environmental component *has experienced* the same impact from activities of similar types of projects.
Moderate	Previous research, knowledge, or experience indicates that the environmental component *may have experienced* the same impact from activities of similar types of projects.
Low	Previous research, knowledge, or experience indicates there is a *small likelihood* that the environmental component has experienced the same impact from activities of similar types of projects.
Unknown	There is *insufficient research, knowledge, or experience* to indicate whether the environmental component has experienced the same impact from activities of similar types of projects.

Source: Based on BHPB 1998.

Timing, Duration, and Frequency

Significance may be based on the timing of an environmental effect—for example, an effect to caribou that occurs during the species migration period may be considered more significant than if it were to occur at other times of the year. Significance may also be based on the duration of the environmental effect or impact—short-term (1 to 5 years after project completion), medium-term (6 to 15 years after completion), or long-term (more than 15 years). This can refer to the amount of time for a VEC or its indicator to return to baseline conditions through mitigation or natural recovery. It is also important to consider whether the impact is continuous, delayed, or immediate. Environmental effects may not become apparent immediately following the activity causing them but still need to be considered. Examples may include effects on human health, which may not occur until several years or decades following project operation. It is also important to consider the frequency of disturbance and the ability of a particular VEC to fully recover between recurring environmental effects.

Spatial Extent

As depicted in Table 8.3, localized or contained effects may not be considered as significant as effects or impacts observed or experienced at locations far removed from the project, such as **acid mine drainage** or atmospheric emissions. Spatial scale provides a frame of reference for EIA and is an important contextual factor in interpreting significance (see Irwin and Rodes 1992). The challenge, however, is that if large boundaries are defined in determining the significance of a proposed activity, then only a superficial assessment of significance may be possible, and uncertainty will increase. If the boundaries are small, then a more detailed examination may be feasible, but an understanding of the broader context and significance of the proposal may be sacrificed. The spatial scale of the proposed project assessment and of the effects should be considered in significance determination, alongside the nature of the proposal, the characteristics of anticipated effects, and whether cumulative or incremental effects are deemed important to consider within the spatial boundary.

Mitigable Effects

It is also important to consider the frequency of disturbance and the ability of a particular VEC to fully recover between recurring environmental effects. Environmental effects are typically considered less significant if they can be mitigated. One source of particular controversy between proponents and regulators, however, is the notion and role of impact mitigation and mitigated FONSIs—Findings of Non-Significant Impacts. A mitigated FONSI refers to a proposed action that has incorporated mitigation measures to reduce any significant negative effects to insignificant ones (Canter 1996). Impact mitigation is inherent in all aspects of EIA systems, and an important issue in significance determination is whether a significant impact can, and will, be sufficiently mitigated. It is common practice under the US NEPA system, for example, for proponents to prepare and submit project proposals and environmental management plans identifying impact mitigation measures that result in mitigated FONSIs. Tinker et al. (2005) explain that a fundamental problem with this approach, unless the environmental management plan is very precise about specific mitigation measures, is that it is not possible to create a valid condition requiring the development to comply with the proposed mitigation. It should not be assumed that conformity with proposed mitigation rules out the need for EIA. In most cases, mitigation is a series of non-binding proposals in a project proposal or preliminary project or impact management plan (Morrison-Saunders et al. 2001). Byron (2000) reports that mitigation measures proposed give no indication as to whether they will actually be implemented or their effectiveness in mitigating adverse effects. Mitigation measures should not be ignored when making decisions about the likely significant effects of proposed development, but they should not form the lead criterion in the decision as to whether an EIA is required. In cases where mitigation measures are considered in the significance determination process, a decision-maker should consider the likelihood that such mitigation will occur, factors to ensure implementation, and proven effectiveness of the mitigation measure (Ross, Morrison-Saunders, and Marshall 2006).

Standards and Regulations

Predicted effects or impacts following mitigation are often compared against environmental standards or regulations—in essence, specified thresholds. A threshold involves a clearly defined performance level, usually applied to distinguish between significant and insignificant effects. Standards and regulations are perhaps the most common approach to thresholds for assessing impact significance. Environmental impacts within specified standards or regulations are often deemed to be insignificant in comparison to impacts that exceed regulations or, for example, those that attain maximum allowable concentrations. Many jurisdictions have standards for drinking water quality or levels of industry emissions. Lawrence (2004) identifies at least three major types of standards or regulatory thresholds:

- *exclusionary:* leads to automatic rejection of a proposal
- *mandatory:* leads to a mandatory finding of significance
- *probable:* normally significant but subject to confirmation

However, with the exception of California and Australia, significance thresholds are not normally included in EIA requirements.

Cumulative Effects

Environmental effects that are cumulative in nature—that is, they add to or interact with an already existing adverse impact to detract additively or synergistically from environmental quality—are more likely to be deemed significant compared to impacts that do not affect already affected environmental components. Significance cannot be avoided by terming an action "temporary" or by breaking it down into small component parts (Canter 1996). Moreover, significance cannot be defined taking into account a proposed development in isolation from other existing and proposed activities in the region. In practice, significance determination in EIA typically focuses on evaluating the effects of a proposed action and the linear, causal effects, direct and indirect, on particular environmental components at the site of the proposed action. Although often individually insignificant, a combined number of small-scale alterations (e.g., a program of development) can lead to significant environmental change. Accordingly, a comprehensive approach to determining the significance of individual proposals that considers cumulative and induced effects is required.

The term "cumulative environmental effects" is generally used to describe the phenomenon of environmental change resulting from numerous, small-scale alterations, whether they are due to the activities of a single development or the combined effects of multiple developments over space and time. Significance determination should involve considering whether the proposed action is related to other actions that may result in individually insignificant but cumulatively significant effects or impacts and whether the proposed action may establish a precedent for future actions or represent a decision, in principle, about future actions. In other words, a piecemeal approach to assessing proposed developments might circumvent the identification and management of potentially significant cumulative effects. Questions concerning the significance of project impacts should be addressed in light of past, existing, and future actions.

Impact Importance

An important aspect of interpreting significance is the consideration of context—that significance should be assessed relative to society, the affected region, the affected interests, and the locality (Canter 1996). The concept of significance in EIA was defined initially by lists of actions regarded as significantly affecting the environment and the identification of suitable assessment thresholds (US Council on Environmental Quality 1973). While these processes considered the nature and scale of effects in significance determination, the context was disregarded (Benson 2004). Significance determination is highly context-sensitive (Kjellerup 1999). What is considered significant is thus highly dependent on a number of contextual factors in addition to the characteristics of the effect itself. In other words, the importance of a particular environmental effect is related to ecological and societal values about the affected components. The importance or value of the affected components is usually based on societal, ecological, economic, political, public, or regulatory concern.

Ecological Value

From an ecological perspective, potential effects on rare or threatened plants, animals, and their habitat, or on particularly vulnerable species or components of the natural environment that are irreplaceable in terms of ecological functioning, are often deemed to be significant impacts (Figure 8.2). The effects of development in highly sensitive areas or on rare or endangered species or features of the environment, for example, are more likely to be considered significant than the same effects on a resilient or "common" landscape (Table 8.7). In the absence of legal or policy instruments, however, or existing land-use designations that clearly identify such features a priori, interpretations of significance based on the sensitivity or uniqueness of the receiving environment should be supported by baseline information, trends analysis, and widespread public consultation or with reference to specified thresholds, criteria, or desired outcomes and objectives for the receiving environment or landscape.

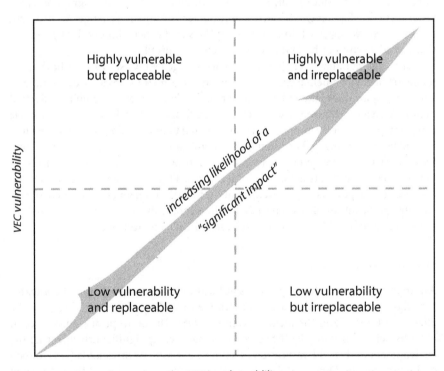

Figure 8.2 ⊙ VEC Vulnerability versus VEC Irreplaceability as a Measure of Significance

Table 8.7 Example Significance Determination for Sensitivity of Landscape Receptors

High sensitivity	Key characteristics and features that contribute to the distinctiveness and character of the landscape. Designated landscapes (e.g., national parks) and landscapes identified as having low capacity to accommodate the proposed form of change.
Medium sensitivity	Other characteristics or features of the landscape that contribute to the character of the landscape locally. Locally valued landscapes that are not designated. Landscapes identified as having some tolerance of the proposed change subject to design and mitigation, etc.
Low sensitivity	Landscape characteristics and features that do not make a significant contribution to landscape character or distinctiveness locally or that are untypical or uncharacteristic of the landscape type. Landscapes identified as being generally tolerant of the proposed change subject to design and mitigation, etc.

Source: Tyldesley and Associates 2005.

Societal Value

Project effects on the biophysical environment often affect factors of importance to society and human life, as well as biophysical components that are valued for aesthetic or sentimental reasons. Societal values that can be affected include:

- human health and safety
- potential loss of green space
- recreational value
- demands on public resources
- demands for infrastructure and services
- demographic effects

Such impacts are likely to be considered significant.

Significance is often measured on the yardstick of values (Lawrence 2005), and any process of identifying and evaluating significance must recognize that the determination of significance is inherently an anthropocentric concept (Rossouw 2003). The values and concerns (or judgments) of decision-makers and of the publics underlie interpretations of significance. As Baker and Rapaport (2005) explain, evaluation of significance based strictly on scientific data is inadequate in many cases, because resources and ecosystems are linked with human values and cultural meaning. As Hilden (1997) explains, a community directly affected by a proposed undertaking may regard any identifiable effect as highly significant, whereas a community outside of the affected area may have little interest in the proposal. On the other hand, if a development is proposed for a sensitive environment (e.g., active sand dunes or native grasslands), stakeholders who are far removed from the direct effects of development may also consider any identifiable impacts as highly significant. In the latter case, such fears may stem from concern over biodiversity or landscape functioning. In that respect, they reflect the same significance concerns held by decision-makers and experts or reflect the need to consider the sensitivity and value of the receiving environment more closely as the main trigger for assessment. Determination of significance and the need for assessment should include consideration of public concerns and the values they represent; however, in Canadian systems it is widely held that significance determination must be based on scientific and credible technical information (Sadler 1996).

Desired Thresholds

Thresholds need not always be based on standards or regulatory requirements. As already discussed, thresholds for adverse impacts are often stated as *maximum allowable effects levels*. These levels are often based on the desired outcomes and are determined by policy, compatibility with existing plans or programs, or societal objectives. In cases where predicted impacts exceed desired thresholds or maximum allowable change for particular environmental components, the impact is considered significant—even though it may not exceed regulatory standards or thresholds. Haug et al. (1984) proposed a set of criteria to help determine the priority of threshold concern and the probability that an effect will cross the threshold of concern. Each is determined by a maximum or minimum value (quantitative or qualitative). Listed from high to low priority for significance determination, the criteria include:

- *Legal thresholds:* established by law or by regulatory limits
- *Functional thresholds:* established for resource use so as to avoid adverse impacts or disturbances that may disrupt ecosystem functioning or destroy resources
- *Normative thresholds:* established by communities or regions based on social norms in relation to environmental, economic, or social concerns
- *Controversial thresholds:* impacts that may have a high profile because of potential conflicts between the affected interest groups but would not otherwise have priority
- *Individual thresholds:* based on the priorities of individuals, groups, or organizations and do not warrant higher priority based on the above reasons

Approaches to Significance Determination

Determining significance is a highly subjective process, yet significance can and should be determined in a consistent and systematic fashion. Several approaches can be used for significance determination, including technical, collaborative, and reasoned argumentation (see Lawrence 2005), and various methods and techniques are available to support such approaches, including Geographic Information Systems, simulation modelling, and **statistical significance** tests; data scaling and screening procedures, such as threshold analysis and constraint mapping; qualitative and quantitative aggregation and evaluation procedures, including concordance analysis, multi-criteria analysis, ranking and weighting, and risk assessment; and formal and informal public interaction procedures, such as open houses, workshops, and advisory committees (Lawrence 2004). As with many other aspects of EIA practice, there is no specific set of methods and techniques for determining significance. A number of guidelines, however, should be adhered to—namely, that significance thresholds, criteria, and methods be explicit, easy to use, traceable, verifiable, readily understandable, relevant to the problem at hand, and capable of facilitating the interpretation of significance (Sadler 1996).

Technical Approach

Under the technical approach, determining significance typically involves adopting one or more standardized scaled or quantified methods, each of which is based on some characterization of the various dimensions of the anticipated impacts of development, such as impact scale, severity, reversibility, probability, and duration. The technical model captures the most widely used set of standardized methods for significance determination (Whitelaw 1997) and currently forms the basis of ISO 14004 guidance for the evaluation of significance.

Under the technical approach, when a predicted effect meets a pre-established standard or threshold, determining significance is relatively straightforward. However, in many cases environmental effects are evaluated only in terms of relative significance. For example, is a short-term effect on human health from factory emissions more significant than a major long-term increase in local employment? It is often the case in EIA that effects and impacts are expressed as "major" or "minor" without any evaluation of the relative importance of the environmental components under consideration (Figure 8.3). Yet without some type of weighting approach to determine the **relative significance** of all the environmental components, comparing project action "i" on environmental components "x" and "y" gives little indication of actual significance (Figures 8.4 and 8.5). Some examples of technical approaches to relative significance determination include fixed-point scoring, rating, and paired comparisons. Hajkowicz, McDonald, and Smith (2000) provide a more detailed discussion of weighting methods and techniques.

Fixed-Point Scoring
In **fixed-point scoring**, a fixed number of values are distributed among all affected environmental components. The higher the point score, the more important the environmental component. In other words, project effects on environmental components assigned high scores are likely to be considered more significant than effects on environmental components with low fixed-point scores. For example, assume four VECs for which a fixed number of points, totalling no more than 1, must be distributed to indicate relative VEC importance:

VEC	Importance ($w_i / 1$)
water quality	0.25
noise	0.10
employment	0.30
human health	0.35

Significance of residual effects

Significance	Symbol
not applicable	blank
negligible	○
negligible/minor	◉
minor	●
minor/moderate	◆
moderate	★
moderate/major	✹
major	⊗
unknown	

Project Period and Activity

Construction Phase

Activity / VECs	air quality	permafrost	eskers	water quality	fish and aquatic habitat	vegetation	wildlife and wildlife habitat	caribou	grizzly bears	wilderness	biodiversity	hydrology	climate	groundwater
blasting	○													
road traffic	○													
roads, dams, infrastructure		○		●	◉	○		●	●	◉		○		
quarrying			◉											
noise								●	◉					
channel diversion				○				●				○		
lake dewatering				●								○	○	○
pre-stripping						●								
diesel power generation						○								
process plant								●	●					
human activity								●	●	◉	○			
tailings disposal								●	●					
flow and stream diversion												◉		

Operations Phase

Activity / VECs	air quality	permafrost	eskers	water quality	fish and aquatic habitat	vegetation	wildlife and wildlife habitat	caribou	grizzly bears	wilderness	biodiversity	hydrology	climate	groundwater
process plant operation	○					○	○	●						
diesel power generation	○					○							○	
roads and road traffic	○			●			○	●	◉					
tailings water disposal		○		○				○						
freshwater supply					●									
waste rock dumping						○							⊗	
excavation of pits					●									
winter roads					◉									
human activity								●	◉	◉	○			
diversion channel								●						
noise								●	◉					
flow and stream diversion												◉		○

Figure 8.3 ◉ Sample Impact Significance Matrix

Source: Based on BHPB 1998.

Project Actions

Environmental parameter	blasting	site clearing	dredging	road construction	waste disposal	equipment transport	Impact Score
Air quality	−1			−1	−1		−3
Water resources	−2	−3	−3				−8
Water quality	−2	−4	−2				−8
Noise	−2		−1	−2		−2	−7
Forests and vegetation		−5		−3			−8
Wildlife	−2	−4		−2			−8
Human health	−2			−1	−4		−7

+ = positive impact
− = adverse impact

no impact =
negligible impact = 1
minor (slight or short term) = 2
moderate impact = 3
major impact (irreversible or long-term) = 4
severe impact (permanent) = 5

Figure 8.4 ◉ Unweighted Impact Assessment Matrix.

Assuming an example of simple additive impacts, the impact of site clearing on forest and vegetation appears to be the most significant individual project impact. In terms of overall project impacts, the impacts of project activities on water resources, water quality, forests and vegetation, and wildlife appear to be equally significant. The relative importance of the affected environmental components have not been determined.

Project Actions

Environmental Parameter	Weight	blasting	site clearing	dredging	road construction	waste disposal	equipment transport	Impact Score
Air quality	0.16	-1(0.16) = -0.16			-1(0.16) = -0.16	-1(0.16) = -0.16		-0.48
Water resources	0.08	-2(0.08) = -0.16	-3(0.08) = -0.24	-3(0.08) = -0.24				-0.64
Water quality	0.22	-2(0.22) = -0.44	-4(0.22) = -0.88	-2(0.22) = -0.44				-1.76
Noise	0.04	-2(0.04) = -0.08		-1(0.04) = -0.04	-2(0.04) = -0.08		-2(0.04) = -0.08	-0.28
Forests and vegetation	0.08		-5(0.08) = -0.40		-3(0.08) = -0.24			-0.64
Wildlife	0.08	-2(0.08) = -0.16	-4(0.08) = -0.32		-2(0.08) = -0.16			-0.64
Human health	0.22	-2(0.22) = -0.44			-1(0.22) = -0.22	-4(0.22) = -0.88		-1.54

+ = positive impact
- = adverse impact

no impact =
negligible impact = 1
minor (slight or short-term) = 2
moderate impact = 3
major impact (irreversible or long-term) = 4
severe impact (permanent) = 5

Figure 8.5 ◉ Weighted Impact Assessment Matrix

Weights are assigned to indicate the importance of the affected environmental components and therefore the relative significance of the impacts. Using the same example as in Figure 8.3, the impact of 'site clearing' on 'forest and vegetation' is no longer the most significant individual project impact. Given the importance of the affected components, the impacts of 'site clearing' on 'water quality' and of 'waste disposal' on 'human health' are now the most significant individual impacts. In terms of overall project impacts, the impacts of project activities on 'water quality' are now the most significant whereas overall impacts on 'forest resources' and 'wildlife' now appear to be relatively less significant. As a result, the unweighted assessment matrix depicted in Figure 8.4 for determining significance may lead to an incorrect allocation of resources for impact management.

Increasing the importance of any one VEC directly affects the relative significance of project effects on that VEC. At the same time, those assigning the weights, whether EIA decision-makers, experts, or public interest groups, are forced to make trade-offs between VECs, because increasing the importance of one VEC requires decreasing the importance of another. While explicit for decision-making purposes, direct trade-offs between VECs may not always be possible.

Rating

With a **rating** approach to assigning impact importance, the importance or significance of each VEC is indicated on a numerical rating scale—for example: 1 = not important; 5 = moderately important; 10 = extremely important; 2, 3, 4, 6, 7, 8, 9 = intermediate values. While rating does not require direct trade-offs in assignment of weights, the disadvantage is that it does not directly indicate the relative importance of one component in comparison to another.

Paired Comparisons

As with fixed-point scoring, **paired comparisons** force the decision-maker to consider trade-offs. However, whereas fixed-point scoring requires multiple, simultaneous, and often complex trade-offs for the entire list of environmental components, paired comparisons involve making trade-offs one at a time, for each pair of VEC, thereby contributing to better overall understanding of the decision problem.

The paired comparison approach is based on the Analytical Hierarchy Process (AHP), a systematic procedure for representing the elements of any problem hierarchically. The AHP organizes the basic rationality by breaking down a problem into its smaller constituent parts and then guides the decision-maker through a series of pair-wise comparison judgments to express the relative importance of the elements in hierarchy in ratio form, from which decision weights are derived based on the principal eigenvector approach (Saaty 1977). An eigenvector is a linear combination of variables that consolidates the variance, or eigenvalues, in a matrix. The eigenvalues indicate the relative strength (weight) of each of the VECs, where the larger the eigenvalue, the larger the role the paired comparison plays in weighting the entire assessment matrix and therefore determining significance.

For any pair of environmental components, or VECs i and j, out of the set of components C, the individual decision-maker can provide a paired comparison Cij of the components under consideration on a ratio scale that is reciprocal, such that $Cji = 1/Cij$. In other words, if environmental component VEC i is "seven times" more important than VEC j, then the reciprocal property must hold—that is, VEC j must be seven times less important than VEC i. When a decision-maker compares any two VECs i, j, one VEC can never be judged to be infinitely more important than another. If such a case should arise where VEC i is infinitely more than j, then no decision tool would be required.

Using the paired-comparison approach, decision-makers are presented with a 9-point decision scale ranging from 1, if both components or VECs are equally preferred, 3 for a weak preference of VEC i over j, 5 for a strong preference, and so forth. If VEC j is preferred to i for any given criteria, the reciprocal values hold true: 1, 1/3, and 1/5 (Box 8.1). The scale is standardized and unit-free; thus, there is no need to transform all measures, for example, into monetary units for comparative purposes, which is a key advantage of paired comparisons over cost-benefit analysis or utility functions. An additional advantage of paired comparisons over other weighting methods is that the paired comparison approach also provides a measure of consistency or of the extent to which weights were assigned purposefully or at random.

While paired comparison generates the most informative weightings for determination of significance, it is also the most complex of the three approaches presented here. In fact, when relying on the public or experts to assign the values, the approach is limited to the number of items that an individual can simultaneously compare—usually 7 +/– 2 (Miller 1956). However, when relying on simulation models or computer-based technology such as Geographic Information Systems, the number of environmental components that can be weighted is limited only by the time and resources available to the EIA practitioner.

Box 8.1 Paired Comparison Process for Assigning Weights to Environmental Components

The relative importance of component Ci in the assessment of project impacts is defined by:

9 = component *i* is extremely more important than *j*
7 = component *i* is very important compared to component *j*
5 = component *i* is strongly more important than component *j*
3 = component *i* is moderately more important than component *j*
1 = component *i* is equally important to component *j*
1/3 = component *j* is moderately more important than component *i*
1/5 = component *j* is strongly more important than component *i*
1/7 = component *j* is very important compared to component *i*
1/9 = component *j* is extremely more important than component *i*

Sample paired comparison assessment matrix for environmental components (VECs)

	VEC A	VEC B	VEC C
VEC A	1	1/3	7
VEC B	3	1	7
VEC C	1/7	1/7	1

For example, VEC B is considered "moderately more important" than VEC A, so a value of 3 is entered in row 2, column 1; thus 1/3 in row 1, column 2.

To derive the weights for each VEC:

1. Divide each cell entry of the matrix by the sum of its corresponding column to normalize the matrix:

	VEC A	VEC B	VEC C
VEC A	0.24	0.23	0.47
VEC B	0.72	0.68	0.47
VEC C	0.03	0.10	0.07

2. Determine the priority vector of the matrix by averaging the row entries in the normalized assessment matrix:

   Priority vector: A = 0.31 B = 0.62 C = 0.07

3. Multiply each column of the initial paired comparison matrix by its priority and sum the results:

$$
0.31 \begin{array}{c} 1 \\ 3 \\ 1/7 \end{array} + 0.62 \begin{array}{c} 1/3 \\ 1 \\ 1/6 \end{array} + 0.07 \begin{array}{c} 7 \\ 7 \\ 1 \end{array} = \begin{array}{c} 1.01 \\ 2.04 \\ 0.21 \end{array}
$$

4. Divide the results by the original priority vector to determine the relative importance or weight of each VEC:

   VEC A = 1.05 / 0.31 = 3.26
   VEC B = 2.04 / 0.62 = 3.29
   VEC C = 0.21 / 0.07 = 3.00

5. The relative importances are used to interpret the relative significance of project impacts on each of the affected VECs.

Collaborative Approach

The collaborative approach, explains Lawrence (2005, 16), "starts from the premise that subjective, value-based judgments about what is important should result from interactions among interested and affected parties." Under this approach, there are no predefined thresholds or criteria for judgment—rather, judgments of what is important, what is acceptable, and what are the limits of allowable change emerge only through consultation with the publics. Issues pertaining to regional and community context are integrated directly into significance determination, and the compatibility of the proposal and its impacts with regard to goals and objectives are central to the decision-making process. Because the approach is collaborative, the key role of regulators is ensuring the involvement of the most directly affected communities or stakeholders and that all stakeholder concerns are assessed in significance determination.

Numerous methods and approaches are available to facilitate collaborative approaches to significance determination, including open houses and community forums, interactive web-based forums, key informant interviews, community or regional profiling, rapid rural appraisal, nominal group processes, and intervener funding. The overall objectives of using such methods are to fully integrate community, technical, and traditional knowledge in significance determination and to forge a stronger relationship with the public throughout the EIA process. It is important to note, however, that public concerns about environmental effects are not always the same as actual environmental effects resulting from project actions.

Reasoned Argumentation

The reasoned argumentation approach views significance determination as based on reasoned judgments supported by evidence. Lawrence (2005, 19) explains that reasoned argumentation starts from the premise that technical and collaborative models are "too narrow to provide an adequate foundation for value-based significance judgments about what is important and what is not important." Usually expressed qualitatively, the reasoned argumentation model is evident at the regulatory level in priorities and objectives of EIA legislation or regulation, often defining "matters of significance," which are used as triggers during the screening process, and further expressed in project-specific guidelines and requirements. At the practical level, reasoned argumentation, similar to a court decision process, involves sifting through information, data, perspectives, and expressed values using structured methods (e.g., decision support aids, matrices, network diagrams) to focus on matters of most importance to decision-making and to build "reasoned" arguments that support significance determination—hence the importance of complete and clear documentation of the evaluation process and reasons for the decision (Kontic 2000). The reasoned argumentation model is flexible and responsive to context; however, a well-reasoned argument for significance does not ensure that full consideration has been given to scientific data, public values, or existing technical information.

Composite Approach

The choice of approach to significance determination will vary with context, taking into consideration the nature of the project, the receiving environment, interests, and the political setting. In a study of methods used to determine significance in EIA practices in England, for example, Wood, Glasson, and Becker (2006) found that professional judgment and experience, consultation, and simple checklists of impacts were the methods most commonly used by local planning authorities and consultants. In other words, a composite model consisting of various combinations of technical, collaborative, and argumentative approaches is desirable. Such a model may consist of technical analysis using conventional significance determination methods, supported by public consultation or traditional knowledge systems, which together, based on existing EIA regulation or land-use plans, compile a reasoned argument for significance. Lawrence (2005) suggests that at minimum, effective impact significance determination relies upon:

- use of a variety of technical methods and analytical techniques;
- a range of public consultation methods, including methods that facilitate stakeholder interaction, to support collaborative significance determination;
- use of existing regional, community, or land-use plans or local social surveys to identify values and aspirations against which to compare the proposed development;
- literature analysis and case study reviews of previous, similar proposals and outcomes in comparable environments and situations;
- exploring the uncertainties and acceptable risks associated with the significance determination.

Toward a Higher Test for Impact Significance

If the underlying objective of EIA is to ensure sustainability, then an important question to ask in determining significance is whether the proposal will make an overall positive. While the definitions and interpretations of sustainability vary considerably, it is generally acknowledged that sustainability includes the desire to maintain, over an indefinite future, necessary and desired attributes of ecological and socio-economic environments that are necessary for system functioning and integrity (Deakin, Curwell, and Lombari 2002). Thus, an alternative approach to determining impact significance is to ask whether the project's impacts make a positive contribution to or detract from sustainability. Gibson (2001), for example, identifies 12 generic sustainability-based questions for evaluating the significance of environmental impacts:

1. Could the effect add to stress that might undermine ecological integrity at any scale in ways, or to an extent, that could damage life-support functions?
2. Could the effect contribute to ecological rehabilitation and/or otherwise reduce stress that might otherwise undermine ecological integrity at any scale?
3. Could the effect provide more economic opportunities for human well-being while reducing material and energy demands and other stresses on socio-ecological systems?
4. Could the effect reduce economic opportunities for human well-being and/or increase material and energy demands and other stresses on socio-ecological systems?
5. Could the effects increase equity in the provision of material security, including future as well as present generations?
6. Could the effect reduce equity in the provision of material security, including future as well as present generations?
7. Could the effect build government, corporate, and public incentives and capacities to apply sustainability principles?
8. Could the effect undermine government, corporate, or public incentives and capacities to apply sustainability principles?
9. Could the effect contribute to serious or irreversible damage to any of the foundations of sustainability?
10. Are the relevant aspects of the undertaking designed for adaptation if unanticipated effects emerge?
11. Could the effect contribute positively to several or all aspects of sustainability in a mutually supportive way?
12. Could the effect in any aspect of sustainability have consequences that might undermine prospects for improvement in another?

These are not new questions, but their direct and explicit focus on sustainability ensures that sustainability considerations are front and centre in significance determinations. In the case of the Voisey's Bay mine project, introduced in Chapter 1, the "sustainability of resources," defined as the capacity of an affected resource to meet present and future needs (Voisey's Bay Nickel Company 1997), was used as a factor to assist in determining the significance of residual impacts based on such concepts as ecosystem integrity, carrying capacity, and assimilative capacity (Table 8.8). In other cases, such as the Kemess North mine project, the sustainability test played a key role in the review panel's determination that the project's impacts were simply unacceptable. See Environmental Assessment in Action: Application of the "Sustainability Test" for Significance: The Case of the Kemess North Mine Project, British Columbia.

Table 8.8 Sustainability Criteria Used in the Voisey's Bay EIS for Significance Determination

Sustainability rating	Sustainable use of renewable resources criteria
High	Previous research/experience indicates that the environmental effect on the VEC would not reduce biodiversity or the capacity of resources to meet present and future needs.
Moderate	Previous research/experience indicates that the environmental effect on the VEC may, to a certain extent, reduce biodiversity or the capacity of resources to meet present and future needs.
Low	Previous research/experience indicates that the environmental effect on the VEC would reduce biodiversity or the capacity of resources to meet present and future needs.
Nil	Previous research/experience indicates that the environmental effect on the VEC would eliminate biodiversity or the capacity of resources to meet present and future needs.
Unknown	There is insufficient research/experience to indicate whether the environmental effect on the VEC would reduce biodiversity or the capacity of resources to meet present and future needs.

Source: Voisey's Bay Nickel Company 1997.

Environmental Assessment in Action

Application of the "Sustainability Test" for Significance: The Case of the Kemess North Mine Project, British Columbia

In 2005, Northgate Mineral Corporation filed an environmental impact statement for development of the proposed Kemess North copper and gold deposit, located approximately 250 kilometres northeast of Smithers and 450 kilometres northwest of Prince George, British Columbia. The proposal represented an expansion of the existing Kemess South mine, located just 6 kilometres south, and would include development of a new open pit mining operation, modification of the existing mill, and related infrastructure. The proposed new mine site would make use of existing infrastructure at the south mine site, including the work camp, airstrip, access roads, and power lines. Current ore milling capacity would be increased from 55,000 tonnes per day to up to 120,000 tonnes per day, with an estimated 397 million tonnes of tailings and 325 million tonnes of waste rock produced. To reduce the risk of acid rock drainage, the proposed undertaking would also involve modification of the adjacent Duncan Lake to increase its capacity for waste rock and tailings disposal.

The project was subject to an environmental assessment under the British Columbia Environmental Assessment Act and the Canadian Environmental Assessment Act. Given the potential of the project to result in significant adverse environmental effects, the use of a lake for mined waste disposal, and the potential effects on Aboriginal people, both the provincial and federal governments determined that an independent review panel be established to carry out a public review of the project, in accordance with the Canada–British Columbia Agreement on Environmental Assessment Cooperation.

The joint review panel was established in May 2005 to conduct an assessment of the potential environmental, economic, social, health, and heritage effects of the project, including such effects on Aboriginal people. In conducting its assessment, the panel consulted mining sector sustainability initiatives, as well as the BC government's 2005 Mining Plan. The panel examined the significance of the project's effects from five key sustainability perspectives: environmental stewardship, economic benefits and costs, social and cultural

continued

benefits and costs, fairness in the distribution of benefits and costs, and present versus future generations.

The panel completed its assessment in 2007 and recommended to the federal and provincial governments that the project not be approved as proposed. In its report, the panel noted that "the economic and social benefits provided by the project, on balance, are outweighed by the risks of significant adverse environmental, social and cultural effects, some of which may not emerge until many years after mining operations cease."

The panel identified several key concerns in relation to the significance of the project's potential environmental effects:

- Environmental stewardship: The creation of a long-term site management legacy was a significant environmental concern, and there were doubts about how much assurance could be provided that Northgate Mineral Corporation's site management regime would remain effective over the long term.
- Economic benefits and costs: Although significant benefits would accrue to mine workers and suppliers, government, and shareholders, there were significant concerns with respect to the short duration of the incremental economic benefits—2 years of construction and only up to 11 years of mining production.
- Social and cultural benefits and costs: The project would likely make a significant contribution to social well-being and community stability in communities where workers live and service suppliers operate, but the socio-cultural implications of the project for Aboriginal people, and obstacles to their participation in project benefits, was a significant concern.
- Fair distribution of benefits and costs: There was concern about the likelihood for inequities in the distribution of benefits and costs between those interests that would receive most of the benefits (workers, suppliers, governments, company shareholders) and those people who would incur most of the costs (locally based, primarily Aboriginal, people).
- Present versus future generations: The creation of a long-term legacy of substantial mine site management and maintenance obligations, lasting for thousands of years, was found to be a major imposition on future generations

This was the first time in Canada that a joint review panel recommended outright rejection of a mining project based on the potential for significant adverse impacts.

Source: Based on the Kemess North Copper-Gold Mine Project Joint Review Panel Report Summary, 17 September 2007.

Guidance for Determining Significance

Sadler (1996) identifies several guiding principles for significance interpretation, suggesting, for example, that significance thresholds, criteria, and methods be explicit, easy to use, traceable, substantiated, readily understandable, and relevant to the problem at hand. Methods should be broadly supported and structured and focus the significance interpretation effort, provision should be made to involve interested and affected stakeholders, and the rationale for significance interpretations should be clearly indicated. Building on the work of Sadler and others, Lawrence (2005) provides a more comprehensive list of guiding principles for significance determination:

1. Focused and efficient: Concentrate on aspects most relevant to decision-making and consistent with regulatory and public concern.
2. Explicit: Values and bases for judgments are clear and understandable.
3. Logical and substantiated: Reasoning behind significance determination is logical, and data and analyses are clearly linked to significance judgments.
4. Systematic and traceable: A coherent procedure exists for integrating information and objectives, and that procedure can be reconstructed.
5. Appropriate: Sensitivity of the context is considered in the significance determination.
6. Consistent: Similar projects and issues are treated in a similar manner.
7. Collective and collaborative: Interested and affected parties are involved.
8. Effective: Outcomes of significance determination help to realize public, policy, or EA goals and objectives.
9. Adaptable: Procedures for significance determination are flexible to different contexts and changing circumstances over space and time.

It is not likely that all of these principles can be adhered to simultaneously; however, when determining whether the effects of a proposed development are significant, consideration should be given not only to the characteristics of the effects but also to the contextual factors of public concern, conditions and sensitivity of the receiving environment, specified thresholds and objectives, significance scale, cumulative change and induced effects, contributions to sustainability, and the nature of proposed mitigation measures. Significance is a highly subjective concept, and differences of opinion are likely to continue as to what constitutes a significant environmental effect, how contextual factors are to be considered, what represents an appropriate level of detail for regulatory requirements and standards, and how to make significance determinations that are transparent and robust. The challenges for significance determination do not lie simply in the realms of improved science and the pursuit of objective expert evaluations but in the clarity of communication of the assessment to decision-makers and the stakeholder community as well (Wood 2008). Significance is not so much the search for objectivity as it is the adequate substantiation of any subjectivity (Lawrence 1993).

Key Terms

acid mine drainage	fixed-point scoring
assimilative capacity	impact significance

paired comparisons
rating
relative significance

residual effects
statistical significance

Review Questions and Exercises

1. Determining the significance of environmental effects is often a subjective process, particularly when issues of social concern are involved. Suppose, for example, that a proposed development is likely to result in the closure of a local outdoor public recreational area. How might you assess the significance of such an impact?

2. Obtain a completed project EIS from your local library or government registry, or access one online. Identify the types of criteria used to determine impact significance. Compare your findings to those of others. Are there noticeable similarities among the significance criteria across impact statements? Is there evidence of "sustainability criteria"?

3. Impact significance in EIA is often classified as "major," "moderate," "minor," or "negligible," without any qualification as to what these terms mean. Following the example illustrated in Figure 8.3, construct a simple impact significance matrix for the construction and operation of a waste incineration project. Identify project actions across the top and VECs down the side. Develop a legend for impact significance similar to that in Figure 8.3, and provide operational definitions for each level of significance based on the significance criteria discussed in this chapter.

4. Discuss the advantages of weighted impact assessment matrices over unweighted impact assessment matrices.

5. Construct a 3 × 2 table to compare and contrast the relative advantages and disadvantages of technical, collaborative, and reasoned argumentation approaches to significance determination.

6. Assume that the following VECs have been identified for a large-scale energy project to be developed in your region:
   - air quality
   - employment
   - forests and vegetation
   - water quality
   - human health

   a) Use fixed-point scoring to assign weights to the affected VECs so that the total of all weights equals 1.

   b) Use a numerical rating scale to assign weights to the affected VECs so that 1 = not important, 3 = moderately important, 5 = important, 7 = very important, and 9 = extremely important.

   c) Construct a paired comparison matrix, and using the scale depicted in Box 8.1, calculate weights for each of the VECs.

   d) For each of the above approaches, standardize each of the weights for each VEC by using the following scaling parameter: $(i - i_{min}) / (i_{max} - i_{min})$, where *i* is the respective weight and i_{min} and i_{max} represent the minimum and maximum values of all weights, respectively, for the set of VECs. This will generate a standardized scale in which the least important VEC = 0 and the most important VEC = 1.

   e) Compare your results across weighting techniques. Are the weights different? Why?

   f) Discuss the advantage of using paired comparisons over the other two weighting techniques.

7. Who should be responsible for determining impact significance? Are there certain criteria that would apply to practically all proposed developments in your area? What is the role of the public in determining impact significance?

References

Baker, D., and E. Rapaport. 2005. "The science of assessment: Identifying and predicting environmental impacts." In K. Hanna, ed., *Environmental Impact Assessment Practice and Participation*. Toronto: Oxford University Press.

Benson, W. 2004. "Determining significant effects within environmental management: A case study of the UK Ministry of Defence." Proceedings of the International Sustainable Development Research Conference, University of Manchester.

BHPB (Broken Hill Proprieties Billiton). 1998. *NWT Diamonds Project Environmental Impact Statement*. Vancouver: BHP Diamonds Inc.

Byron, H. 2000. *Biodiversity and Environmental Impact Assessment: A Good Practice Guide for Road Schemes*. Sandy, UK: The RSPB, WWF-UK, and English Nature and Wildlife Trusts.

Canter, L.W. 1996. *Environmental Impact Assessment*. 2nd edn. New York: McGraw-Hill.

———, and G.A. Canty. 1993. "Impact significance determination—Basic considerations and a sequenced approach." *Environmental Impact Assessment Review* 13: 275–97.

Deakin, M., S. Curwell, and P. Lombari. 2002. "Sustainable urban development: The framework and directory of assessment methods." *Journal of Environmental Assessment Policy and Management* 11 (3): 171–97.

Duinker, P.N., and G.E. Beanlands. 1986. "The significance of environmental impacts: An exploration of the concept." *Environmental Management* 10 (1): 1–10.

FEARO (Federal Environmental Assessment Review Office). 1994. "Reference guide: Determining whether a project is likely to cause significant adverse environmental effects." In Canadian Environmental Assessment Agency, ed., *The Canadian Environmental Assessment Act Responsible Authority's Guide*. Ottawa: Supply and Services Canada.

Gibson, R.B. 2001. *Specification of Sustainability-Based Environmental Assessment Decision Criteria and Implications for Determining "Significance" in Environmental Assessment*. A report prepared under a contribution agreement with the Canadian Environmental Assessment Agency Research and Development Program. Gatineau, QC: CEAA.

Hajkowicz, S.A., G.T. McDonald, and P.N. Smith. 2000. "An evaluation of multiple objective decision support weighting techniques in natural resource management." *Journal of Environmental Planning and Management* 43 (4): 505–18.

Haug, P.T., et al. 1984. "Determining the significance of environmental issues under the

National Environmental Policy Act." *Journal of Environmental Management* 18: 15–24.

Hilden, M. 1997. "Evaluation of the significance of environmental impacts." Report of the EIA Process Strengthening Workshop, 4–7 April, Canberra, Australia.

Irwin, F., and B. Rodes. 1992. *Making Decisions on Cumulative Environmental Impacts: A Conceptual Framework.* Washington: World Wildlife Fund.

Kemess North Copper-Gold Mine Project Joint Review Panel. Report Summary. 17 September 2007.

Kjellerup, U. 1999. "Significance determination: A rational reconstruction of decisions." *Environmental Impact Assessment Review* 19: 3–19.

Kontic, B. 2000. "Why are some experts more credible than others?" *Environmental Impact Assessment Review* 20: 427–34.

Lawrence, D.P. 1993. "Quantitative versus qualitative evaluation: A false dichotomy?" *Environmental Impact Assessment Review* 13 (1): 3–12.

———. 2000. "Significance in environmental assessment." Research and Development Monograph Series. Gatineau, QC: CEAA.

———. 2004. *Significance in Environmental Assessment.* Research supported by the Canadian Environmental Assessment Agency's Research and Development Program for the Research and Development Monograph Series, 2000. Gatineau, QC: CEAA.

———. 2005. "Significance criteria and determination in sustainability-based environmental impact assessment." Report to the Mackenzie Gas Project Joint Review Panel. Langley, BC: Lawrence Environmental.

Lyhne, I., and L. Kørnøv. 2013. "How do we make sense of significance? Indications and reflections on an experiment." *Impact Assessment and Project Appraisal* doi: 10.1080/14615517.2013.795694.

Miller, G.A. 1956. "The magical number seven plus or minus two: Some limits on our capacity for processing information." *Psychological Review* 63: 81–97.

Morrison-Saunders, A., et al. 2001. "Roles and stakes in environmental impact assessment follow-up." *Impact Assessment and Project Appraisal* 19 (4): 289–96.

Ross, W.A., A. Morrison-Saunders, and R. Marshall. 2006. "Common sense in environmental impact assessment: It is not as common as it should be." *Impact Assessment and Project Appraisal* 24 (1): 3–22.

Rossouw, N. 2003. "A review of methods and general criteria for determining impact significance." *African Journal of Environmental Assessment* 6: 44–61.

Saaty, T.L. 1977. "A scaling method for priorities in hierarchical structures." *Journal of Mathematical Psychology* 15: 243–81.

Sadler, B. 1996. *Environmental Assessment in a Changing World: Evaluating Practice to Improve Performance.* Final report of the International Study of the Effectiveness of Environmental Assessment. Fargo, ND: IAIA.

Shell Canada. 2002. *Application for Approval of the Jack Pine Mine—Phase 1.* Environmental Impact Statement submitted to the Alberta Energy and Utilities Board and Alberta Environment. Edmonton.

Sippe, R. 1999. "Criteria and standards for assessing significant impact." In J. Petts, ed., *Handbook of Environmental Impact Assessment,* vol. 1, *Environmental Impact Assessment: Process, Methods and Potential.* London: Blackwell Science.

Tinker, L., et al. 2005. "Impact mitigation in environmental assessment: Paper promises or the basis of consent conditions?" *Impact Assessment and Project Appraisal* 23 (4): 265–80.

Tyldesley and Associates. 2005. *Environmental Assessment Handbook: Guidance on the Environmental Assessment Process.* Doc Ref: 1477 4th edition, Issue: 04A.

US Council on Environmental Quality. 1973. *Preparation of Environmental Impact Statements: Guidelines*. Washington: Council on Environmental Quality.

———. 1987. "Regulations for implementing NEPA." Section 1508.27.40 Code of Federal Regulations, 1987. http://ceq.eh.doe.gov/nepa/regs/ceq/toc_ceq.htm.

Voisey's Bay Nickel Company. 1997. *Voisey's Bay Mine/Mill Project Environmental Impact Statement*. St John's: Voisey's Bay Nickel Company.

Westman, W.E. 1985. *Ecology, Impact Assessment and Environmental Planning*. New York: John Wiley & Sons.

Whitelaw, K. 1997. *ISO 14001 Environmental System Handbook*. Oxford: Butterworth-Heinnemann.

Wood, C. 1995. *Environmental Impact Assessment: A Comparative Review*. London: Longman Scientific and Technical.

———, and J. Becker. 2004. "Evaluating and communicating impact significance in EIA: A fuzzy set approach to articulating stakeholder perspectives." Paper presented at the annual meeting of the International Association for Impact Assessment, 26–29 April.

Wood, G. 2008. "Thresholds and criteria for evaluating and communicating impact significance in environmental statements: See no evil, hear no evil, speak no evil." *Environmental Impact Assessment Review* 28: 22–38.

———, J. Glasson, and J. Becker. 2006. "EIA scoping in England and Wales: Practitioner approaches, perspectives and constraints." *Environmental Impact Assessment Review* 26: 221–41.

Follow-up

Much of what has been covered in this book thus far has focused on what happens before a decision is made to approve or not to approve a project. In this context, EIA is a linear, predictive process with no mechanism to verify impact predictions or evaluate the effectiveness of measures proposed to manage actual project impacts. Follow-up occurs after a project's approval and involves determining whether a project has had or is continuing to have environmental effects (Kilgour et al. 2007). Follow-up is the element that transforms EIA from a static to a dynamic process—the link between EIA and effective **life-cycle assessment** (Arts, Caldwell, and Morrison-Saunders 2001). The rationale for follow-up is similar to that of EIA itself—to come to grips with the uncertainties intrinsic to a particular activity. Some EIA practitioners interpret follow-up strictly as ensuring that mitigation measures identified in the assessment have been implemented, while others view follow-up as an umbrella that covers the activities, such as routine monitoring or auditing, undertaken during the post-decision stages of the environmental assessment process.

Follow-up Requirements

In Canada, under the Canadian Environmental Assessment Act, 2012, a follow-up program means a program for:

a) verifying the accuracy of the environmental assessment of a designated project;
b) determining the effectiveness of any mitigation measures.

Regarding the first point, one question that emerges is whether there is much value in verifying the accuracy of impact predictions when environmental conditions are constantly changing. Thus, it is the second aspect of follow-up, its management function, that is of greater value (Box 9.1). This is the perspective adopted in this chapter, which views follow-up as largely an adaptive process of mitigation performance evaluation, state-of-the-art environmental monitoring, and ongoing revision of mitigation programs and project impact management measures (Figure 9.1).

In many cases, mitigation in EIA is a series of non-binding proposals (Morrison-Saunders et al. 2001), and as such, conditions and recommendations need to be

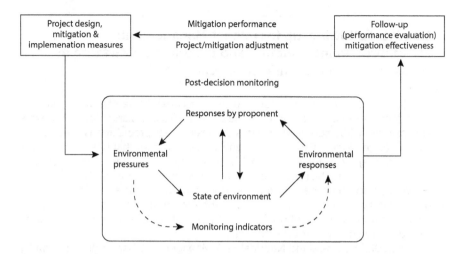

Figure 9.1 ◉ Follow-up and Monitoring in EIA

Source: Based on Ramos, Caeiro, and de Melo 2004.

monitored and enforced to ensure their implementation and effectiveness (Tinker et al. 2005). The lack of commitment to, and enforcement of, follow-up has been a long-standing concern in Canadian EIA practice, particularly at the federal level. A significant improvement under the Canadian Environmental Assessment Act, 2012 is the introduction of an enforceable decision statement at the conclusion of the review process. Under section 54(1)(b), for example, a proponent is required to comply with the conditions defined in the minister's decision statement about the assessment, and under section 89–102 the minister can, among other actions, designate persons or authorities to carry out enforcement of the act.

That said, the scope of follow-up under the act is relatively restrictive in comparison to the Canadian Environmental Assessment Agency's Operational Policy Statement (CEAA 2007) for follow-up programs under the former Canadian Environmental Assessment Act. The Operational Policy Statement adopts a broader perspective on follow-up than the current act itself, stating that the purposes of a follow-up program are to:

- verify the predictions made in an environmental assessment;
- determine the effectiveness of mitigation measures in order to modify or implement new ones if required;
- support the implementation of adaptive management measures;
- provide information on environmental effects and mitigation measures that can be used to support future assessments;
- support project environmental management systems.

At the provincial level, follow-up is required under the laws and regulations of the provinces, with each jurisdiction placing more or less emphasis on monitoring and post-decision impact management activities.

Box 9.1 Following-up for Impact Management:
EIA of Bison Reintroduction and Experimental Grazing in
Grasslands National Park, Saskatchewan

Grasslands National Park is located in southwest Saskatchewan, near the Saskatchewan–Montana border in the Great Plains grassland biome. It covers approximately 906 square kilometres of mixed-grass prairie and is divided into two separate blocks. The park is a semi-arid ecosystem that has evolved over time with migratory and sedentary bison grazing, a disturbance considered to have contributed to the heterogeneity and ecological integrity of the landscape. However, bison have been extirpated from the grasslands area since the turn of the nineteenth century, and most of the region outside the park has been cultivated for crop production. The remaining native vegetation has been subjected to grazing by domestic animals.

In 2002, the Grassland National Park Management Plan identified grazing as an important ecosystem function for maintaining mixed-grass prairie and necessary for restoring ecological integrity. The management plan proposed implementing a grazing prescription to represent regimes more consistent with historical patterns of migratory and sedentary bison herds. In 2005, the park prepared environmental impact assessment screening reports for two large-scale experiments: a grasslands grazing experiment and the reintroduction of plains bison. The grasslands grazing experiment was to involve the introduction of livestock to the park as part of a broader adaptive management process for restoring ecological integrity. The primary objective of the grazing experiment was to determine how grazing intensity alters spatial and temporal heterogeneity of the mixed-grass prairie community. Over a 10-year period, approximately 2664 hectares of native prairie was to be set aside for the experiment, which was based on a "before-after-control-impact" design process. The bison reintroduction project would involve an initial introduction of 50 to 75 animals to the park, transported from Elk Island National Park in Alberta. The long-term stocking-rate objective was to develop a bison herd size of up to 358 animals over approximately 185 square kilometres.

Both EIAs were screening assessments prepared under the requirements of the former Canadian Environmental Assessment Act. As the proponent of

Follow-up Components

Follow-up in EIA consists of three interrelated components: monitoring, auditing, and ex-post evaluation. Generally speaking, **monitoring** is an activity designed to identify the nature and cause of change. More specifically, it is a data collection activity undertaken to provide specific information on the characteristics and functioning of environmental and social indicators (Bisset and Tomlinson 1988). Such indicators can be indicators of human- or project-induced stress (e.g., changes in road densities, measures of habitat fragmentation) or indicators of environmental effects (e.g., changes in water sediment/chemistry). Monitoring usually consists of

both projects, Grasslands National Park was also responsible for the design and implementation of a follow-up program. The two projects were different from most development initiatives in that their overall purpose was to introduce disturbance to the environmental system in order to learn how it responds to particular management actions. At the same time, however, a number of unintended, potentially adverse environmental effects were also recognized.

A total of 45 potentially adverse impacts were identified in the impact statements, as well as 64 prescribed mitigation measures. Impacts and mitigation concerned a range of biophysical and socio-economic issues, including visitor experience, vegetation, wildlife, aquatics, and cultural resources. An integrated follow-up program, encompassing both projects, was designed in 2005 and focused on following-up for mitigation effectiveness. The objectives of the follow-up program were to: (i) ensure that mitigation measures identified in the impact statements were implemented; (ii) establish systems and procedures for this purpose; (iii) monitor the effectiveness of mitigation; and (iv) take any necessary action when unforeseen impacts occurred or when mitigation measures were not performing as expected.

The follow-up framework consisted of four core components: (i) an implementation audit or one-time check or verification that proposed mitigation measures had been implemented; (ii) compliance monitoring and regular inspection to ensure that agreed-upon impact management procedures and best-management practices identified in the impact statements were being adhered to; (iii) effects monitoring and evaluation to track and evaluate changes in a range of biophysical, economic, and social variables so as to compare data collected during project implementation and operation against baseline conditions; and (iv) the development of effectiveness indicators, targets, and thresholds so as to determine whether the mitigation measure was working to avoid or minimize potentially adverse effects. The first year of the follow-up program was implemented in 2007.

Source: Noble 2006.

a program of repetitive observation, measurement, and recording over a period of time for a defined purpose (Arts and Nooteboom 1999), the objective of which is to detect whether change in a particular indicator or set of indicators has occurred and to understand the magnitude of that change. Of course, monitoring requires much more than simply determining whether change has occurred; determining the causes of change and whether such changes, and how much change, are the consequence of the project is an essential part of monitoring for environmental management.

Auditing refers to the objective examination or a comparison of observations with predetermined criteria. There are a variety of types of audits (Box 9.2), but auditing generally is a periodic activity that involves comparing monitoring observations against a set of criteria, such as standards or expectations, and reporting the results. Whereas monitoring is often a frequent or continual process, auditing is usually a periodic or single event. There is little point in collecting monitoring data unless those data are subject to some form of comparative analysis or audit (Arts and Nooteboom 1999).

A third and related concept is **ex-post evaluation**, which refers to the collection, structuring, analysis, and appraisal of information concerning project impacts and making decisions on remedial actions and communication of the results of this process (Arts 1998, 75). The differences and relationships among these three components are illustrated in Figure 9.2.

Box 9.2 Types of EIA Audits

Draft EIS audit: Review of the project EIS according to its terms of reference

Project impact audit: Determination of whether the actual environmental impacts of the project were those that were predicted

Decision-point audit: Examination of the role and effectiveness of the EIS based on whether the project is allowed to proceed and under what conditions

Implementation audit: Determination of whether the recommendations included in the EIS were actually implemented

Performance audit: Examination of the proponent's environmental management performance and ability to respond to environmental incidents

Predictive technique audit: Comparison between actual and predicted effects of the project

Source: Tomlinson and Atkinson 1987.

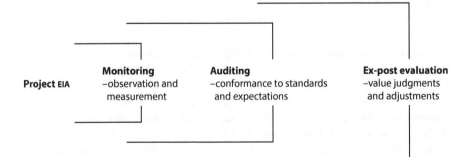

Figure 9.2 ◉ Follow-up Components
Source: Based on Arts 1998.

Rationale for Post-decision Monitoring

There are several, often overlapping reasons for undertaking monitoring activities. However, three broad reasons for post-decision monitoring can be identified: compliance, progress, and understanding.

Compliance Monitoring

The primary purpose of **compliance monitoring** is to determine project compliance with regulations, mitigation commitments, agreements, or legislation. In this sense, follow-up has a *control* function—to ensure that a project is operating within specified guidelines. Compliance monitoring alone does not fulfill the requirements of an EIA follow-up program. It is simply a means of ensuring that what a proponent said would be done in the EIS has actually been done once the project has been implemented. There are several types of compliance monitoring, including inspection monitoring, regulatory permit monitoring, and agreement monitoring.

Inspection monitoring, the simplest form of monitoring, is site-specific and involves checking to ensure that operating procedures are being followed and that environmental degradation is not occurring. Inspection monitoring typically involves on-site visits and regular reporting of relevant activities and is mostly used for regular checking for compliance with agreed-upon procedures and operation within acceptable standards of safety. As mentioned above, provisions in the Canadian Environmental Assessment Act, 2012 provide the authority to undertake both compliance and inspection monitoring of commitments or decision recommendations emerging from an EIA, including property searches.

Regulatory permit monitoring is also site-specific and involves the regular documentation of conditions required for maintenance or renewal of a permit, such as permits for the operation of waste disposal systems.

The **monitoring of agreements** between project proponents and affected groups, such as impact benefit agreements, are becoming commonplace, particularly in Canada. They sometimes include a monitoring component to track changes in population, housing, and other infrastructure demands in order to assign costs associated with the project and to ensure compliance with stated commitments.

Progress Monitoring

The purpose of progress monitoring is to confirm anticipated outcomes and to alert managers to unanticipated outcomes. In this sense, follow-up has a *watchdog* function and allows managers to measure project and environmental progress and to respond to adverse environmental change in a timely fashion when necessary. Common approaches to progress monitoring include ambient environmental quality monitoring, monitoring for management, cumulative effects monitoring, and project evaluation monitoring.

Ambient environmental quality monitoring is concerned with the effects of the project on its surrounding environment. Information collected prior to project approval and implementation at and near the site and at control sites can provide

baseline data against which to compare data collected during the development, operation, and post-project phases. While most ambient environmental quality monitoring is associated with measures of the status of biophysical phenomena such as
air and water quality, socio-economic changes such as quality of life or health and
well-being can also be the focus.

Monitoring for management may include tracking and evaluating changes in
a range of environmental, economic, and social variables. This type of monitoring is
usually associated with high-profile projects with uncertain outcomes, which have
the potential to result in significant adverse outcomes unless prompt action is taken
to address issues as they emerge. The size and scope of the monitoring requirements
may be such that the co-ordination of the monitoring program is undertaken by a
formally constituted and funded monitoring organization, such as the Independent
Environmental Monitoring Agency (IEMA), established to oversee outcomes from the
Ekati Diamond Project in northern Canada (www.monitoringagency.net) and to ensure
that the project meets the requirements set forth in IEMA's environmental approval.
Monitoring for management is the primary focus of its mandate, and finding solutions
to environmental management issues arising from the project is a primary objective.

Cumulative effects monitoring is less site- and project-specific and attempts
to monitor the accumulated effects of developments within a particular region. The
broad range of interests involved and the need for co-ordination mean that cumulative effects monitoring is best achieved by an organization mandated with monitoring responsibilities. The nature of cumulative environmental effects is discussed
in greater detail in Chapter 11.

Also referred to as productivity measurement, **project evaluation monitoring**
is concerned with measuring a project's performance with respect to established
goals or objectives such as overall efficiency. The focus is often social or economic
in nature and includes measures of performance of programs designed to provide
job training to address social concerns.

Monitoring for Understanding

A third reason for monitoring is to better understand the complex relationships
between human actions and environmental and social systems. In this sense, follow-up
has a *learning* function by increasing knowledge and understanding that can be applied
to the science of assessment of future projects or related policy decisions. Monitoring
for understanding includes experimental monitoring and monitoring for knowledge.

The purpose of **experimental monitoring** is to generate information and knowledge about environmental systems and their impacts through research methodologies
that test specific hypotheses. Whereas the approaches discussed above, while designed
to monitor changes, do not take anticipated consequences into account, experimental
monitoring is guided by questions to test specific hypotheses. In this sense, experimental
monitoring is science-driven rather than motivated by impact management per se.

Monitoring for knowledge is a type of data collection and reporting that often
takes place well after impacts occur. Rather than being used for impact management
purposes, the data are used to provide insights for the management of future projects.

Effective Follow-up and Monitoring

In Canada and elsewhere, post-decision follow-up has been described as less than satisfactory. While weak or non-existent legal requirements and institutional support mechanisms may in part be contributing to the current state of practice, a number of substantive and procedural elements are also required to facilitate effective post-decision EIA follow-up programs.

Objectives and Priorities Identified

Perhaps the most important requirement for effective follow-up is the clear articulation and identification of follow-up and monitoring program objectives and priorities at the outset of program design. If a clear set of objectives for follow-up and monitoring does not exist, whether for monitoring for compliance, performance, or knowledge, then one cannot adequately determine whether the follow-up or monitoring program itself is effective. The specific objectives of follow-up and monitoring programs vary from one project to another (Box 9.3) and may include such objectives as verifying impact predictions or ensuring regulatory compliance for project licensing. Failure to state clear program objectives and priorities leads to dissatisfaction with and confusion regarding what the monitoring program is actually trying to

---

Box 9.3 Follow-up Program Objectives for Two Canadian Mining Projects

Voisey's Bay nickel mine-mill project

1. To continue to provide baseline data so that project activities can be scheduled or planned to avoid or reduce conflict with VECs

2. To verify earlier predictions and evaluate the effectiveness of mitigation to lower uncertainty or risk

3. To identify unforeseen environmental effects

4. To provide an early warning of undesirable change in the environment

5. To improve understanding of environmental cause-and-effect relationships

Ekati diamond mine project

1. To ensure regulatory compliance

2. To measure operational performance and the effectiveness of mitigation strategies

3. To monitor natural environmental changes as well as those caused by the project (environmental effects monitoring)

4. To assess the validity of impact predictions

5. To trigger response to and mitigation of unexpected adverse effects

achieve. Consequently, consideration should be given to follow-up and monitoring objectives and priorities from the outset of EIA—during the scoping process, when key environmental issues are first identified, since that will allow pre-development baseline monitoring. Specific monitoring program design can be postponed until the post-decision stage when project design elements are finalized and priorities for management purposes can be better determined.

Targeted Approach to Data Collection

Not everything of interest or importance in an EIA can be monitored and effectively followed up. Further, because of the complexity of most stressor–response relationships, it is impossible to characterize all of the variables. Monitoring programs, particularly for the biophysical environment, must be targeted and focused on the indicators that are most useful for understanding, or at least correlating, stressor–response relationships. Such indicators must be comparable over space and time, and in order to differentiate project-induced change from natural change, these variables and indicators and the data derived from them must be comparable to previous, current, and forecasted baseline conditions.

One approach to monitoring is to focus on early warning indicators. Measurable parameters of either a non-biological or a biological nature, **early warning indicators** serve to indicate stress on particular VECs before these VECs themselves are adversely affected. Early warning indicators might include, for example, changes in worker productivity and number of sick days taken as warnings of worker stress, changes in nitrogen and phosphorous concentrations as early warning of adverse effects to water quality, or changes in benthic organisms as early warning of adverse effects to fish communities (Table 9.1). Benthic community surveys are a core part of Environment Canada's Environmental Effects Monitoring programs for pulp and paper and mining operations and are required as part of their licences to operate if they are releasing effluent to surface waters (Kilgour et al. 2007). Early warning indicators must be:

- directly or indirectly related to the VEC;
- physically possible to monitor;
- amenable to quantitative analysis;
- indicative of change in VEC condition before the onset of change.

Hypothesis-Based or Threshold-Based Approaches

For each indicator to be monitored, significance levels and probability levels must be specified as part of the monitoring program protocol. Monitoring to inform management action requires that monitoring programs be formulated as testable hypotheses whereby analyses of significance are made against an a priori null hypothesis (Bisset and Tomlinson 1988) or formulated with specific thresholds in mind. Squires and Dubé (2012) suggest that "benchmark" is the more acceptable terminology, particularly in the regulatory environment, because "threshold" implies that sufficient

Table 9.1 Selected Environmental Components, VECs, and Monitoring Indicators for the Cold Lake Oil Sands Project

Environmental component	Issues of concern	Valued components	Warning indicators
Air systems	Acidic deposition, odours, greenhouse gas emissions	Air quality	Emitted gases transported over long distances (NO_x, SO_2)
Surface water	Lowering of lake water levels, contamination of water	Water quality and quantity	Combined water volume withdrawals, water quality constituents affecting drinking-water standards
Groundwater	Depletion of aquifers	Potable well water	Combined water volume withdrawals
Aquatic resources	Contamination of fish, increased harvest pressures	Sport fish species	Northern pike populations/health
Vegetation	Loss of vegetation through land clearing, effects of airborne deposition	Vegetation ecosites	Low bush cranberry, aspen, white spruce
Wildlife	Loss, sensory alienation, and fragmentation of habitat, direct mortality due to increased traffic and hunting harvest	Hunted and trapped species	Moose, black bear, lynx, fisher populations/health

Source: Based on Hegmann et al. 1999.

knowledge exists to understand when the assimilative capacity of an environmental system has been exceeded. Refer to Chapter 5 for further discussion on benchmarks and cautionary, target, and critical thresholds.

For each indicator, appropriate thresholds (or benchmarks) should thus be established as a measure of environmental effects. Such thresholds may be established on the basis of:

- consultation with regulatory agencies;
- levels of acceptable change;
- scientific research or recommendation;
- range of natural variability;
- project goals or objectives.

The approach, which is common to many biophysical environmental effects monitoring programs, including the Alberta Regional Aquatic Monitoring Program (RAMP) (Box 9.4), is often based on testing to determine whether changes in specific indicators exceed stated threshold levels. Thresholds need to be linked to actions or decision-making processes. Monitoring of these indicators helps to verify that the management measures implemented are effective and that adverse effects are not occurring.

Box 9.4 Regional Aquatic Monitoring Program (RAMP)

The Athabasca River originates in the Columbia Icefield in the Rocky Mountains of Alberta and drains into Lake Athabasca in northeast Alberta and northwest Saskatchewan, covering an area of approximately 157,000 square kilometres. In addition to human settlement, the Athabasca River basin is home to a range of land-use activities, including large-scale agriculture, forestry, and petroleum extraction. The Athabasca is among the most stressed river systems in Canada. In addition to point source sewage discharge from urban settlement and non–point source urban and agricultural runoff, there are two bleached kraft pulp mill operations discharging to the river system. Most notably, the Athabasca River basin is home to the Alberta oil sands, with proven reserves of 170 billion barrels of oil sands bitumen and up to 315 billion barrels should favourable economic conditions prevail and new technologies become available.

RAMP is an industry-funded environmental monitoring program initiated in 1997. It was established to achieve a better understanding of the potential effects of oil sands development on aquatic systems so that long-term trends and issues related to oil sands and other developments in the Athabasca River basin would be identified and addressed. The focus of RAMP is on the Athabasca River and Peace-Athabasca Delta. The purposes of RAMP are to:

- monitor aquatic environments in the oil sands region to detect and assess effects and regional trends;
- collect baseline data to characterize variability in aquatic indicators;
- collect and compare data against predictions contained in EIAs; and
- collect data that assist with the monitoring required by regulatory approvals.

An effects-based monitoring program, RAMP is focused on monitoring the aquatic environment, including water quality (e.g., dissolved organic carbon, pH, total alkalinity, total dissolved solids, suspended sediments, major ions, nutrients, total and dissolved phosphorus, biological oxygen demand, total

---

Combined Stressor- and Effects-Based Monitoring

The focus of EIA is often on predicting the potential effects of a proposed development on specified VECs using a stressor-based approach. In other words, the nature and magnitude of changes in a particular VEC or VEC indicator are predicted on the basis of the conditions imposed on the local environment by project development (Kilgour et al. 2007). Under such an approach, it is assumed that there is an adequate level of understanding of the properties of the affected components to identify and predict the most likely outcomes. For simple projects and environmental systems, the use of a stressor-based approach and subsequently monitoring the stressors may be an appropriate means of effectively assessing the actual effects of development on the receiving environment.

Ball et al. (2012), for example, explain that where data on indicators (e.g., water quality) are limited, there is the potential to use surrogate indicators, such as land-use

phenolics, total and dissolved metals); benthic invertebrates (e.g., abundance, taxon richness, diversity and evenness); sediment quality (e.g., physical variables, total inorganic carbon, benzene, xylene, hydrocarbons by size class, total metals, polycyclic aromatic hydrocarbons); and fish populations (e.g., relative abundance, length/age frequency, condition factors, nutritional health).

Monitoring is designed to answer several key questions, including:

- What are the baseline conditions and range of natural variability?
- Are monitored conditions outside the range of natural variability or baseline conditions?
- Do fish measurement endpoints vary significantly between areas or water bodies exposed and unexposed to development?
- Do fish measurement endpoints from exposed areas exhibit time trends reflective of effects associated with increasing development?

The significance of monitoring results are then examined against the range of natural variability or baseline variability for the indicator in question or against specified thresholds for metal toxicity in fish tissue, such as lethal thresholds in whitefish for Al, Ba, Cu, Fe, and Pb.

RAMP is currently in its sixteenth year of operation. The 2012 annual report of monitoring activities reported localized changes in some indicators and watershed when compared to pre-development conditions but did not detect large-scale changes related to oil sands development (see http://www.ramp-alberta.org). However, the Office of the Auditor General of Canada (2010) and the Oil Sands Advisory Panel (2010) report that monitoring programs in the region are currently inadequate to detect any measurable change. Schindler (2010), for example, reports on inconsistent sampling and methodology and a lack of accessibility to data, which make it impossible to determine the extent to which mining has increased the concentrations of contaminants in the river over natural background levels.

and land-cover metrics (e.g., riparian zone habitat, stream crossing density, percent impervious surfaces), which can act as indicators of responses by affected systems to environmental change and can be used in regression and correlation analyses to provide an indication of cause–effect relationships (Seitz, Westbrook, and Noble 2011). Such stressor-based indicators can often be monitored efficiently and cheaply with the assistance of remote sensing and Geographic Information Systems tools. There are, however, a number of cautions to monitoring stressors:

- The utility of the indicators (i.e., landscape metrics) will depend on the response variable, the strength of the relationship between the indicator and response, and the relative importance of other controlling variables (e.g., topographic relief, bedrock, or climate variability) (Gergel et al. 2002).
- The relationship between the stressor (e.g., road density) and the effect (e.g., water quality) is an assumed or statistical association rather than

cause–effect per se. Such analysis is thus indicative, but not descriptive, of cause–effect.

- Projects and environmental systems are rarely that simple, and focusing monitoring programs solely on stressor–VEC indicator response relationships may mask broader systemic environment changes caused by the project if no measurable changes are detected in the VECs themselves (Kilgour et al. 2007).

Effects-based monitoring, such as in the case of RAMP (see Box 9.4), focuses on the performance of biological indicators. Effects-based monitoring is based on the premise that measuring change in environmental (biological) indicators is the most direct and relevant means of assessing change (Munkittrick et al. 2000). Exceedances of stress-based thresholds or benchmarks, such as habitat loss or stream-crossing density, are considered insignificant if there are no biological effects in the receiving environment. In Canada, effects-based monitoring programs are used by many agencies to determine whether environmental quality has been compromised, and they are currently a legal component under the Fisheries Act for monitoring and evaluating the effects associated with discharges to aquatic systems from metal mines and pulp mills. Among the current challenges in developing and implementing such effects-based monitoring systems in EIA practice, however, is that EIA and environmental effects monitoring practitioners have generally operated independently and have evolved relatively distinct jargon that is not easily transferred between the disciplines (Kilgour et al. 2007).

As with stressor-based monitoring, there are a number of challenges:

- Indicators selected for monitoring must be responsive and provide some early warning of a potentially adverse effect.
- Indicators selected for monitoring must be responsive to human actions and, in particular, be associated with different types and/or levels of human-induced stress. Benthic invertebrates and indices of biological integrity, for example, are often advocated for monitoring in aquatic systems, because the biota integrates the effects of multiple stressors over time (see Jensen 2006). The difficulty, however, is in establishing a relationship between the monitored state and the specific project or project actions of interest.
- Historical monitoring data needed to understand temporal trends are often scarce, particularly in remote regions.

Considering the relative strengths and limitations of stressor- and effects-based monitoring, good follow-up and monitoring programs focus on a complementary stressor- and effects-based approach that monitors stressors due to project actions, the performance of the environmental system, and attempts to establish a relationship between the two. That said, not all VECs identified and assessed in an EIA are easily tied to specific measurable parameters or effects-based indicators that can be easily monitored. If an effect cannot be tied to either a stressor or a VEC identified in a project assessment, challenges emerge concerning roles and responsibilities for mitigation.

Control Sites

Where possible, particularly for effects-based monitoring, **control sites** should be established as reference monitoring locations to compare with the treatment (project-affected) locations. This will help in correctly differentiating between project impacts and natural change and will facilitate the spatial and temporal consistency of the effects monitoring program with initial project impact predictions. In cases where a well-established control site is not readily available, it may be possible to use the range of natural variability or create an artificial one using a **gradient-to-background monitoring** approach.

The gradient-to-background approach assumes that there is a well-defined, localized source of impact, such as pollution, and that effects can be monitored at increasing distances from the point source. With increasing distance from the point source, effects should decrease and eventually reach background or ambient conditions. The point on this gradient at which background levels are attained is considered the control site (Figure 9.3).

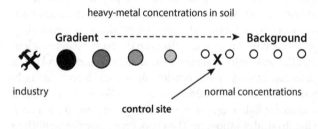

Figure 9.3 ⊚ Gradient-to-Background Approach

Continuity in Data Collection

There should be continuity in data collection and handling procedures to ensure that monitoring data are transferable and comparable. Failure to achieve continuity may lead to problems in comparing monitoring results and the significance of those results from one monitoring period to the next. Such quality control problems in the continuity of data collection were evident in the Rabbit Lake uranium mine project in Saskatchewan, the province's oldest operating uranium mining facility. In 1993, following a joint federal–provincial panel review of the environmental, health, and socio-economic effects of uranium mining in northern Saskatchewan, the panel noted that while monitoring that met regulatory requirements was ongoing, there was concern over the quality of the monitoring data, consistency of the methods used to test for radionuclides and trace elements in fish, and the effectiveness of monitoring programs in determining the impacts of mining activities. Monitoring and testing procedures had changed several times throughout the 1980s, and data collected during 1989 and 1990 were discarded because of quality problems. After decades of biophysical monitoring and data collection, the proponent was unable to make a direct connection between project actions and the impacts of project-induced environmental change.

Ensuring Adaptability, Flexibility, and Timeliness

Care should be taken to ensure that as project or environmental conditions change, data quality is consistent and comparable over time. While objectives and indicators need to be established at the outset, project changes, unanticipated effects, emerging project concerns, and indicator choices may warrant changes in elements of the monitoring as it proceeds (Davies and Sadler 1990). This requires that monitoring and reporting be completed in a timely fashion so that those using the results can respond promptly. While data collected late in the process are still valuable for reflection and learning, such monitoring processes are often a waste of time from the perspective of impact management.

Socio-economic-Inclusive Monitoring

Follow-up programs should address not only traditional biophysical impacts but socio-economic and other human and cultural impacts as well. In practice, however, social issues have been given less than sufficient attention in post-decision monitoring. While the assessment of socio-economic and other effects on the human environment are often given as much attention as biophysical effects in impact statements and panel reports, such attention is often not carried over into follow-up—and this is particularly true for the "social" components of socio-economic impacts. When social and other human issues do carry over to the follow-up stage, they tend to be treated with considerably less rigour than biophysical components (Box 9.5). Proponents often wish to distance themselves from socio-economic follow-up, partly because of the complexity of the pathways that link project actions to socio-economic impacts. In the case of the Cluff Lake uranium mine in northern Saskatchewan, the final panel report of the Cluff Lake Board of Inquiry (1978, 174) explicitly recognized the difficulty of assessing the social and other human impacts associated with uranium mining activities, stating:

> there now exists in the north (and it has nothing to do with uranium mining) a social disorder . . . To superimpose upon that kind of society a project such as a uranium mine and mill which has the potential of exacting additional social costs and then try and measure those additional costs presents a near impossible task.

In principle, according to discussions with an industry representative, there are often so many confounding factors in the communities that it is impossible to tell whether or not there has been an effect. Further, many communities are too far away to permit monitoring direct effects based on scientific risk and pathways monitoring.

Box 9.5 Canadian Experiences with Socio-economic Monitoring: Ekati Diamond Mine

In 1994, the Department of Indian Affairs and Northern Development initiated an environmental review of Canada's first diamond mine, 300 kilometres northeast of Yellowknife in the Northwest Territories. The proposal involved the development of a diamond mine in an area of unsettled and overlapping Aboriginal land claims where there had been little previous industrial development. The proponent, now BHP Billiton (BHPB), submitted its assessment documents in 1994, and a full panel review followed.

The importance of follow-up was recognized, and a variety of suggestions were made to ensure that it was undertaken. The panel, based on information from the government of the Northwest Territories (GNWT) that it had already identified and collected data on a number of health and wellness indicators, recommended a partnership approach, primarily between the proponent and the GNWT.

In 1996, a Socio-economic Agreement was signed between the GNWT and BHPB for the Ekati project. The agreement is intended to promote the development and well-being of the people of the Northwest Territories, particularly people in communities close to the mine. The agreement focuses on monitoring and promoting social, cultural, and economic well-being. The GNWT is responsible for the establishment and maintenance of the monitoring program. The objective is to look for ways to strengthen opportunities and to mitigate negative effects associated with the project. Public statistics and surveys of mine employees are the primary data sources used to track a series of indicators chosen to reasonably match the possible effects identified during the assessment phase for the project. In addition to data on social stability and community wellness, BHPB issues its own reports describing realization of business and employment opportunities and also reports directly to the Aboriginal groups with whom it has signed impact benefits agreements.

Public statistics for 14 indicators are currently used to monitor and assess the effects of the Ekati project. Data are monitored and reported for the small communities in the West Kitikmeot Slave area, Lutselk'e, Rae-Edzo, Rae Lakes, Wha Ti, Wekweti, Dettah, and Ndilo, as well as for Yellowknife. These data are then compared with information for the rest of the Northwest Territories. The results are presented in an annual report. While these data provide indications of overall changes within the region for the indicators tracked, most of the results are at best inconclusive in linking social and economic changes in the communities to the diamond project activities. Cause and effect are always difficult to determine for social variables, but the "coarseness" of the indicators used to measure change, the small size of some communities, and the quality of some of the available data make it particularly difficult to do so in this case.

Key socio-economic monitoring indicators for the BHPB Ekati project include:

continued

Social stability and community wellness	Non-traditional economy
• number of injuries	• average income of residents
• number of potential years of life lost	• employment levels and participation
• number of suicides	• number of income assistance cases
• number of teen births	• high school completion
• number of children in care	• cultural well-being
• number of complaints of family violence	
• number of alcohol/drug-related crimes	
• number of property crimes	
• number of communicable diseases	
• housing indicators	

Sources: GNWT 2001; Storey and Noble 2004.

If sustainable development is the objective of EIA, then biophysical and socio-economic components must be given equal consideration throughout all phases of EIA, including post-decision monitoring. That said, an important question is: Who is responsible for **socio-economic monitoring**? Often, proponents suggest that they are willing to cooperate with governments by sharing project information but maintain that socio-economic follow-up is primarily the responsibility of other parties. Mobil Oil held to this view in connection with the Hibernia offshore oil project, Newfoundland and Labrador. Similarly, in the case of the Voisey's Bay mine-mill project, the proponent took the position that the financial provisions in the impact benefits agreements to be signed with the Labrador Inuit Association and the Innu Nation were in part intended to provide these groups with the resources to carry out any socio-economic follow-up studies that they deemed necessary (Voisey's Bay Mine and Mill Environmental Assessment Panel 1999).

Participatory Monitoring

Active engagement of the local community in follow-up and monitoring programs can lead to increased capacity at the local level to deal with environmental change and the impacts of project development (Austin 2000). Noble and Birk (2011) report that the benefits of involving communities in the EIA process have long been recognized and range from ensuring that the project itself carries more legitimacy to enhancing the effectiveness of monitoring programs and impact management measures. In practice, however, meaningful community involvement in monitoring has not frequently occurred. Follow-up programs are often criticized for not effectively engaging communities in the monitoring and management of project impacts or building trust and capacity among stakeholders. Hunsberger, Gibson, and Wismer (2005) note that it is important for communities to play a role in determining the purpose, scope, and priorities of follow-up and monitoring activities, but practice has traditionally been weak in the areas of community engagement in monitoring programs. As reported by Lawe, Wells, and Mikisew Cree First Nations Industry Relations Corporation (2005), although local communities are becoming more involved in resource management processes, "they often have limited influence over EIA follow-up including long-term monitoring programs that determine effectiveness of mitigation." One exception to this may be community-based monitoring programs in the Saskatchewan uranium mining industry (see Environmental Assessment in Action: Community-Based Monitoring in the Saskatchewan Uranium Mining Industry).

Monitoring Methods and Techniques

As with impact prediction, there is no single set of monitoring techniques for all projects and components. The type of monitoring technique selected to provide data and to evaluate environmental change depends on the nature of the environmental components, the purpose of the data, and particular monitoring program objectives. Thus, for both biophysical and socio-economic monitoring, a variety of quantitative and qualitative techniques, or combinations thereof, are available. For example, socio-economic impacts can be monitored using annual surveys of residents' perceptions of quality of life and local police and hospital records. Selected examples of techniques for monitoring biophysical components are listed in Table 9.2. When selecting a technique for monitoring, as for any other EIA procedure, it is important to keep in mind the nature and resolution of the data required.

Table 9.2 Biophysical Monitoring Components, Parameters, and Techniques

Components	Parameters	Techniques
wildlife habitat	fragmentation	remote sensing aerial photography
rangeland health	vegetation cover vegetation composition soil structure	remote sensing field transects trend transects
water quality	metals nutrients physical parameters	stream gauges and go-flow samplers chemical analysis
lake biology	lake benthos	Eckman Grab stratified depth sampling and species counts
stream biology	stream benthos fish	Hester-Dendy sampler and frequency counts tagging electro fishing age-length keys
hydrology	water level	staff gauges pressure transducers

Environmental Assessment in Action

Community-Based Monitoring in the
Saskatchewan Uranium Mining Industry

The Athabasca basin in northern Saskatchewan is one of the world's most important sources of uranium. The basin is also home to three remote Aboriginal reserves and settlements and four small hamlets and communities. The region is largely "fly-in," and local economies are based primarily on traditional hunting, trapping, fishing, and guiding activities and, in recent years, uranium mining. Uranium was first discovered in the Athabasca basin in the 1930s, but high-grade uranium production did not begin until 1975 at the Rabbit Lake mine site. Currently, three uranium mines are operating in the basin, owned by Cameco Corporation and Areva Resources Canada.

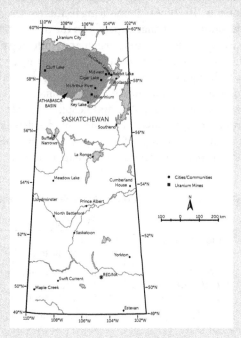

In 1991, in response to six new proposals for uranium mine development and expansion in the region, the governments of Canada and Saskatchewan appointed an independent joint federal–provincial panel to review the proposed developments and evaluate the future of uranium mining in the Athabasca basin. The panel's recommendations on follow-up and monitoring varied by project and focused on such aspects as monitoring the effects of uranium mining on air, soils, terrestrial plans and wildlife, and aquatic resources (see Joint Federal–Provincial Panel on Uranium Mining Developments in Northern Saskatchewan 1991). The joint panel also noted that because of the

proximity of mining operations to northern communities, community involvement in monitoring and managing the impacts of uranium operations should
extend beyond consultation to include participation.

In 1993, in response to the recommendations of the joint panel, the uranium industry established the Athabasca Working Group (AWG). The AWG is
a private partnership between the two uranium mining companies, Cameco
and Areva, and the seven communities of the Athabasca basin. As part of the
agreement's commitment to environmental protection, and in part because
of the joint panel's recommendation for more direct community involvement
in mining and impact management activities, a community-based environmental monitoring program was established in 2000 to monitor and follow
up on the "off-site" impacts of uranium mining operations within the vicinity
of AWG communities.

The monitoring program originated at the request of AWG community
members to provide a means for communities to generate environmental
monitoring data independent of industry and regulatory-based follow-
up programs. The program is funded by the industry; the community is
responsible for appointing AWG members and local residents to participate
in monitoring. Many of the monitoring samples are collected during the
course of normal hunting and trapping activities. Samples are processed
and analyzed by independent research laboratories in Saskatchewan and
the United States.

One of the benefits of the community-based monitoring program, and a
primary benefit of the environmental agreement itself, has been the increase in
communication between industry and community. Specifically, the agreement
and monitoring partnership has provided an opportunity for communities to
become involved in impact management and to discuss their environmental
concerns directly with industry. The program has been described by industry
as a good example of mixing traditional knowledge and Western science in
a co-ordinated monitoring program. Communication of community-based
monitoring program results, however, has been problematic. Monitoring
results are made available in English in the form of a technical report, a format
not necessarily sensitive to the Aboriginal community.

Perhaps more challenging is that although methods for data analysis for
the community-based monitoring program and industry's EIA monitoring program are similar, the results are not directly comparable. Many of the monitoring samples are taken biannually, often limited in sample size, sparse in
coverage, and far removed from industry activities and do not meet scientific
standards for effects detection. In this regard, the community-based monitoring program does not directly contribute to industry impact monitoring for the
purposes of identifying unexpected impacts or verifying the effectiveness of
impact mitigation measures. More accurately, the community-based monitoring program can be described as "comfort monitoring," a monitoring approach
to alleviating community concerns while providing communities with a sense
of ownership and an opportunity for technology transfer.

In their analysis of the uranium industry community-based monitoring
program, Noble and Birk (2011) conclude that comfort monitoring may be
valuable in its own right and of benefit to industry in ensuring community

continued

cooperation for future development in the region; however, if community-based monitoring is to be credible over the long term, then the results generated must be useful for, and integrated with, mainstream project effects monitoring and impact management practices.

Source: Based on Noble and Birk 2011.

Key Terms

ambient environmental quality
      monitoring
auditing
compliance monitoring
control site
cumulative effects monitoring
decision-point audit
draft EIS audit
early warning indicators
experimental monitoring
ex-post evaluation
gradient-to-background monitoring
implementation audit

inspection monitoring
life-cycle assessment
monitoring
monitoring for knowledge
monitoring for management
monitoring of agreements
performance audit
predictive technique audit
project evaluation monitoring
project impact audit
regulatory permit monitoring
socio-economic monitoring

Review Questions and Exercises

1. What provisions exist for post-decision monitoring and auditing under your national, provincial, or state EIA system? Are these provisions mandatory or voluntary?
2. What is the value added to EIA from follow-up and monitoring activities?
3. Who should be responsible for monitoring the environment after project approval?
4. Why are socio-economic effects difficult to monitor post-project implementation? How might we address these difficulties?
5. What is the role of the public in environmental monitoring?
6. Obtain a completed project EIS from your local library or government registry, or access one online. Is there a monitoring component to the impact statement? What environmental aspects are included in the monitoring program? Are methods and techniques for monitoring identified? Compare your findings with those of others.

References

Arts, J. 1998. *EIA Follow-up: On the Role of Ex-post Evaluation in Environmental Impact Assessment.* Groningen, Netherlands: Geo Press.

———, P. Caldwell, and A. Morrison-Saunders. 2001. "Environmental impact assessment follow-up: Good practice and future directions." *Impact Assessment and Project Appraisal* 19 (3): 175–85.

Arts, J., and S. Nooteboom. 1999. "Environmental impact assessment monitoring and auditing." In J. Petts, ed., *Handbook of Environmental Impact Assessment.* London: Blackwell Science.

Austin, E. 2000. "Community participation in EIA follow-up." Paper presented at the 2000 International Association for Impact Assessment annual meeting. Hong Kong: IAIA.

Ball, M.A., et al. 2012. "Scale, assessment components, and reference conditions: Issues for cumulative effects assessment in Canadian watersheds." *Integrated Environmental Assessment and Management* doi: 10.1002/ieam.1332.

Bisset, R., and P. Tomlinson. 1988. "Monitoring and auditing of impacts." In P. Wathern, ed., *Environmental Impact Assessment: Theory and Practice.* London: Unwin Hyman.

CEAA (Canadian Environmental Assessment Agency). 2007. *Follow-up Programs under the Canadian Environmental Assessment Act.* Operational Policy Statement. Ottawa: CEAA.

Cluff Lake Board of Inquiry. 1978. *Cluff Lake Board of Inquiry Final Report.* Regina.

Davies, M., and B. Sadler. 1990. *Post-project Analysis and the Improvement of Guidelines for Environmental Monitoring and Audit.* Report EPS 6/FA/1 prepared for the Environmental Assessment Division, Environment Canada. Ottawa.

Gergel, S.E., et al. (2002) "Landscape indicators of human impacts to riverine systems." *Aquatic Science* 64: 118–28.

GNWT (Government of the Northwest Territories). 2001. *Communities and Diamonds: Socio-economic Impacts on the Communities of Lutselk'e, Rae-Edzo, Rae Lakes, Wha Ti, Wekweti, Dettah, Ndilo and Yellowknife.* Yellowknife: Annual Report of the Government of the Northwest Territories under the BHP Socio-economic Agreement.

Hegmann, G., et al. 1999. *Cumulative Effects Assessment Practitioners Guide.* Prepared by AXYS Environmental Consulting Ltd. and the CEA Working Group for the Canadian Environmental Assessment Agency. Hull, QC: CEAA.

Hunsberger, C., R. Gibson, and S. Wismer. 2005. "Citizen involvement sustainability-centred environmental assessment follow-up." *Environmental Impact Assessment Review* 25: 609–27.

Jensen, E.V. 2006. *Cumulative Effects Monitoring of Okanagan Streams Using Benthic Invertebrates, 1999 to 2004.* Penticton, BC: Environmental Protection Division, Ministry of Environment.

Joint Federal–Provincial Panel on Uranium Mining Developments in Northern Saskatchewan. 1991. *Uranium Mining Developments in Northern Saskatchewan.* Regina: Saskatchewan Environment and Resource Management.

Kilgour, B., et al. 2007. "Aquatic environmental effects monitoring guidance for environmental assessment practitioners." *Environmental Monitoring and Assessment* 130 (1–3): 423–36.

Lawe, L., J. Wells, and Mikisew Cree First Nations Industry Relations Corporation. 2005. "Cumulative effects assessment and EIA follow-up: A proposed community-based monitoring program in the oil sands region, northeastern Alberta." *Impact Assessment and Project Appraisal* 25 (3): 191–6.

Morrison-Saunders, A., et al. 2001. "Roles and stakes in environmental impact assessment follow-up." *Impact Assessment and Project Appraisal* 19 (4): 289–96.

Munkittrick, K., et al. 2000. *Development of Methods for Effects-Driven Cumulative Effects Assessment Using Fish Populations: Moose River Project*. Pensacola, FL: Society of Environmental Toxicology and Chemistry.

Noble, B. 2006. *Environmental Impact Assessment Follow-up Prescription and Reporting Framework: Grasslands Grazing Experiment and Reintroduction of Plains Bison, Grasslands National Park of Canada*. Val Marie, SK: Grasslands National Park.

———, and J. Birk. 2011. "Comfort monitoring? Environmental assessment follow-up under community–industry negotiated environmental agreements." *Environmental Impact Assessment Review* 31: 17–24.

Office of the Auditor General of Canada. 2010. *Report of the Commissioner of the Environment and Sustainable Development*. Chapter 2: "Monitoring water resources." Ottawa: Minister of Public Works and Government Services Canada.

Oil Sands Advisory Panel. 2010. *A Foundation for the Future: Building an Environmental Monitoring System for the Oil Sands*. A report submitted to the minister of environment.

Ramos, T., S. Caeiro, and J. deMelo. 2004. 'Environmental indicator frameworks to design and assess environmental monitoring programs'. *Impact Assessment and Project Appraisal* 22 (1): 47–62.

Schindler, D.W. 2010. "Tar sands need solid science." *Nature* 468: 499–501.

Seitz, N.E., C.J. Westbrook, and B.F. Noble. 2011. "Bringing science into river systems cumulative effects assessment practice." *Environmental Impact Assessment Review* 31 (3): 172–9.

Squires, A., and M.G. Dubé. 2012. "Development of an effects-based approach for watershed scale aquatic cumulative effects assessment." 9 (3) *Integrated Environmental Assessment and Management* doi: 10.1002/ieam.1352.

Storey, K., and B. Noble. 2004. *Toward Increasing the Utility of Follow-up in Canadian EA: A Review of Concepts, Requirements and Experience*. Report prepared for the Canadian Environmental Assessment Agency. Gatineau, QC: CEAA.

Tinker, L., et al. 2005. "Impact mitigation in environmental assessment: Paper promises or the basis of consent conditions?" *Impact Assessment and Project Appraisal* 23 (4): 265–80.

Tomlinson, P., and S. Atkinson. 1987. "Environmental audits: Proposed terminology." *Environmental Monitoring and Assessment* 8 (3): 187–98.

Voisey's Bay Mine and Mill Environmental Assessment Panel. 1999. *Report on the Proposed Voisey's Bay Mine and Mill Project/Environmental Assessment Panel*. Hull, QC: Canadian Environmental Assessment Agency.

Involving the Publics

Public engagement in impact assessment is required in most EIA systems around the world. Broadly defined, public participation refers to the involvement of individuals and groups that are positively or negatively affected by a proposed intervention subject to a decision-making process or are interested in it (André et al. 2006). Public participation initiatives began to develop in the 1960s, and have had an influence on environmental decision-making in the United States. The public was a critical driving force behind the initial development of EIA, and since NEPA in 1969, public involvement has increasingly been recognized as a key element in EIA practice. Interestingly, however, under the NEPA system, requirements for public involvement in the EIA process are limited to public scoping of the issues to be addressed in assessment, review of the draft EIS, and court challenge.

There are international provisions with respect to public participation in EIA, including the 1991 Espoo Convention on Environmental Impact Assessment in a Transboundary Context and the 1998 Aarhus Convention on Access to Information, Public Participation in Decision-Making and Access to Justice in Environmental Matters. Many national EIA systems also have specific requirements for public involvement in project evaluation and decision-making, with some more stringent than others. In the UK, for example, Article 6 of Directive 85/337/EEC provides for public participation in EIA through the opportunity for a public review of the project's EIS before the project is initiated; there is no requirement for public involvement either before or during project planning.

Revisions to the Canadian Environmental Assessment Act in 2003 strengthened the requirements for early public involvement in federal EIA in Canada and provided for the incorporation of public and traditional knowledge during the assessment process. In 2012, the new act provided further support for participation, particularly for Aboriginal consultation, but restricted public participation in hearing processes to written submissions, unless the individual or organization is an "interested party"— one that is directly affected by the designated project. One of the stated purposes of the Canadian Environmental Assessment Act, 2012 is to ensure that opportunities are provided for meaningful public participation. In practice, arguably, sustained involvement of the public in any substantive way, other than a review of the final EIS, remains limited and is still largely at the discretion of the project proponent.

Rationale for Involving the Public

It is recommended that public involvement commence as early as possible in the EIA process. The Canadian Environmental Assessment Agency's (CEAA 2012, 9) *Guide to Preparing a Description of a Designated Project* under the federal act, for example, notes:

> Experience has shown that engagement by proponents with Aboriginal groups early in the planning and design phase of a proposed project can benefit all concerned. By learning about Aboriginal interests and concerns and identifying ways to avoid or mitigate potential impacts, proponents can build these considerations into their project design, reducing the potential for future project delays and increased costs.

In practice, however, public involvement is frequently limited to the submission of written public comment. Some proponents have argued that it can be more efficient to exclude the public from EIA, given that the public often lacks the project-specific expertise necessary to contribute to development decision-making, and argue instead for "educating the public" rather than involving them. As a result, far too often the common answer to the question "why involve the public in EIA?" is simply "because it is required" (Shepherd and Bowler 1997). However, public involvement can be beneficial at all stages of the EIS process, from initial project design to post-decision analysis and monitoring (Table 10.1).

The lack of public involvement during an EIS process, or inappropriate and tokenistic involvement, can be detrimental to the success of an EIS and to project approval (Box 10.1). However, by involving the public in the decision-making process, it is possible to:

- define the problem more effectively;
- access a wider range of information, including traditional knowledge;
- identify socially acceptable solutions;
- ensure more balanced decision-making;
- minimize conflict and costly delays;
- facilitate implementation;
- reduce the possibility of legal challenge;
- promote social learning.

Public involvement in EIA can be seen as meeting a number of ends. While public involvement may extend the time needed during the initial project planning and scoping phases, this initial investment is usually returned later in the process because it minimizes or avoids conflict and facilitates project approval and implementation (Noble 2004). From a corporate perspective, Eckel, Fisher, and Russell (1992) suggest that consultation with different groups—regulators, shareholders, governments, and communities—will help to clarify expectations about environmental performance and boost a firm's reputation with the public.

Table 10.1 Objectives of Public Involvement throughout the EIA Process

EIA stage	Public involvement objectives
Initial project design	Early identification of affected interests and values
	Identification of values relevant to site selection for conflict minimization
Screening	Public review of decisions concerning EIA requirements
	Early notification of affected interests of potential development
Scoping	Further identification of active and inactive publics
	Learning about public interests and values
	Elicitation of local knowledge for baseline assessment
	Identification of potentially significant impacts
	Identification of other areas of public concern and suggestions for management
	Identification of alternatives
	Establishment of credibility and trust between proponent and publics
Impact prediction and evaluation	Elicitation of values and knowledge to assist impact predictions
	Identification of criteria for evaluation of project impacts
	Development of public's technical understanding of project impacts
Reporting and review	Informing public of project details, baseline conditions, likely impacts, and proposed management measures
	Obtaining public feedback on key concerns, outstanding issues, and suggestions for improved management
	Identification of errors or omissions in the EIS
	Providing public an opportunity to challenge EIS assumptions and predictions
Decision-making	Resolution of potential conflicts
	Final integration of responses from EIS review in project approvals or conditions
Follow-up	Maintenance of trust and credibility
	Identification of management effectiveness
	Elicitation of local knowledge in data collection and ongoing monitoring of environmental change

Source: Based on Petts 1999.

Nature and Scope of Public Involvement

Petts (1999, 147) defines public involvement in EIA as:

> a process of engagement, where people are enlisted into the decision process to contribute to it . . . provide for exchange of information, predictions, opinions, interests, and values . . . [and] those initiating the process are open to the potential need for change and are prepared to work with different interests to develop plans or amend or even drop existing proposals.

Box 10.1 The Power of Public Involvement and Nuclear Fuel Waste Disposal in Canada

Current debates about the disposal of nuclear fuel waste in Canada date back more than 30 years when in 1978, under the Canada/Ontario Nuclear Fuel Waste Management Program, the governments of Canada and Ontario directed Atomic Energy of Canada Limited (AECL) to develop the concept of deep geological disposal of Canada's nuclear fuel waste (NFW). A subsequent joint statement issued in 1981 established that a disposal site would not be selected until the concept itself was publicly reviewed and approved by both governments.

In 1988, the concept, along with related issues concerning NFW, was referred for public review under the Federal Environmental Assessment and Review Process Guidelines Order, and in 1989 an independent environmental assessment panel, which would later become known as the Seaborn Panel, was appointed to develop guidelines for the assessment and to conduct a public review of the assessment document upon completion.

Terms of reference defining the review were released in 1989, and responsibility for the review process was given to an independent review panel administered initially by FEARO and later by CEAA. The panel's mandate was to review the disposal "concept" rather than a specific project and location, the implementing agency of the concept was not defined, a broad range of policy issues were to be considered, and the public review was to span five provinces. The terms of reference directed the review in four areas: evaluation of the acceptability of the nuclear fuel waste disposal concept, including the burden on future generations, and a future course of action; comparison of the Canadian disposal concept to the approaches for nuclear fuel waste management adopted by other countries; a focus explicitly on nuclear fuel waste and the disposal concept; prohibition against discussing energy policy, nuclear plant operation and construction, and military applications. The terms of reference were not publicly negotiated; in essence, the scope of the assessment was predetermined.

The panel embarked on a series of scoping meetings to gather public input on the concept and to set the scope of AECL's impact statement. Public meetings were held in 1990 in 14 communities. Draft guidelines were released for public review, and the final guidelines were presented to AECL in 1992. The panel directed AECL to consider ethical, moral, and social perspectives as equally important as the scientific and technical information. In a review of the terms of reference and guidelines for the AECL concept review and impact statement, Murphy and Kuhn (2001) note considerable debate among AECL and government and non-government agencies and an array of publics as to what could and should be considered in an assessment of a NFW disposal facility. Some contended that the panel was hampered by a very specific mandate and terms

---

Identifying the Publics

There is no such thing as "the public" in EIA; rather, there are many "publics"—some of whom may emerge at different times during the EIA process depending on

of reference; others suggested that the terms of reference were clear and that the entire process required further streamlining to restrict hearing submissions to those that conformed to the official mandate of the review.

AECL submitted its EIS in 1994, and a series of public hearings followed throughout 1995. The hearings were co-ordinated by the review panel and addressed three major areas of concern: the management of NFW within a broad societal context; a technological review of the assessment concept itself; and local perspectives on public safety and acceptability of the concept. The panel conducted its review in 16 communities across Saskatchewan, Manitoba, Ontario, Quebec, and New Brunswick and received more than 500 written submissions.

The public's response was for a broader consideration of alternatives, broader in scope than what was considered in the assessment. The terms of reference for the AECL NFW assessment were focused on a technical review of the proposed concept to the near exclusion of alternative definitions of and perspectives on what constitutes NFW management and the social and ethical aspects that NFW management might involve (Murphy and Kuhn 2001). The review panel, however, went beyond the scope of the terms of reference to ensure consideration of broader social issues and concerns regarding NFW management in Canada. Interestingly, the panel's report to government found that AECL's concept was technically safe but not publicly acceptable, partly because only one plan option, AECL's proposed concept, was considered. The panel's report contained numerous recommendations, including the creation of an independent organization to manage and co-ordinate all activities dealing with nuclear fuel waste in the long term and that such an organization would be subject to regulatory control and regular public review. In 2001, An Act respecting the Long-Term Management of Nuclear Fuel Waste was introduced. It passed in 2002 as the Nuclear Fuel Waste Act and resulted in the establishment of the Nuclear Waste Management Organization—a private industry organization led by the major owners and producers of NFW, including provincial Crown nuclear power utilities and AECL. The Nuclear Waste Management Organization has the mandate to review and select a preferred option for long-term NFW management. It was not the independent organization recommended by the panel, and the legislation and process have been criticized by some for lacking transparency and accountability.

Source: Noble and Bronson 2007.

their particular concerns and the issues involved. Mitchell (2002) makes a distinction between **active publics** and **inactive publics**. The active publics are those who affect decisions, such as industry associations, environmental organizations, quasi-statutory bodies, and other organized interest groups. Inactive publics are those who do not typically become involved in environmental planning, decisions, or issues and may include the "average" town citizen (Diduck 2004). When involving the publics in EIA, it is important not to overrepresent the active publics and to ensure adequate representation of the inactive publics. Particular attention should thus be given to those who reside in the area where the project will be implemented and who may be directly affected by the project.

One way to approach identification of the publics for involvement in EIA is to consider the "influence" of the public group versus the group's "stake in the outcome" (Figure 10.1). Different publics may need to be involved in different capacities and at different points in the EIA process. For example, publics with little stake in the outcome (i.e., they will not be directly affected) and with limited power and influence over the project decision and EIA process may be considered spectators and involved in the EIA process only indirectly through public communications, news releases, and education about the project. Individuals with a high stake in the outcome, such as the affected local population, but with limited power and influence should be intimately involved throughout the EIA process so as to ensure that their concerns are addressed at the time of project decision; they have no influence over that decision. This second group is often referred to as the "victims" of project development in that although they may experience some benefits from development, they also have the most to lose if project impacts are not properly managed. It is for this group that funding to participate, such as through the Canadian Environmental Assessment Agency's participant funding program, is important in order to facilitate participation and to ensure that the public has the necessary resources and capacity to become involved.

At the other end of the spectrum are publics with a limited stake in the outcome but who are highly influential. This group might include quasi-regulatory bodies, the media, and other special interest groups. Caution must be taken to ensure that the voice of this highly influential group is not overrepresented relative to that of the "victims." The final group of publics is those with both high stakes in the outcome

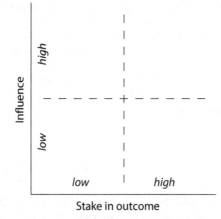

Figure 10.1 ◉ Influence versus Stake in Outcome when Identifying Publics

and a high degree of influence over the process. For regulators and proponents, this is the most complex group of publics, since they have a potential for considerable gains and losses from project development and at the same time are highly influential over the EIA process and project success. In the Canadian context, some of the more influential Aboriginal groups and quasi-government bodies could be classified as high-stake and high-influence interests.

Levels of Involvement

Public involvement, while an important concern in EIA, rarely consists of highly participatory approaches in which proponents are willing to significantly alter project design or implementation plans. Arnstein (1969), for example, observed that different levels of public involvement could be identified, ranging from manipulation of the public to citizen control (Box 10.2). At one end of the spectrum is "non-participation"—involvement of the publics in a way that does not include direct participation. This approach consists of what Arnstein labels "rubberstamp committees" and efforts to inform or "educate" the public rather than to genuinely seek their involvement. At the opposite end of the spectrum is citizen power in which the affected public is granted full control and authority in development decision-making—rarely, if ever, the case in EIA. In practice, public involvement in EIA has focused primarily on *consulting* or *informing* the public and sometimes negotiating trade-offs, with limited options for greater participation. In other words, public participation in EIA typically involves *providing information* to the public about the proposed project and, at best, degrees of tokenism. Whether this is a sufficient level of public involvement in development impact assessment and decision-making is a matter of debate.

Box 10.2 Arnstein's Ladder of Citizen Participation

Participation	Nature of involvement	Degree of power
1. Manipulation	Rubberstamp committees	Non-participation
2. Therapy	Power-holders educate or cure citizens	
3. Informing	Citizens' rights and options are identified	Degrees of tokenism
4. Consultation	Citizens are heard but not always heeded	
5. Placation	Advice is received but not acted on	Degrees of citizen power
6. Partnership	Trade-offs are negotiated	
7. Delegated power	Citizens are given management power	
8. Citizen control	For all or parts of projects or programs	

Source: Arnstein 1969.

Provisions and Requirements for Public Participation in Canada

Provisions for public participation in Canadian EIA vary from one jurisdiction to the next. However, there are five minimum provisions for public participation in EIA. In the sections below, these provisions are explored at the federal level and within the context of EIAs conducted primarily under the responsibility of the Canadian Environmental Assessment Agency.

Adequate Notice

There is no requirement under Canadian federal EIA for project proponents to notify or consult with either the public or Aboriginal peoples in the preparation of a project description for submission to the federal government for a determination of the need for an EIA; however, the Canadian Environmental Assessment Agency's *Guide to Preparing a Description of a Designated Project under the Act* strongly encourages them to do so. When a project description is submitted, the agency conducts a 20-day public comment period on the designated project and seeks the views of other federal authorities and public and Aboriginal groups before making a determination as to whether an EIA is required. If an EIA is required, the agency posts a "notice of commencement" of the EIA on its website. Notice is also posted for opportunities for the public to comment on the draft EIS guidelines and the submitted EIS itself.

Box 10.3 Environmental Assessment Registries: Promise and Reality

A 2009 report of the Commissioner of the Environment and Sustainable Development found that project files were complete for all the comprehensive studies and panel reviews contained in a sample of studies and reviews. As Hanna and Noble (2011) report, in terms of compliance it would seem that agencies are mostly doing what is officially required of them under the Canadian Environmental Assessment Act. These same requirements exist under the Canadian Environmental Assessment Act, 2012. However, Hanna and Noble caution that it is one thing to have information available in the form of a registry; it is another to have it available in a language, format, and style that are easily accessible and understood by diverse audiences.

- Findlay (2010) reports that in a sample of 30 road EIAs for which decisions had been posted, obtaining documents was neither easy nor often successful. In 23 instances, requests for reports were met with no reply or "endless redirections." In only one instance was the requested material available through the Internet site, and in the remainder the wait was one to two weeks. Reports were not uncommonly copyrighted by

Access to Information

Documents relevant to a project and the assessment process, including the project registration and description, various ministerial decision, the EIS, and often public submissions, are made available to the general public and interested parties through online registries. Each responsible authority, the Canadian Environmental Assessment Agency, the Canadian Nuclear Safety Commission, and the National Energy Board, maintains its own web-based registry of EIA documentation from the time of EIA commencement until completion of all follow-up programs (Box 10.3).

Public Comment

There are several opportunities for public comment during the assessment process. When a project description is submitted, the public, Aboriginal groups, and other interested federal authorities have a 20-day window to submit written comment. The agency also posts a draft of the EIS guidelines for a 30-day public comment period. An opportunity for public comment is also available during the conduct of an EIA by the responsible authority. For example, section 19(1) of the Canadian Environmental Assessment Act, 2012 notes that the environmental assessment of a designated project must take into account comments from the public or, under section 54 of the National Energy Board Act, any interested party. Section 19(3) states that the EIA of designated projects may take into account community knowledge and Aboriginal traditional knowledge. But the proponent is required to incorporate into the EIS the community and Aboriginal knowledge or comments to which it has access or that is acquired through EIA processes. A summary of the EIS is also posted on the agency's

consultants, and thus several approvals were required to obtain material. Older EIAs were often not available in electronic form.

- Sinclair and Diduck (2009) report cognitive inaccessibility as a persistent problem in EIA registries. In particular, they point to the "overly technical language and general lack of readability" of EISs and related project documents—an obstacle for most members of the public.
- Ball, Noble, and Dubé (2012) report on an analysis of a sample of 35 EISs from the South Saskatchewan watershed, including assessments under federal and Saskatchewan and Alberta provincial jurisdiction. Approximately 25 per cent of the EISs initially selected for review could not be accessed because of the incompleteness of public registries, limited logistical support from some responsible authorities to locate the documents, and the reluctance of some proponents to share documents. It required three months of active soliciting to collect the sample of "publically available" EISs. Among the EISs that were reviewed, many referred to technical studies and to data that were not included in the registry.

website for public comment, and, in the case of a review panel assessment, there is an opportunity to submit written comment to the panel.

Public Hearings

Typically, public hearings are associated with review panel assessments and managed by the review panel itself; however, the minister of environment does have the authority to call a public hearing for any assessment. In the case of a review panel assessment, the public must be notified and provided access to documents relating to the environmental assessment. Any member of the public may attend review panel public hearings and present written comment to the review panel. Under the former Canadian Environmental Assessment Act, any member of the public could apply to present information orally at a public hearing. Under the new act, only "interested parties" (a person who is directly affected by the project or has relevant information or expertise) may be granted intervener status and present information orally at a public hearing. Public hearings provide a formal opportunity for directly affected interests to be heard, for public information and knowledge to be collected, and for public concerns to be formally documented. The information gathered through a hearing process is used to support a review panel's recommendations. Public hearings and review panels are not decision-making bodies.

Participant Funding

Also referred to as "intervener funding," participant funding is available to support the participation of individuals, non-profit organizations, and Aboriginal groups interested in participating in federal EIAs. Each responsible authority, the Canadian Environmental Assessment Agency, the Canadian Nuclear Safety Commission, and the National Energy Board, manages its own participant funding program. Eligibility for participant funding varies by responsible authority, but generally speaking, individuals, non-profit organizations, and Aboriginal groups can apply for participant funding if they have a direct, local interest in the project (e.g., living on or owning property in the project area), have community awareness of Aboriginal traditional knowledge relevant to the assessment, or have expert information relevant to the anticipated environmental effects of the project. Recipients can use the funds to collect their own data or conduct their own research, to convene or participate in public meetings or hearings, and to support their review of the EIS. Participant funding programs are not intended to cover the full costs for individuals, non-profit organizations, or Aboriginal groups to participate in an EIA.

Aboriginal Consultation and Engagement

In Canada, Aboriginal peoples are acknowledged to have particular rights under the Canadian Constitution and, where applicable, through treaty land entitlements. Aboriginal consultation is required in environmental assessment in Canada. Federally, for example, several sections of the Canadian Environmental Assessment Act, 2012 make reference to Aboriginal consultation and engagement or acknowledgment of Aboriginal peoples' concerns. Among the purposes of the act is to "promote communication and cooperation with Aboriginal peoples with respect to environmental assessments." The definition of an "environmental effect" under the act also makes direct reference to Aboriginal peoples with respect to health and socio-economic conditions, physical and cultural heritage, and use of lands and resources for traditional purposes.

Ensuring meaningful Aboriginal engagement in EIA requires more than meeting the requirements of legislation. Much has been written about the failure of EIA with regard to Aboriginal engagement, and similar concerns about the EIA process have been raised across federal, provincial, and territorial jurisdictions. Issues and concerns about the process from the perspective of legal experts, scholars, and Aboriginal communities are well documented. But what lessons can be learned from those engaged in day-to-day practice?

Booth and Skelton (2011) examined industry and government perspectives on First Nations' participation in the British Columbia environmental assessment process. They report in particular on the lessons and perspectives of consultants and project proponents concerning what works to ensure successful engagement and what issues cause difficulty for consultants and proponents. The lessons and perspectives are synthesized below.

Lessons for successful engagement:

- Early development of personal relationships with the First Nations affected by the project. Frequent on-site visits are important and allow proponents to identify potential problems and concerns and to have time to work together on developing solutions.
- A successful relationship can be established through multiple visits to the First Nation community and attendance at various First Nations social and community events.
- Inviting the First Nations on site visits of related existing projects can provide a grounded perspective in terms of what they can expect to see from the proposed project in terms of infrastructure and related developments.
- Couching initial discussions in terms of treaty rights as the framework for understanding First Nation concerns can open up communication.
- It is important for the proponents to be actively involved "on the ground," not just the consultants.
- Reciprocal studies are important, giving First Nations an opportunity to participate through employment and access to new data.
- If a project or assessment is controversial, a third-party review can help to ensure trust.

Observed constraints to successful engagement:

- Limitations on the EIA process in terms of addressing non-quantifiable issues, such as spirituality and cultural issues, can lead to misunderstanding.

- Absence of trust and lack of genuine willingness to work together can mean a lack of understanding of the issues by all parties and inability to resolve them.
- Lack of provisions at all stages on the EIA process to address First Nations concerns can lead to tensions and lack of trust.
- Poor relationships between First Nations and governments can negatively affect a proponent's progress through the EIA process.
- The EIA process can be "hijacked" by deeply rooted issues concerning First Nations and governance that are not necessarily within the scope of a proponent's ability to address but emerge because such issues have not been addressed by government.
- First Nations communities may have limited capacity to engage in EIA and respond to the requests of proponents and governments in a timely fashion.
- Consultants employed by industry, including those conducting traditional use studies, are often not adequately skilled or trained in the area, resulting in conflicts with First Nations concerning the credibility of the results.
- Government and the First Nations often hold different ideas about which Nations are impacted by the project and therefore should be included in an EIA.
- First Nations governments can have high staff turnover rates, resulting in a lack of consistency and delays in the engagement process.
- There may be communication challenges internal to the First Nation—between staff and community members and political leaders.

Duty to Consult

> Stakeholder input has generally improved in Canada in the last decade . . . but true meaningful involvement is difficult and has not frequently occurred from a community/First Nations perspective (Lawe, Wells, and Mikisew Cree First Nations Industry Relations Corporation 2005, 207).

The **duty to consult** is a formal, legal obligation for government to consult with Aboriginal peoples. This legal obligation is separate from and in addition to any requirements for public participation under various federal and provincial environmental assessment systems. In 2004, the Supreme Court of Canada announced two important decisions concerning the *Taku River Tlingit First Nation v. British Columbia* and *Haida Nation v. British Columbia* (Box 10.4). Together, these two decisions would change Aboriginal rights law by specifying the "duty to consult" with Aboriginal peoples on the part of federal and provincial governments in cases where Aboriginal rights, claims, or titles are known and may be affected by a development or decision but where those rights, claims, or titles have not yet been proven in court. When there is a possibility that Aboriginal or treaty rights may be infringed, the government involved has a duty to consult with the affected First Nations group. Such consultation must be carried out with the goal of addressing the concerns of the affected First Nation, but the extent of consultation is determined on a case-by-case basis according to the severity of the potential impact; the extent to which there is an asserted claim or treaty right; the status, merit, or strength of that claim; and whether the Aboriginal or treaty right potentially affected is already established or claimed but not yet established (Potes, Passelac-Ross, and Bankes 2006). The duty to consult applies even before the Aboriginal claim, right, or title is established conclusively by the courts (Brackstone 2002). There is no duty to reach an agreement during consultation; rather, the intent is to substantially address the concerns of the affected Aboriginal peoples. Consent during consultation is not a requirement. For further information on the duty to consult, including consultation requirements, constitutional duty, and an analysis of court decisions and policies, see Newman (2014).

Traditional Ecological Knowledge

Section 19(3) of the Canadian Environmental Assessment Act, 2012 states that the environmental assessment of a designated project may take into account community knowledge and Aboriginal traditional knowledge. The duty to consult should not be confused with the integration of **traditional ecological knowledge** (TEK) into the EIA process. Indeed, the duty to consult can be met in the absence of any consideration of TEK. Often used interchangeably with "local knowledge" or "indigenous knowledge," TEK is defined as "a cumulative knowledge, practice, and belief, evolving by adaptive processes and handed down through generations by cultural transmission, about the relationship of living beings (including humans) with one another and with their environment" (Berkes 1999, 8 and Berkes, Berkes and Fast 2007). In other words, TEK is associated with societies with historical continuity in resource use in a particular region. At its foundation is knowledge of the land, animals, and the local environment, followed by knowledge of management systems, values, social institutions, and a particular world view. Although TEK is not attached to any particular group in society, it is used largely in an indigenous or Aboriginal context.

TEK and Western Science
Perhaps the best way to understand TEK is to contrast it with Western science. According to Peters (2003), TEK is embedded in local culture and communities bounded by the local environment, has a significant moral and ethical context, and emphasizes the absence of separation between nature and culture. Western science, in contrast, is highly analytical, hierarchical in structure, and compartmentalized (e.g., VECs) and authoritative and bureaucratic in nature (Table 10.2). The two different knowledge systems are highlighted in the recent Keeyask hydroelectric generating project EIS, Manitoba.

In July 2012, the Keeyask Hydropower Limited Partnership submitted an EIS to Manitoba Conservation and Water Stewardship and to the Canadian Environmental Assessment Agency for development of the Keeyask hydroelectric generating project. The Keeyask project is a collaborative effort between Manitoba Hydro and four Manitoba First Nations, referred to as the Keeyask Cree Nations (KCN). Keeyask is located in northern Manitoba, 725 kilometres northeast of Winnipeg on the lower Nelson River. The EIS makes explicit reference to the KCN's world view as being "interconnected and/

Box 10.4 Duty to Consult and the Haida Case Ruling

In 2004, the Supreme Court of Canada released its decisions in *Haida Nation v. British Columbia (Minister of Forests)* and *Weyerhaeuser*. The Haida case involved a judicial review, pursuant to the British Columbia Judicial Review Procedure Act, of the minister's decision to replace and approve the transfer of a tree farm licence. In 1961, tree farm licences were issued to MacMillan Bloedel, a forest harvesting company, permitting the company to harvest in an area of Haida Gwaii, the Queen Charlotte Islands. The licences were replaced in 1981, 1995, and 2000; the tree farm licence was transferred to Weyerhaeuser in 1999. The Haida challenged these replacements and the transfer, arguing that they were made without the consent of the Haida and over their objections. The Haida's case was dismissed by the British Columbia Supreme Court, which noted that the law could not presume the existence of Aboriginal rights based only on their assertion and concluded that the government had only a "moral duty" to consult. The decision was appealed, and the court of appeal found that the government had fiduciary obligations of good faith to the Haida with respect to their claims to Aboriginal title and right and that both the province and Weyerhaeuser were aware of the Haida's claims to the area covered by the licence. The court concluded that both the Crown and Weyerhaeuser had a legally enforceable duty to consult with the Haida in an attempt to address their concerns. Weyerhaeuser's appeal of the decision to the Supreme Court of Canada was allowed, but the government's appeal was dismissed. The Supreme Court of Canada ruled that the duty to consult rests with government and government must consult and accommodate when there is knowledge of the potential existence of an Aboriginal right or title, regardless of whether that right or title has been legally established.

Source: Bergner 2005.

or interrelated with all living things of the ecosystem, with emphasis on relationships, harmony and balance" and notes that this holistic world view "does not consider individual parts of components as does a technical science approach; rather it focuses on the relationships among all elements and functions of an ecosystem" (Keeyask Hydropower Limited Partnership 2012, vol. 6, 6-427). As such, Manitoba Hydro and the KCN agreed that the EIS would contain a KCN evaluation process, as well as a government regulatory assessment process. The KCN's evaluation process would provide their own assessment of the effects of the project on their community and would include Aboriginal traditional knowledge relevant to the EIS guidelines.

The contrasting views and understandings of two different knowledge systems are reflected throughout the EIS. For example, Manitoba Hydro reports that project construction is expected to alter water quality in the immediate reservoir construction area causing an increase in total suspended solids, with the largest increase occurring immediately downstream of the construction site, but that any increase is unlikely to have a measureable effect on biota. The KCN's evaluation reports that water quality is expected to be affected upstream of the construction site and water quality is expected to be poor in all areas affected by the Keeyask generating station. Manitoba Hydro reports that adverse effects to fish—namely, walleye and lakefish—will occur because of the loss of spawning habitat during construction but that populations are expected to remain the same or increase over the long term because of an increase in the amount of foraging habitat created by reservoir flooding. The technical reports prepared in support of the EIS indicate that no adverse effects to fish populations are expected from the project outside the reservoir and local study area and that any adverse effects would be restricted to the construction period and localized. The KCN's evaluation reports an expected decline in the numbers and health of most fish species as a result of the project and that adverse effects will extend well beyond the construction site. The KCN's view is that effects will be of a larger spatial and temporal extent than indicated in the EIS technical reports.

The Keeyask project is a good example of the often-contrasting views of TEK and Western science and the challenges this can pose to the EIA process. At the same time, however, there are similarities between TEK and Western science knowledge systems. Berkes, Berkes, and Fast (2007), for example, in a study of bowhead whales in the Canadian North, found that TEK was very similar to science in many respects in that many indigenous peoples recognize and monitor various environmental indicators or "signs" and "signals" of change.

Table 10.2 TEK and Western Science–Based Knowledge Systems

	TEK	Western science
Basic characteristics	Holistic Experimental Intuitive Subjective	Reductionist Positivist Analytical Objective
Structural governance	Non-hierarchical Egalitarian	Hierarchical Centralized
Communication	Oral record Teaching	Written record Instructive
Decision-making Effectiveness	Consensus	Authoritative
Data creation	Slow, inclusive	Fast, selective
Data type	Long time series, single location	Short time series, larger area
Prediction	Short-term cycles	Short-term linear
Explanation	Spiritual and cumulative knowledge	Hypotheses and theory
Biological class	Ecological	Genetic and hierarchical
Experts	Elders Hunters and trappers Communities	Scientists Academics Regulators

Sources: Table compiled by Jasmine Birk, University of Saskatchewan. Based on Berkes, Berkes, and Fast 2007; Houde 2007; Whitye 2006; Peters 2003; Usher 2000; Berkes 1999; Nadasdy 1999; Johannes 1993; Johnson 1992; Wolfe et al. 1992.

Benefits and Challenges to TEK Integration

The integration of TEK in Canadian EIA is a policy requirement, and various international organizations, such as the United Nations and the World Bank, recognize the value of TEK as an important means of integrating indigenous communities in environmental assessment and decision-making processes. Some believe that incorporating TEK in EIA will help to improve communication between proponents and governments and the local Aboriginal community and create a better, overall understanding of the environment, thus providing an opportunity to improve impact assessment and management (Nadasdy 1999) (see Environmental Assessment in Action: Integrating Traditional Ecological Knowledge in Impact Assessment in the Mackenzie Valley). Others, however, such as Howard and Widdowson (1996), argue that efforts to integrate TEK in EIA may be less than beneficial to decision-makers and to the EIA process in general and hinders rather than enhances the ability of governments to more fully understand ecological processes, since there is no mechanism, or will, by which spiritually based knowledge claims can be challenged or verified.

While Howard and Widdowson's view does not represent mainstream thinking on the value of TEK, it may represent the "unspoken" view reflected in the current approach to TEK in EIA. Integrating science and TEK is often an exercise in combining two alternative sets of data, but in most instances the science remains relatively unchanged. Houde (2007), on the basis of a review of TEK and co-management boards, observed that "schemes to involve First Nations in decision-making processes have been created for equating TEK to a collection of data about the environment that could complement and be integrated within the existing data sets used by state management systems and for failing to acknowledge the value and cosmological context within which this traditional knowledge was generated and makes sense."

Environmental Assessment in Action

Integrating Traditional Ecological Knowledge in Impact Assessment in the Mackenzie Valley

The Mackenzie Valley Environmental Impact Review Board (MVEIRB) is one of several co-management boards established following the settlement of comprehensive land claims in the Northwest Territories. As a condition of these claims, the Mackenzie Valley Resource Management Act (MVRMA) was enacted, transferring responsibility for many resource management decisions, including EIA, to Aboriginal–government co-management boards. The MVEIRB is responsible for all environmental assessments and panel reviews in the Mackenzie Valley, and the MVRMA mandates an assessment of direct social and cultural impacts and in "regard to the economic well-being of the residents of the Mackenzie Valley" as well as "the importance of conservation to the well-being and way of life of the aboriginal peoples of Canada."

Development Proposal

In 2004, Imperial Oil Resources Ventures Limited (Imperial) proposed to undertake geotechnical investigations in support of a possible future pipeline along an approximately 450-kilometre route, north to south, in the Dehcho region of the Mackenzie Valley. About 3000, mostly Aboriginal, people live in eight Dehcho communities. The region supports a variety of ungulates, large carnivores, fur-bearers, and small mammals, including rare species like wood bison, woodland caribou, grizzly bears, and wolverines. Moose and caribou are of special importance to local communities, particularly as traditional foods.

The purpose of the proposed development was to investigate sub-surface conditions along the likely pipeline route in order to assess the feasibility of any subsequent engineering and construction. The project was to be carried out over one or two winters. Access to the work sites would be primarily via a gravel highway, an existing winter road, and an existing pipeline right-of-way. For access beyond these areas, Imperial proposed to open 406 kilometres of existing seismic lines and trails and 45 kilometres of new access. The proposed operations involved setting up rig camps, clearing work sites, and drilling holes to study potential gravel sources as well as permafrost conditions. At larger river crossings, the developer planned to undertake in-river drilling. Equipment included bulldozers, graders, backhoes, drill rigs on tracked carriers or sleighs, and helicopters as well as various trucks and smaller vehicles.

Environmental Assessment Process

Imperial applied for a land-use permit and a water licence from the Mackenzie Valley Land and Water Board, and following a preliminary screening, it was found that the development would not have a significant impact. However, the MVEIRB received several letters from communities in the Dehcho region, describing concerns about adverse impacts on fish and wildlife as well as wild-life harvesting and spiritually significant areas. The MVEIRB decided to conduct an assessment on its own motion.

The EIA process consisted of terms of reference for Imperial's assessment, Imperial's assessment report documenting their analysis and conclusions on the likelihood and significance of any adverse impacts, and public hearings held by the MVEIRB in the three communities closest to the development to hear directly from the developer, government experts, and the potentially affected communities. Following its deliberation, the MVEIRB recommended that the proposed development be allowed to proceed subject to 15 mitigation measures. The MVEIRB also made 11 suggestions to further reduce the effects of the development.

Traditional Knowledge Contributions

Traynor Lake watershed: The Sambaa K'e First Nation, based in Trout Lake, views the Traynor Lake watershed to be of significant cultural, spiritual, and ecological importance and contended that the proposed development was unacceptable. Imperial stated that they had already moved the corridor for the future pipeline in response to the communities' concerns and were investigating alternative gravel sources with the help of community members.

continued

The MVEIRB accepted the importance of the Traynor Lake watershed to the Sambaa K'e First Nation, stating that the area is vital to the cultural and spiritual well-being of the people of Trout Lake. The MVEIRB also found, however, that the majority of concerns were related to the future pipeline and that the proposed project under consideration had a lesser risk of affecting the watershed than the pipeline, which was expected to happen sometime in the future. The MVEIRB stopped short of disallowing the proposed activities but imposed several measures, such as having community members, including elders, monitor the activities; locating alternative gravel sources as far away from the lake as possible; and for government to initiate a process to seek protected area status for the watershed.

Caribou and moose: The communities and Imperial differed in their evaluation of potential impacts on moose and woodland caribou. Imperial concluded that the proposed activities would not have any significant impact. The First Nation pointed out that Imperial's conclusions were drawn from limited baseline data and that the First Nation's own observations showed a considerable change of behaviour in moose already from the limited activities involved in designing the project, such as helicopter traffic. The MVEIRB rejected Imperial's analysis and concluded that while in isolation the proposed development might not be likely to cause significant impacts, it would contribute to significant cumulative impacts. As a result, the MVEIRB imposed three measures on the development, ranging from collecting more baseline data to the development of policies to minimize impacts from aircraft over-flights.

Blackwater River area: The Blackwater River is important spiritually, culturally, and ecologically to the Pehdzeh Ki First Nation of Wrigley. Unlike in the Traynor Lake case, the MVEIRB found no evidence that the developer had changed their proposal or otherwise incorporated traditional knowledge into the project design. The MVEIRB found that Imperial's commitments to minimizing impacts on the environment were greatly outweighed by the evidence from the Pehdzeh Ki First Nation that the development would cause significant adverse impacts. To mitigate these cultural and social impacts, the MVEIRB imposed a 3-kilometre-wide exclusion zone along a 15-kilometre stretch of the Blackwater River. This measure disallowed several sites the developer had proposed for their field investigation but fell short of the 15-kilometre-wide exclusion zone the Pehdzeh Ki First Nation had requested.

Other sites near Wrigley: In addition to the Blackwater River, the Pehdzeh Ki First Nation identified several sites in the vicinity of their community as important and that would be adversely impacted by the proposed development. The MVEIRB found in this case that the developer had not provided sufficient site-specific evidence to conclude that impacts were not likely. The MVEIRB also found that several families in the area still maintained a subsistence lifestyle and were more vulnerable to impacts from development than the community at large and thus worthy of extra protection. The MVEIRB imposed measures requiring Imperial to conduct site visits with community elders and resident families to identify heritage resources and, further, to consult directly with three individual families when conducting activities near their homes and main harvesting areas about how to minimize or prevent impacts in each specific location.

Conclusion

Traditional knowledge played a role in subjecting the proposed development to an environmental assessment. According to the developer, traditional knowledge also played a role in designing the project. In several instances, the MVEIRB placed more weight on traditional knowledge than on the conclusions drawn by Imperial based on Western scientific study. A number of mitigation measures imposed by the MVEIRB (e.g., exclusion zones) were largely based on traditional knowledge, since the MVEIRB explicitly rejected Imperial's conclusions and accepted those of the community. What is unclear, however, is to what extent the submissions made by community leaders were informed by traditional knowledge. Traditional knowledge undoubtedly played an important role in the environmental assessment, but it remains impossible to quantity. Also, since the proposed development was delayed for several years in subsequent consultation processes between the Canadian government and the MVEIRB, as well as between government and the First Nations, it also remains unclear whether the MVEIRB balance between Western scientific analysis and traditional knowledge satisfied any of the stakeholders.

Source: The above case is a summary of a case study contributed by Martin Haefele, Manager, Environmental Impact Assessment, Mackenzie Valley Environmental Impact Review Board. The case study was contributed as part of curriculum development for the Geography 386 Introduction to Environmental Impact Assessment course at the University of Saskatchewan for an online course offering for the University of the Arctic.

Methods and Techniques

There are many different methods and techniques for public involvement and communication, and the capability and capacity of these methods and techniques vary considerably. Many different methods and techniques may be used in any single EIA, and at each stage of the EIA process, for different publics and for different purposes (Table 10.3). For example, a proponent may hold information seminars early in the project design stage for the communities or interest groups most likely to be affected, whereas other communities or publics outside of the direct project impact zone may be informed at this point only through the mass media. The specific methods and techniques selected for public involvement in any EIA depend on:

- the proponent's objectives;
- the proponent's commitment to public involvement;
- the nature of legal requirements for participation;
- the sensitivity of the receiving environmental and socio-economic environment;
- the availability of time and resources;
- the magnitude of potential project impacts;
- the level of public interest.

Table 10.3 Selected Techniques for Public Involvement and Communication

Capability meets the criterion = ✓ Capacity high = ◉ medium = ⊙ low = O	Capability				Capacity		
	Provide information	Obtain feedback	Resolve conflict	Identify problems and values	Two-way communication	Caters to special interests	Number of people involved
Public meetings	✓	✓		✓	⊙	O	⊙
Public displays	✓	✓			⊙	O	◉
Presentations to small groups	✓	✓		✓	⊙	⊙	O
Workshops		✓	✓	✓	◉	◉	O
Advisory committees		✓		✓	◉	◉	O
Public review of EIS	✓	✓		✓	O	◉	⊙
Information brochures	✓				O	⊙	⊙
Press release inviting comments	✓	✓			O	O	◉
Public hearings		✓		✓	O	O	⊙
Task forces				✓	◉	◉	O
Site visits	✓			✓	◉	◉	O
Information seminars	✓	✓			◉	◉	O
Mass-media information	✓				O	O	◉

Sources: Based on Sadar 1996; Westman 1985.

Meaningful Participation

Public participation in the EIA process does not necessarily ensure meaningful participation. Based on the International Association for Impact Assessment's (André et al. 2006) international best-practice principles for public participation in EIA, good-practice participation requires adherence to a number of basic and operating principles.

Basic Principles

The basic principles are fundamental to public participation in EIA and apply to all stages of the EIA process and to all tiers, from project-level impact assessment to applications at the broader planning and policy tiers. They are summarized below, based on the IAIA:

Adapted to context: Understanding the context (social institutions, values, culture, history, politics) of the places and communities affected by the proposed development

Informative and proactive: Early and meaningful provision of information to communities or populations that may be affected by or have an interest in the proposal

Adaptive and communicative: Recognition of the heterogeneity of affected populations based on differences in values, demographics, knowledge, and interests

Inclusive and equitable: Ensuring that represented and unrepresented groups and interests are included in participation, including the concerns of future generations

Educative: Contributing to mutual understanding and respect

Cooperative: Promoting cooperation, convergence, and consensus-building

Imputable: Improving the proposal under consideration and reporting back to participants on how their involvement has contributed to decision-making

Operating Principles

The various operating principles concern how the above basic principles should be applied in the EIA process. They are summarized below, based on the IAIA:

Initiated early and sustained: Participation should commence early in the EIA process, before major decisions are made, and be sustained throughout the process.

Well planned and focused on negotiable issues: Emphasis should be placed on negotiable issues relevant to decision-making, and the objectives, organization, and procedures of participation should be clear to all those involved.

Supportive to participants: There should be adequate information and financial support to facilitate participation, and capacity-building support should be provided to interested groups who lack the capacity to participate.

Tiered and optimized: Participation should occur at the most appropriate level of decision-making.

Open and transparent: Those interested in participating should have access to all relevant information, and that information should be available in an understandable format.

Context-oriented: Methods and procedures for participation should be adapted to the local community, social, cultural, and political context.

These basic and operating principles should be applied in all participatory contexts in EIA. However, based on Mezirow (2000), in order to ensure meaningful participation in EIA, or in any other context, a number of conditions must be met. Namely, participants must have: complete and accurate information; openness to alternative points of view; freedom from coercion; the ability to weigh evidence and assess arguments objectively; equal opportunity to participate; and a willingness to seek understanding and agreement and to accept best judgment until new information is presented and validated. These are, of course, ideal conditions. No practitioner or regulator, for example, can ensure that participants are able to weight evidence and assess arguments objectively or are willing to seek understanding and agreement. In some cases, particularly when dealing with controversial projects, conflicts are deeply rooted in values and past experiences. Conflict during the EIA process should not be seen as negative; one of the purposes of EIA is to ensure open and public debate about often controversial resource development projects.

Key Terms

active publics

duty to consult

inactive publics

traditional ecological knowledge

Review Questions and Exercises

1. Suppose the development of a new waste disposal facility was proposed for your area. Identify who the "active" and "inactive" publics might be. Based on "Arnstein's ladder," at what level would you suggest each of these different groups be involved?
2. Identify a recent controversial development project in your region, and assess how public involvement was conducted. Do you think the outcome of the project proposal or the current situation might have been different if a different degree of public involvement had been incorporated?
3. Discuss the advantages and disadvantages to the proponent of involving publics early in the EIA and project design process.
4. What are some of the benefits and challenges of integrating local or traditional ecological knowledge in the EIA process?

References

André, P., et al. 2006. *Public Participation: International Best Practice Principles*. Special Publication Series no. 4. Fargo, ND: IAIA.

Arnstein, S. 1969. "A ladder of citizen participation." *Journal of the American Institute of Planners* 35: 216–24.

Ball, M., B.F. Noble, and M. Dubé. 2012. "Valued ecosystem components for watershed cumulative effects: An analysis of environmental impact assessments in the South Saskatchewan River watershed, Canada." *Integrated Environmental Assessment and Management* doi: 10.1002/ieam.1333.

Bergner, K. 2005. *The Crown's Duty to Consult and Accommodate*. Paper presented at the Canadian Institute's Fifth Annual Advanced Administrative Law and Practice, Ottawa.

Berkes, F. 1999. *Sacred Ecology: Traditional Ecological Knowledge and Resource Management*. Philadelphia: Taylor and Francis.

———, M. Berkes, and H. Fast. 2007. "Collaborative integrated management in Canada's North: The role of local and traditional knowledge and community-based monitoring." *Coastal Management* 35: 143–62.

Booth, A.L., and N.W. Skelton. 2011. "Industry and government perspectives on First Nations' participation in the British Columbia environmental assessment process." *Environmental Impact Assessment Review* 31: 216–25.

Brackstone, P. 2002. *Duty to Consult with First Nations*. Victoria, BC: Environmental Law Centre.

CEAA (Canadian Environmental Assessment Agency). 2012. *Guide to Preparing a Description of a Designated Project under the Canadian Environmental Assessment Act, 2012*. Ottawa: CEAA.

Diduck, A. 2004. "Incorporating participatory approaches and social learning." In B. Mitchell, ed., *Resource and Environmental Management in Canada*, 3rd edn., 497–527. Toronto: Oxford University Press.

Eckel, L., K. Fisher, and G. Russell. 1992. "Environmental performance measurement." *CMA Magazine* (March): 16–23.

Findlay, S. 2010. "The CEAA registry as a tool for evaluating CEAA effectiveness." Presentation to the Ontario Association for Impact Assessment annual general meeting and conference, Ottawa.

Hanna, K., and B.F. Noble. 2011. "The Canadian environmental assessment registry: Promise and reality." *UVP-report* 25 (4): 222–5.

Houde, N. 2007. "The six faces of traditional ecological knowledge: Challenges and opportunities for Canadian co-management arrangements." *Ecology and Society* 12 (2): 34.

Howard, A., and F. Widdowson. 1996. "Traditional knowledge threatens environmental assessment." *Policy Options* (November): 34–6.

Johannes, R. 1993. "Integrating traditional ecological knowledge and management with environmental impact assessment." In J. Inglis, ed., *Traditional Ecological Knowledge: Concepts and Cases*. Ottawa: International Program on Traditional Ecological Knowledge, International Development Research Centre.

Johnson, M. 1992. "Research and traditional environmental knowledge: Its development and its role." In M. Johnson, ed., *Lore: Capturing Traditional Environmental Knowledge*. Yellowknife: Dene Cultural Institute and International Development Research Centre.

Keeyask Hydropower Limited Partnership. 2012. *Keeyask Generation Project Environmental Impact Statement: Response to EIS Guidelines*. Canadian Environmental Assessment Registry Reference 11-03-64144. Winnipeg: Keeyask Hydropower Limited Partnership.

Lawe, L.B., J. Wells, and Mikisew Cree First Nations Industry Relations Corporation. 2005. "Cumulative effects assessment and EIA follow-up: A proposed community-based monitoring program in the oil sands region, northeastern Alberta." *Impact Assessment and Project Appraisal* 23 (3): 205–9.

Mezirow, J. 2000. *Learning as Transformation: Critical Perspectives on a Theory in Progress*. San Francisco: Jossey-Bass.

Mitchell, B. 2002. *Resource and Environmental Management*. 2nd edn. New York: Prentice Hall.

Murphy, B., and R. Kuhn. 2001. "Setting the terms of reference in environmental assessments: Canadian nuclear fuel waste management." *Canadian Public Policy* 27 (3): 249–66.

Nadasdy, P. 1999. "The politics of TEK: Power and the integration of knowledge." *Arctic Anthropology* 36 (1–2): 1–18.

Newman, D.G. 2014. *Revisiting the Duty to Consult Aboriginal Peoples*. Saskatoon, SK: Purich Publishing

Noble, B.F. 2004. "Integrating strategic environmental assessment with industry planning: A case study of the Pasquai-Porcupine Forest Management Plan, Saskatchewan, Canada." *Environmental Management* 33 (3): 401–11.

———, and J. Bronson. 2007. *Models of Strategic Environmental Assessment in Canada*. Ottawa: Canadian Environmental Assessment Agency.

Peters, E. 2003. "Views of traditional ecological knowledge in co-management bodies in Nunavik, Quebec." *Polar Record* 39 (208): 49–60.

Petts, J. 1999. "Public participation and environmental impact assessment." In J. Petts, ed., *Handbook of Environmental Impact Assessment*. London: Blackwell Science.

Potes, V., M. Passelac-Ross, and N. Bankes. 2006. "Oil and gas development and the Crown's duty to consult: A critical analysis of Alberta's consultation policy and practice." Paper no. 14 of the Alberta Energy Futures Project. Calgary: Institute for Sustainable Energy, Environment and Economy, University of Calgary.

Sadar, H. 1996. *Environmental Impact Assessment*. 2nd edn. Ottawa: Carleton University Press.

Shepherd, A., and C. Bowler. 1997. "Beyond the requirements: Improving public participation in EIA." *Journal of Environmental Planning and Management* 40 (6): 725–38.

Sinclair, J., and A. Diduck. 2009. "Public participation in Canadian environmental assessment: Enduring challenges and future directions." In K. Hanna, ed., *Environmental Assessment: Practice and Participation*, 2nd edn., 58–82. Don Mills, ON: Oxford University Press.

Usher, P. 2000. "Traditional ecological knowledge in environmental assessment and management." *Arctic* 53 (2): 183–93.

Westman, W. 1985. *Ecology, Impact Assessment, and Environmental Planning*. New York: John Wiley & Sons.

Whitye, G. 2006. "Culture in collision: Traditional knowledge and Euro-Canadian governance processes in northern land claim boards." *Arctic* 59 (4): 401–14.

Wolfe, J., et al. 1992. "The nature of indigenous knowledge and management systems." In J. Wolfe et al., eds., *Indigenous and Western Knowledge and Resource Management Systems*. Guelph, ON: School of Planning and Development, University of Guelph.

III

Advancing Principles and Practices

Cumulative Environmental Effects

Describing the loss of coastal wetlands along the east coast of the United States between 1950 and 1970, Odum (1982, 728) explains:

> No one purposely planned to destroy almost 50% of the existing marshland along the coasts of Connecticut and Massachusetts . . . However, through hundreds of little decisions and the conversion of hundreds of small tracts of marshland, a major decision in favour of extensive wetlands conversion was made without ever addressing the issue directly.

It is not possible to determine the true significance of a project's effects without the consideration of cumulative environmental effects. Each additional disturbance or impact, regardless of its magnitude, can represent a high marginal cost to the environment. Cumulative effects can be characterized as "progressive nibbling," "death by a thousand cuts," or the "tyranny of small decisions." In other words, cumulative effects are the culmination of effects—many of which can be individually small and seemingly insignificant, such as seismic lines, pipelines, water withdrawals, or the incremental filling of wetlands. Such characterizations are based on the notion that a significant adverse effect can result over space or over time because of the culmination of seemingly small and insignificant actions. For each action, the effects are deemed marginal or relatively insignificant when compared to other types or scales of change or disturbances. But over time, such seemingly insignificant effects can result in significant cumulative environmental change (Gunn and Noble 2012).

Definition of Cumulative Effects

The terms "cumulative environmental change," "cumulative effects," and "cumulative impacts" are often used interchangeably. Generally speaking, these terms all refer to effects of an additive, interactive, synergistic, or irregular (surprise) nature, caused by individually minor but collectively significant actions that accumulate over space and time (Canter 1999). There is no universally accepted definition of cumulative effects, and various definitions have been proposed in the literature, for example:

- the accumulation of human-induced changes in VECs across space and over time that occur in an additive or interactive manner (Spaling 1997);
- the impact on the environment [that] results from the incremental impact of the action [under review] when added to other past, present, and reasonably foreseeable future actions (US Council on Environmental Quality 1978);
- changes to the environment caused by an action in combination with other past, present, and future actions (Hegmann et al. 1999).

Perhaps the most commonly used definition of "cumulative environmental effect" is the one provided by the US Council on Environmental Quality (1997), which characterizes cumulative environmental effects as:

- the total effect, including direct and indirect, on a given resource, ecosystem, or human community of all actions taken;
- effects that may result from the accumulation of similar effects or the synergistic interaction of different effects;
- effects that may last for many years beyond the life of the action that caused them;
- effects that must be analyzed in terms of the specific resource, ecosystem, or human community affected and not from the perspective of the specific action that may cause them;
- effects that must be approached from the perspective of carrying capacity, thresholds, and total sustainable effects levels.

Gunn and Noble (2014), in a report commissioned by the Canadian Council of Ministers of the Environment, indicate that jurisdictional interpretations of a "cumulative effect" vary in scope—some being inclusive of social, cultural, and economic aspects and others more restrictive in scope and focused only on biophysical aspects. They argue that ensuring a standard definition of "cumulative effect" requires that the term be defined independently of the focus of concern, be it biophysical, social, or economic or any combination thereof. In other words, whether a social or cultural effect is included in the definition of a cumulative effect is a matter of jurisdictional preference. Adding or removing any one or more of these components does not change the core definition of what constitutes a "cumulative effect"—it changes only what must be considered within the scope of assessment in that particular jurisdiction. Based on a review of Canadian and international legislation, policies, regulations, and guidelines, Gunn and Noble (2014) suggest a set of mutually supportive definitions for cumulative effects concepts, with each definition nested within the context of the previous:

- **Cumulative effect**: a change in the environment caused by multiple interactions among human activities and natural processes that accumulate across space and time.

- **Cumulative effects assessment**: a systematic process of identifying, analyzing, and evaluating cumulative effects.
- **Cumulative effects management**: the identification and implementation of measures to control, minimize, or prevent the adverse consequences of *cumulative effects*.

Sources of Cumulative Effects

The Canadian Environmental Assessment Research Council (1988) suggests that cumulative effects can occur when impacts on the biophysical or human environments take place frequently in time or densely in space to such an extent that they cannot be assimilated or when the impacts of one activity combine with the activities of another in a synergistic manner. This suggests that a variety of different *sources* of change contribute to cumulative environmental effects (Table 11.1). Consider, for example, the total downstream effects on water quality and fish resulting from upstream **point-source** and **non-point-source stress** in a watershed (Figure 11.1), including:

Table 11.1 Sources of Change That Contribute to Cumulative Environmental Effects

Source of change	Characteristics	Example
Space crowding	High spatial density of activities or effects	Multiple mine sites in a single watershed
Time crowding	Events frequent or repetitive in time	Forest harvesting rates exceeding regeneration and reforestation
Time lags	Activities generating delayed effects	Human exposure to pesticides
Fragmentation	Changes or interruptions in patterns and cycles	Multiple forest access roads cutting across wildlife habitat
Cross-boundary movement	Effects occurring away from the initial source	Acid mine drainage moving downstream to community water supply systems
Compounding	Multiple effects from multiple sources	Heavy metals, chemical contamination, and changes in dissolved oxygen content resulting from multiple riverside industries
Indirect	Second-order effects	Decline in recreational fishery caused by decline in fish populations due to heavy-metal contamination from industry
Triggers and thresholds	Sudden changes or surprises in system behaviour or system structure	Collapse of a fish stock when persistent pressures from harvesting and environmental stress result in a sudden change in population structure

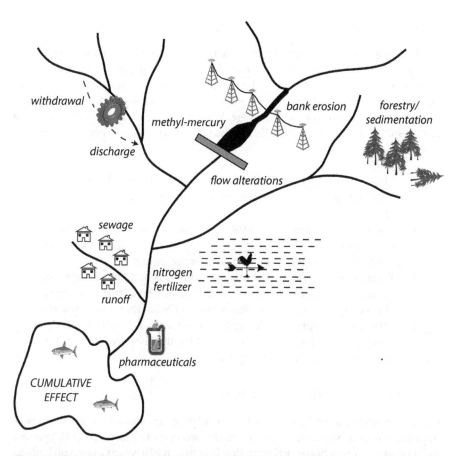

Figure 11.1 ⊚ Sources of Cumulative Environmental Effects in a Watershed

- increased sedimentation due to forestry activity;
- alterations in flow at a hydroelectric facility;
- increased methyl-mercury concentrations caused by reservoir flooding;
- bank erosion at a transmission line crossing;
- water withdrawal and discharge from heavy industry;
- septic leakage from residential areas;
- urban storm water runoff as a result of surface imperviousness;
- nutrient loadings from agricultural runoff;
- pharmaceuticals and other chemicals from industry and manufacturing.

The total environmental effect of all of these activities, combined with larger-scale stress caused by climate change and transboundary effects acting on a single VEC, such as fish or water quality, is a cumulative environmental effect. The problem is that not all of these point and non-point sources would be subject to EIA, and certainly few assessments ever would consider non-point sources of stress or capture the point sources from headwater to mouth.

Types of Cumulative Effects

While multiple types of activities and impacts can lead to cumulative environmental change, it is often useful to characterize the different types of cumulative effects. Such characterizations can help in the communication of cumulative effects issues and in identifying cumulative effects management measures—such as setting limits or maximum levels of allowable change. Based on Peterson et al. (1987), Sonntag et al. (1987), and Hegmann et al. (1999), four broad types of cumulative effects can be identified:

1. **Linear additive effects**. Incremental additions to, or deletions from, a fixed storage where each increment or deletion has the same individual effect.
2. **Amplifying effects**. Incremental additions to, or deletions from, an apparently limitless storage or resource base where each increment or deletion has a larger effect than the one preceding.
3. **Discontinuous effects**. Incremental additions that have no apparent effect until a certain threshold is reached, at which time components change rapidly with very different types of behaviour and responses.
4. **Structural surprises**. Changes that occur as a result of multiple developments or activities in a defined region. They are often the least understood and most difficult to assess.

Pathways of Cumulative Effects

Cumulative environmental effects result from different combinations of actions or pathways that consist of both additive and interactive processes. Peterson et al. (1987) present a classification of functional pathways that lead to cumulative environmental effects (Figure 11.2); each pathway is identified and differentiated according to the sources of change and type of impact accumulation. An example of pathways that lead to cumulative effects is illustrated by the Cold Lake oil sands project in Alberta (Box 11.1).

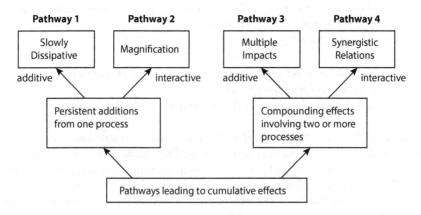

Figure 11.2 ⊚ Pathways Leading to Cumulative Effects

Box 11.1 Cold Lake Oil Sands Project Cumulative Effects Pathways

The Cold Lake oil sands project is a heavy oil facility in northern Alberta. Approximately 2500 wells are currently operating in the region. The Cold Lake facility is at present the second largest producer of oil in Canada. In 2003, the Cold Lake operations accounted for 10 per cent of Canada's crude oil production. Oil deposits are located in sand deposits approximately 400 metres below the surface and are extracted by a steam recovery process that injects high-pressure steam into the reservoir to separate the sand and oil. Wells are drilled and steam injected via clusters of vertical and directional-drilled wells, organized onto large surface pads. In 1997, the proponent, Imperial Oil Resources Limited, proposed to expand its operation in the Cold Lake area with the development of a central plant and additional production wells. A total of 35 impact models were contained in the EIA to assess the cumulative effects of the project on surface water quality, including the additive effects of roads and facilities (well pads) on sediment and contaminant levels in nearby water bodies.

Cumulative impact statement:

- Operation and maintenance of roads and facilities will result in the generation of sediment and transport of contaminants to receiving waters.

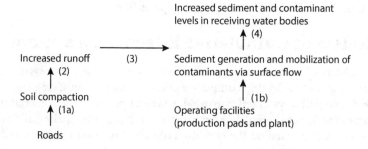

Pathways:

1a. The operation and maintenance of roads will lead to compaction of the roadbed.

1b. Operation and maintenance of pads and plant facilities will result in the generation of sediment and mobilization of contaminants via overland flow from these facilities.

2. Compaction will cause an increase in surface runoff from the road.

3. Increased runoff from roads will result in erosion of exposed soils, resulting in an increase in sediment generation and transport. Soluble contaminants from the road and the roadbed will be transported along with the sediment.

4. Increased sediment and contaminant transport will result in higher levels of these parameters in receiving waters, which will result in a decline in surface water quality.

Sources: Based on Hegmann et al. 1999; Imperial Oil Resources Ltd. 1997.

Single-Source Perturbations
Pathway one results from the persistent effects of a single project on a particular environmental component, such as repeated changes in water temperature resulting from a reservoir development. When any single activity has multiple effects, potential interactions between them may create cumulative effects. Pathway two is characterized by a single activity, but the effects accumulate synergistically. For example, the creation of a reservoir can change water temperature, lower dissolved oxygen content, and lead to heavy-metal contamination. While each of these effects can individually affect aquatic life, they can also accumulate in such a way that the toxicity of certain contaminants is multiplied because of high water temperatures and low dissolved oxygen content (Bonnell 1997).

Accumulation of Effects from Two or More Projects
Pathway three occurs when the environmental effects of multiple actions accumulate in an additive manner, as would be the case with the development of multiple reservoirs in a river basin. Although no interaction occurs between the effects of individual projects, they collectively result in significant impacts on aquatic resources. Pathway four occurs when these multiple effects do interact in a synergistic manner. For example, each project may alter water temperature, change dissolved oxygen content, and introduce heavy metals, thereby contaminating aquatic life, but the impacts from the interaction of these effects across all projects would be greater than the sum of the individual project impacts (Bonnell 1997).

Models of Cumulative Effects Assessment

Cumulative effects assessment (CEA) refers to the systematic process of identifying, analyzing, and evaluating cumulative effects—that is, identifying environmental effects and pathways in order to avoid, wherever possible, the potential triggers or sources that lead to cumulative environmental change (see Spaling and Smit 1994). Good CEA is focused on the condition of environmental receptors and whether the total effects via all stressors in a project's regional environment are acceptable, including the potential additional stress caused by the proposed project. There are, however, two broad and often competing models of how this should be achieved: effects-based and stressor-based models.

Effects-Based CEA

Effects-based CEA is focused on assessing existing environmental conditions relative to a reference condition and is typically retrospective in design—*what has happened*. Examples include environmental effects monitoring programs (see Environment Canada 2010; 2011) and ecological modelling and baseline studies (see Culp, Cash, and Wrona 2000; Munkittrick et al. 2000; Dubé et al. 2006; RAMP 2010). The strength of effects-based approaches is in measuring the accumulated environmental state of a system and identifying whether performance indicators are at or below an acceptable level (Dubé and Munkittrick 2001). Doing so can inform the identification of thresholds and help to inform risk assessment processes that, in principle, support decision-making about the impacts of development. Emphasis is on understanding the total effects on a particular VEC from all sources of stress (point, non-point, direct, indirect) and comparing these effects to some reference condition in order to determine an actual measure of cumulative change, irrespective of the number and nature of the impacts causing that change. Under this model, the focus of cumulative effects shifts away from the individual project and its localized stressors to allow for questions of a broader nature related to ecological thresholds and synergistic effects. The underlying premise of this approach is that cause–effect relationships can be established through long-term monitoring, which can then be used to predict cumulative impacts. According to Spaling et al. (2000), however, rarely under this sort of framework is there authority to implement recommendations or to carry forward CEA findings to specific project-based assessments. Many effects-based CEA studies are "one-offs," disconnected from regulatory-based development decision-making (see Sheelanere, Noble, and Patrick 2013).

Stressor-Based CEA

Stressor-based CEA is prospective in design—*what might or could happen*. The focus is typically on quantifying current (and, in some cases, past) levels, types, and distributions of human disturbance in the project's environment (e.g., industrial footprint, road densities, habitat fragmentation) and then projecting disturbances, caused by the project and other sources of human actions, into the future under different scenarios of resource use or development. Attention is placed on predicting the cumulative stress associated with particular agents of change—such as different projects or types of disturbances. This involves an analysis of the distribution and rates of change in disturbance in the baseline and predictive modelling of future disturbance patterns. The assumption is that stressors and VEC response can be correlated or that stressors are a good proxy for threats to VEC sustainability.

Good CEA

Both the effects-based and the stressor-based approaches are useful, but each offers a different type of understanding of cumulative effects—the first from the perspective of change in the receiving environment, the second from the perspective of change in human disturbance or stress to the environment. If the role of CEA is solely to understand the accumulated state and set thresholds through monitoring, then further development of effects-based models is required. If the role of CEA is to guide decisions about the potential implications of proposed land and resource use, then further development of stressor-based models is required. Arguably, good CEA requires both effects- and stressor-based approaches. As Duinker and Greig (2006) report, dwelling on the past is useful but only in the sense of possible learning about interactions, knowledge that can be used to sharpen predictive analysis for the future. At the same time, focusing solely on the future is useful only if we are able to understand the implications of future environmental change, which is often based on learning from the past and understanding thresholds or limits of change.

Good CEA must focus on understanding the accumulated environmental state and human stressors—past, present, and future.

Framework for the Application of CEA

The notion of cumulative environmental effects is not new to EIA, and the terms "cumulative impacts" and "cumulative effects" actually appeared in many national EIA guidelines and laws during the early 1970s. The US Council on Environmental Quality (1978), for example, suggested that project impacts on the environment could interact with other past, present, and reasonably foreseeable actions to generate collectively significant environmental change. It was not until the late 1980s, however, that cumulative effects started to receive any real attention in EIA. In Canada, CEA emerged on the scene in the early to mid-1980s as a priority of the Canadian Environmental Assessment Research Council (CEARC). Federally and provincially, CEA is now an accepted part of most assessment systems and is mandatory at the federal level for all EIAs conducted under the Canadian Environmental Assessment Act, 2012.

Section 19(1)(a) of the Canadian Environmental Assessment Act, 2012 requires that an EIA under the act take into account the environmental effects, including cumulative environmental effects, that are likely to result from the project in combination with other physical activities that have been or will be carried out. Cumulative environmental effects must be considered when determining the significance of a project's impacts and in the design of mitigation and follow-up programs. The Canadian Environmental Assessment Agency's (2013) *Operational Policy Statement* outlines the general requirements and approach to CEA under the Canadian Environmental Assessment Act, 2012 and suggests that CEA include an initial scoping, analysis, identification of mitigation measures, determination of significance, and follow-up.

There are a variety of frameworks that present steps or phases for CEA (e.g. Ross 1998; Hegmann et al. 1999; European Commission 1999; Canter and Ross 2010; Gunn and Noble 2012), including the Canadian Environmental Assessment Agency's (2013) *Operational Policy Statement*. Regardless of the number of steps identified or their labels, a review of standards and practices for CEA by Gunn and Noble (2012) suggests that good CEA can be distilled to four necessary components. These four components are scoping, retrospective analysis of cumulative effects, prospective or futures analysis of cumulative effects, and the management of cumulative effects (Figure 11.3). In the absence of any one of these components, CEA is incomplete.

Scoping

Scoping, or context setting, establishes all that will be included and all that will be excluded when evaluating cumulative effects and subsequent impacts to VECs. When conducting CEA as part of the regulatory EIA process, the Canadian Environmental Assessment Agency's (2013) *Operational Policy Statement* recommends that the CEA consider those VECs for which residual environmental effects are predicted after consideration of project mitigation measures, *regardless of whether those residual environmental effects are predicted to be significant*. Good CEA thus adopts ecosystem

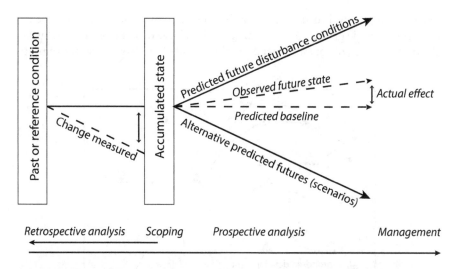

Figure 11.3 ⊛ Conceptualization of Cumulative Effects Assessment

Source: Adapted from Gunn and Noble 2012. Developed originally based on Dubé et al., "Developing cumulative impacts assessment and management strategies" project, funded by the Canada Water Network and depicted in the Lower Athabasca Water Quality Monitoring Program Phase I (Environment Canada, Catalogue no. En14-42/2011E-PDF.)

health and functioning as a core determinant of VEC selection; thus, effective CEA must be spatially and temporally bound based on the distribution of the VECs affected by both the project(s) in question and the effects of other projects and disturbances—past, present, and future.

Spatial Boundaries

The spatial boundaries for CEA vary considerably and are defined by a combination of factors, including: (i) the specific land uses or industrial activity of interest; (ii) planning or management jurisdictions; and (iii) the characteristics or distribution of the VECs or indicators of concern (Noble 2013). Cumulative effects occur over a large spatial scale and over a long time period. Determining the spatial boundaries for a CEA is thus critical to its success in effectively managing the cumulative impacts associated with development. Boundaries in CEA delimit the spatial extent of the assessment and thus the environments and VECs that are considered.

It is generally acknowledged that in order to assess cumulative effects effectively, the spatial boundaries of assessment must be extended well beyond the project site. The Canadian Environmental Assessment Agency's (2013) *Operational Policy Statement*, for example, indicates that the spatial boundaries for CEA need to encompass the potential environmental effects on VECs of the project being assessed in combination with other physical activities that have been or will be carried out. However, if the boundaries identified are too large, only a superficial assessment may be possible, and uncertainty will increase. Moreover, the incremental addition of a single project may seem less and less significant—only a small drop in a large bucket. If the boundaries are too small, a more detailed examination may be feasible, but an understanding of the broad context may be sacrificed. In addition, the incremental impacts of a single project may be exaggerated—a large drop in just a small bucket (Figure 11.3). This, of course, depends on the nature of the project and the impacts.

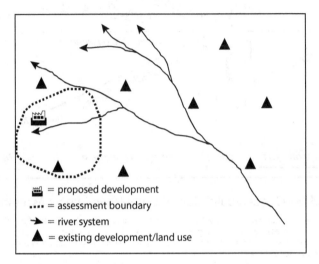

Restrictive bounding may result in the project's additive effects on water quality being perceived as quite *significant* when considered together with the surrounding development and land-use activities.

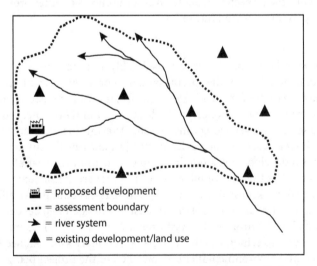

Ambitious bounding may result in the project's additive effects on water quality being perceived as quite *insignificant* when considered in light of the total effects of all development and land-use activities on the watershed.

Figure 11.4 ◉ Restrictive and Ambitious Spatial Bounding

For example, by selecting a relatively small spatial boundary, the impacts of emissions from a proposed smelting operation might seem quite insignificant, especially if emission stacks are relatively high. The choice of spatial boundaries for CEA, then, can be to the proponent's advantage or disadvantage.

Establishing the appropriate boundaries for CEA requires consideration of three types of scale.

Spatial scale refers to the actual geographic extent of the assessment and is typically based on one or both of natural boundaries, such as watersheds, or administrative boundaries, such as townships or landownership. This is usually the most common interpretation of "scale" in CEA; however, it is certainly not the most functional with regard to the actual analysis of cumulative effects.

Analysis scale is used to examine VECs and impacts across space and is represented by such ideas as data resolution, detail, and granularity. In the Cold Lake oil sands project discussed in Box 11.1, for example, the geographic boundaries for wildlife and vegetation were restricted to local township areas, based on the availability of historical and current information on vegetation composition and wildlife habitat as well as on the extent of available aerial photo coverage. Spatial boundaries were determined on the basis of data availability and desired analysis scale.

Phenomenon scale is perhaps the most important type of scale in CEA, since it refers to the spatial units within which various processes operate or function. Thus, in any single assessment, different spatial boundaries may be appropriate for different cumulative effects and for different VECs. The boundaries selected for cumulative effects on air quality might be quite different from those chosen for cumulative effects on soil quality or sedentary versus migratory wildlife.

While there is no best method for determining the spatial boundaries for any particular CEA, the CEA literature offers a number of guiding principles to assist the practitioner:

- *Adequate scope.* Boundaries must be large enough to include relationships between the proposed project, other existing projects, and the VECs. This means crossing jurisdictional boundaries if necessary to account for interconnections across systems.
- *Natural boundaries.* Natural boundaries such as watersheds, airsheds, or ecosystems are perhaps the best reflection of the natural components of a system and should be respected.
- *VEC differentiation.* Different VECs and VEC processes operate at different spatial scales, and boundaries must therefore reflect spatial variations in the VECs considered.
- *Maximum zones of detectable influence.* Impacts related to project activities typically decrease with increasing distances; thus, boundaries should be established where impacts are no longer detectable.
- *Multi-scaled approach.* Multiple spatial scales, such as local and regional boundaries, should be assessed to allow for a more in-depth understanding of the scales at which VEC processes and impacts operate.
- *Flexibility.* CEA boundaries must be flexible enough to accommodate changing natural and human-induced environmental conditions.

In principle, CEA should focus on ecological units, such as watersheds or eco-regions (Seitz, Westbrook, and Noble 2011). In practice, administrative boundaries play an important role in the success of CEA programs. Ambitious ecological boundaries often need to be tempered by institutional arrangements and the administrative authority to implement CEA, including mitigation and monitoring programs. Squires and Dubé (2012) suggest that the spatial scale of CEA be determined by the spatial scale of the processes (i.e., industry, land uses) that most affect or control the resources of concern in a region.

Temporal Boundaries

The Canadian Environmental Assessment Agency's (2013) *Operational Policy Statement* indicates that the spatial boundaries for CEA should take into account future activities and the degree to which the environmental effects of these activities will overlap those predicted from the proposed project. However, good-practice temporal bounding for CEA also requires asking "how far into the past" should cumulative environmental change resulting from other actions and activities be considered in the assessment. The extent of temporal boundaries depends on the amount of information desired, the amount of information available, and what the assessment is trying to accomplish. Examining past conditions may be as simple as examining land-use maps, and in certain cases it may be feasible to incorporate 50 years of historical data if deemed necessary.

The Canadian Environmental Assessment Agency's *Cumulative Effects Assessment Practitioners' Guide* (Hegmann et al. 1999), developed under the former act, outlines several options for establishing how far into the past a CEA should extend. The first two options have limited historical perspective and are based on the temporal characteristics of the proposed project itself:

- temporal bounds established only on the basis of existing environmental conditions; or
- when impacts associated with the proposed action first occurred.

Other options are based on more historical perspectives of land use and conditions of environmental change and include:

- the time when a certain land-use designation was made (for example, the establishment of a park or the lease of land for development);
- the time when effects similar to those of concern first occurred; or
- a time in the past representative of desired environmental conditions or pre-disturbance conditions, especially if the assessment includes determining to what degree later actions have affected the environment.

Identifying which potential future actions and activities to include in CEA can be much more uncertain. The Canadian Environmental Assessment Agency's (2013) *Operational Policy Statement* recommends that CEA for future actions include those that are certain and reasonably foreseeable. Actions that are certain are those that will proceed or for which there is a high probability that they will proceed (e.g., the proponent has received the necessary authorizations or is in the process of obtaining those authorizations). Actions that are reasonably foreseeable are those that are expected to proceed (e.g., the proponent has publicly disclosed its intention to seek the necessary EIA or other authorizations to proceed). Arguably, good CEA is not limited to certain and reasonably foreseeable actions but also gives some consideration to hypothetical actions—those for which there is considerable uncertainty as to whether they will proceed but that are of potential concern for cumulative environmental effects should they proceed. These may include actions or activities discussed only on a conceptual basis or speculated to proceed, based on current information. **Scenario analysis** is a common tool used to identify such hypothetical actions.

These actions lie on a continuum from most likely to least likely to occur. For each assessment, the practitioner or the regulatory agency will have to decide how far into the future the assessment should reach. Often, a major criterion is whether the future action or actions are likely to affect the same VECs as the proposal under consideration. While practical, this criterion may detract from these projects, creating "nibbling" effects that, while they may not directly affect the same VECs, contribute to overall decline in environmental quality.

Retrospective Analysis

Regardless of the temporal boundary selected, it is important to consider the significance of past changes to the VECs of concern—and not treat past changes in, or effects to, VEC conditions simply as the "new normal." The latter approach, whereby the magnitude of the cumulative effects of past projects is discounted and treated as part of the current baseline condition, misses important opportunities for impact management—particularly for those VECs that might be nearing, or already beyond, a critical sustainability threshold (see Chapter 5, Box 5.4).

The concept of retrospective analysis, as part of baseline assessments, was introduced in Chapter 5. Retrospective analysis involves assessing past VEC conditions and analyzing trends and changes in conditions over time and against thresholds. Good CEA requires an understanding of how VEC conditions have changed over time and whether that change is significant in terms of the sustainability of the VEC. An attempt should be made to identify relationships between indicators of change in VEC conditions (e.g., caribou population, water quality indices) and measures of human or natural disturbance so as to determine trends and associations that can be used to predict and monitor VEC conditions or responses to future cumulative change (Gunn and Noble 2012).

Examples of disturbance measures of interest in cumulative effects analysis may include the density of linear features per unit area on the landscape (e.g., road or trail density—km / km^2), percentage disturbed landscape (e.g., cleared area), edge density or perimeter area ratio, the rate of land conversion (e.g., rate and area of change from forested to non-forested), the number or density of river crossings (e.g., number of crossings per river kilometre in a river reach), the density of impervious or hard surfaces in a watershed (e.g., road surfaces and parking lots have been linked to contaminant transfer and measurable responses in water quality [see Brydon et al. 2009]), and broader natural processes of change such as flood or fire frequency (Gunn and Noble 2012). In many cases, cause-and-effect relationships between disturbances and VEC responses may not be known, but correlations or qualitative associations can be relied upon. The objective is to identify measures of the drivers of change in the region, characterize VEC or indicator responses over space and time, and iden-tify—when and where appropriate—thresholds, management targets, or maximum allowable limits of change.

Prospective Analysis

Using knowledge gained and models developed from the scoping phase and retro-spective analysis, prospective analysis is about predicting and evaluating how VECs or their indicators (e.g., caribou population, water quality index) might respond to additional stress in the future—stress caused by the project and by other projects and actions in the regional environment (e.g., fragmentation, river crossings). The focus of analysis is on the VEC conditions and understanding potential VEC response to cumulative disturbance. Gunn and Noble (2012) explain that prospective analysis for CEA might involve "summing up" individual effects such that the total effects on VECs are evaluated and summarized into trend information, focusing on regional environmental issues and whether they will grow worse or better, and assessing the effects on VECs of broad regional change agents such as "surface disturbance" that are, by definition, cumulative and provide a measure of ecosystem health.

Predicting such future conditions is often uncertain, and data are often incom-plete. Greig and Duinker (2007) suggest the use of scenario analysis, particularly for large projects, to address the range of possible future VEC conditions under different development/disturbance regimes. Other methods that support CEA, many of which are discussed in Chapter 3, include landscape metrics, correlation and statistical modelling, and more complex simulation tools such as ALCES and MARXAN. The Canadian Environmental Assessment Agency's (2013) *Operational Policy Statement* also recommends that scientific data can often be supplemented with knowledge from other areas with comparable conditions, from community knowledge, and from Aboriginal traditional knowledge.

Management

The best way to manage cumulative effects is to avoid them. However, this not always feasible—most projects are proposed in areas that have already been subject to some level of human disturbance. The management phase of CEA involves the identifica-tion of potentially significant cumulative effects, identifying impact management measures, and developing follow-up and monitoring programs for cumulative effects. Of particular importance is determining whether the incremental or cumulative effects caused by the project under consideration are significant. This requires an evaluation of the total effects on each VEC of concern, including the effects of the project plus the effects caused by other sources. Gunn and Noble (2012) suggest that important to this determination is assessing how much more change in VEC conditions is acceptable. This requires some assessment against thresholds, management targets, or maximum allowable limits of change identified during the retrospective or scoping stages of the assessment or set out in regulations or broader environmental policy objectives. Viable management measures are then proposed, considering the range of possible future outcomes or VEC conditions. The objective is to minimize, if not eliminate, the cumulative contribution of the project to an adverse effect on VEC conditions. In those cases where a VEC is already unhealthy or unsustainable, or nearing such levels, the only acceptable management action may be rectification or restoration of VEC conditions—i.e., no additional cumulative effect caused by the project is acceptable.

State of CEA in Canada

Notwithstanding the recognized need for CEA and the implications of not doing it, there are constant and consistent messages that CEA is either not being done or not being done well when it is done. In a 1998 report of the Auditor General of Canada, the auditor noted that on a sample of 159 environmental assessments conducted by federal authorities, excluding Parks Canada, only 48 indicated that cumulative effects had been considered. In Canada's western prairie watersheds, Schindler and Donahue (2006) suggest an impending water crisis, arguing that policy decision-makers and planners have seldom, if ever, considered the cumulative effects of climate warming, drought, and human activity. Rather, the focus of attention has been on project-by-project decision-making, while cumulative environmental change and broader regional and non-point sources of stress have been ignored. In a more recent panel review process for the Manitoba Hydro's Bipole III transmission line project, witnesses for the Consumers' Association of Canada (Manitoba) reported that the project's CEA largely ignored cumulative effects on VECs of other, past projects, including the proponent's own past projects. For most VECs, impacts were measured against, rather than in addition to, the effects of other future disturbances (see Environmental Assessment in Action: Review of the Bipole III Transmission Line Project Cumulative Effects Assessment).

Environmental Assessment in Action

Review of the Bipole III Transmission Line Project Cumulative Effects Assessment

The Bipole III Transmission Line project was introduced in Chapter 5, Box 5.4. The project, proposed by Manitoba Hydro, involves the construction of an approximately 1400-kilometre transmission line from northern Manitoba, near Gillam, south to Winnipeg. The transmission line will traverse boreal forest and

continued

caribou habitat in the north and agricultural land in the south, including several river and stream crossings along the route. The project is to help improve the reliability of electricity supply to Manitoba and reduce the risk of supply interruptions due to ice storms, fires, and other events. Currently, more than 70 per cent of the province's electricity is transmitted via a single corridor on the Bipole I and II transmission lines. Construction is planned for 2013, with a project operation date set for 2017.

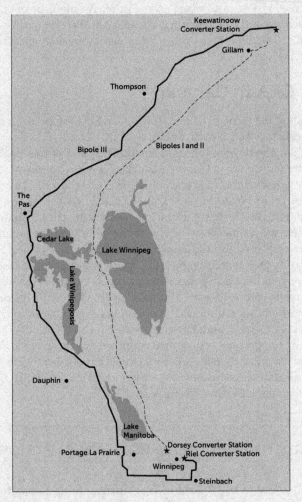

The project was subject to EIA under the Environment Act of Manitoba. As part of its EIA, the proponent submitted a CEA of its project. In section 9.1 of the EIS, the proponent noted that its CEA was conducted based on guidance from the project's scoping document, the Canadian Environmental Assessment Act, and review of other guidance documents for cumulative effects assessment. In 2005, the minister of Manitoba Conservation and Water Stewardship requested that the Manitoba Clean Environment Commission (CEC) conduct a public hearing into the proposal. The CEC is an arms-length provincial agency mandated to provide advice to the minister and ensure public participation in environmental matters. The CEC held its public hearings between October and November 2012 and in March 2013. Several interveners participated in the hearing process, including the Consumers' Association of Canada (Manitoba), which focused on, among other issues, the nature and quality of the proponent's CEA.

An analysis of the quality of the proponent's CEA was undertaken by Gunn and Noble (2012) on behalf of the Consumers' Association of Canada. Gunn and Noble applied several criteria to guide their analysis of the CEA, as follows:

A. Scoping practices:

i.   Is the CEA methodology distinct from the project impact assessment?
ii.   Does the CEA consider all types of activities and stresses that may interact with the project's effects?
iii. Does the CEA adopt "ambitious," ecologically based scoping?
iv.   Is an explicit rationale for VEC selection documented?
v.   Do the spatial boundaries reflect the natural distribution patterns (present and historical) of VECs selected for the cumulative effects assessment?
vi.   Does the CEA adopt "pre-disturbance" conditions as the historic temporal limit and capture other certain and reasonably foreseeable future projects and activities?

B. Retrospective analysis:

i.   Does the baseline analysis delineate past and present cumulative effects in the study area?
ii.   Does the baseline analysis establish trends in VEC conditions and known or suspected relationships between changes in VEC conditions and the primary drivers of change?
iii. Are thresholds specified against which cumulative change and the significance of effects can be assessed?

C. Prospective analysis:

i.   Is the time scale of cumulative effects predictions/analysis sufficient to capture the scope of impacts associated with the project's life cycle?
ii.   Is there sufficient analysis/evidence to support conclusions about potential cumulative effects?
iii. Are the tools and techniques used capable of capturing the complexities of cumulative effects pathways and the uncertainties of future developments?
iv.   Are trends and linkages established between VEC conditions and disturbances in the baseline analysis used to inform predictions about cumulative impacts?
v.   Is the analysis centred on the total effects on VECs in the project's regional environment?

continued

D. Cumulative effects management:

i.   Is the significance of the project's cumulative effects measured against a past reference condition and not simply the current, cumulative, or disturbed condition?

ii. Is the significance of cumulative effects adequately described and justified and based on VEC sustainability, defined by a desired or healthy condition or threshold as opposed to the magnitude of the individual project stress on that VEC?

iii. Are the incremental impacts of the proposed initiative "traded off" against the significance of all other disturbances of activities in the region (i.e., minimized or masked)?

iv. Are mitigation measures identified that help offset significant cumulative environmental effects, and if so, is consideration is given to multi-stakeholder collaboration to develop joint management measures?

v.   Is adaptive management identified for significant cumulative effects contingent upon future and uncertain developments and impact interactions?

Gunn and Noble concluded that the Bipole III CEA fell short of good practice and significantly short of the standard identified in the EIS scoping document, which commits to a cumulative effects assessment based on best and current practices. Several significant deficiencies were identified in the report, including:

- The baseline against which cumulative effects are assessed largely ignored the cumulative effects on VECs of past actions and changing VEC conditions over time.
- There was a lack of supporting analysis of cumulative effects to support many of the conclusions.
- The baseline was descriptive; few trends or condition changes were identified and analyzed, and thus there was little means of predicting or modelling cumulative effects into the future.
- The temporal scope of analysis was insufficient and inconsistent with the lifetime of the project. For example, the CEA adopted only a five-year horizon for what is one of the VECs of most concern, caribou.
- The majority of VEC conditions were not examined within the context of regional ecological health but rather from the perspective of absorbing the project's stress.
- Much of the effects analysis was restricted to the transmission line right-of-way, ignoring the effects of the Bipole I and II projects.
- The CEA often assessed the magnitude of the project's impacts against or "compared to" the effects of other actions, versus "in addition to" past changes in VEC conditions and "in addition to" the effects of other current and future actions. As a result, the total or cumulative effects were rarely addressed or analyzed.

Among the recommendations of their report was that the project not proceed until the government of Manitoba undertakes a regional and strategic environmental assessment of the cumulative effects of current and future land uses, particularly in the northern portion of the Bipole III study area.

The Manitoba Clean Environment Commission issued its final panel report on the public hearings in July 2013. In its report, the CEC noted that "The cumulative effects analysis should be the most important section of an environmental assessment report" (p. 11), but the CEC also indicated that "The Commission has a long history of being less than satisfied with the nature of cumulative effects assessments conducted by proponents in Manitoba" (p. 11). The CEC reported that it was "simply inconceivable—given the 50-plus-year history of Manitoba Hydro development in northern Manitoba and given that at least 35 Manitoba Hydro projects have been constructed in the north in that time—that there are few, if any, cumulative effects identified in this EIS" (p. 112). The CEC recommended, as a non-licensing requirement, that Manitoba Hydro implement a CEA approach that would go beyond the minimal standard and would be more in line with current "best practices" (p. 129). The panel also recommended that Manitoba Hydro, in cooperation with the Manitoba government, conduct a regional CEA for all Manitoba Hydro projects and associated infrastructure in the Nelson River sub-watershed and that this be undertaken prior to the licensing of any additional projects in the region after the Bipole III project. The CEC concluded, however, that it was prepared to concede that the proponent had met the minimum standards and that the project be approved.

The Bipole III CEA was a significant step forward in publicly recognizing the limits of CEA as currently practised and established the need for a better standard. It was also a significant step forward in recognizing that effective CEA requires a regional approach and that a regional CEA be undertaken in northern Manitoba prior to approving further development proposals. The Bipole III CEA was a also a step backward in advancing the practice of CEA, sending a message via project approval that the current practice of CEA is still "good enough" to secure EIA approval.

Enduring Challenges and Concerns

Parkins (2011) suggests that "thinking cumulatively and regionally does not emerge naturally from a project-based perspective." In a review of the state of cumulative effects assessment in Canada, Duinker and Greig (2006) conclude that "continuing the kinds and qualities of CEA currently undertaken may be doing more harm than good." Baxter, Ross, and Spaling (2001), Duinker and Greig (2006), Canter and Ross (2010), Noble (2010), and Seitz, Westbrook, and Noble (2011) point to several enduring challenges and concerns with the current practice of CEA in Canada. Some of the main ones are synthesized below.

The first problem concerns the context of CEA as currently required in Canada—situated within project-based EIA. As explained at the outset of this chapter, cumulative environmental effects concern the total effects of human activities on a VEC. Project EIA, in contrast, is concerned about project-induced stress and making sure that the impacts of a project are acceptably small rather than understanding the total effects of all stressors, project and non-project, on any single VEC. Parkins (2011) reports that cumulative effects simply as additive impacts from multiple projects on indicators such as water use or pollution does not facilitate broader discussions about regional limits to development and change and the ways in which specific projects and impacts are aligned or misaligned with regional development goals and objectives.

Second, and closely related, is that project EIA is concerned primarily with minimizing project stress to a level of acceptability. The objective of proponents is to ensure that their project meets regulatory and public approval—this usually means minimizing any efforts regarding CEA and paying little attention to understanding VEC quality and longer-term sustainability. The result is often findings of "non-significance" when, in reality, the project is contributing to incremental, if not synergistic, cumulative environmental change (Box 11.2)

A third concern relates to thresholds. Understanding the cumulative effects of human activity on a VEC, and the implications of such effects, requires some understanding of thresholds and carrying capacities. The challenge, however, is that thresholds are not easily determined, particularly within the spatial and temporal confines of a project assessment. There is often reluctance to set thresholds or to limit development when our understanding of natural variability and adaptability within the system is poor. However, in general, for any assessment it is useful to have a management target or benchmark against which to assess condition change (either

---

Box 11.2 Individually Insignificant Actions

In southwest Saskatchewan, a 1940 km² ecologically rich land base, consisting of active sand dunes, rare and endangered species, and plants of Aboriginal cultural importance is subject to the pressures of approximately 1500 natural gas wells, cattle grazing, and more than 3000 kilometres of access roads and trails. The landscape is significantly fragmented, and biodiversity, in a once native grassland ecosystem, is at risk. Cattle grazing and roads and trails in the region have not been subject to EIA. Of the 1500 wells in the area, only five proposals were subject to assessment—none of which was deemed to have significant environmental effects (GSH SAC 2007). Nasen, Noble, and Johnstone (2011), however, found that the ecological footprint of petroleum and natural gas wells in southwest Saskatchewan grasslands has an effect on soils and range health up to 25 metres from the well head—well beyond the physical footprint of the infrastructure and with a duration of at least 50 years (see Chapter 4, Environmental Assessment in Action).

The Athabasca River basin, Alberta, is exposed to a wide range of land-use activities, including agriculture, forestry, pulp and paper operations, and petroleum extraction. Roads, power lines, pipelines, and other disturbances have fragmented forests, and the amount of old-growth forest has been significantly reduced. Between 1966–76 and 1996–2006, the number of pulp mills discharging into the Athabasca basin increased from one to five; total farm area increased from 47.2 million acres to 52.1 million acres; the number of operating oil sands leases increased from 2 to 3360; water withdrawals increased from

effects-based change or stressor-based change); otherwise, it is difficult to determine when to take action and what action to take when undesirable change occurs. When thresholds are addressed, they are usually defined within the context of the project as opposed to the total effects on a VEC and, further, typically defined on the basis of public acceptability as opposed to ecological knowledge.

Fourth, CEA and management are ultimately about the future and demand, looking far enough into the future to capture the full array of human activities and natural changes that may affect the sustainability of VECs of concern. This is a highly uncertain environment and one that is about possible futures and outcomes—a view that stands in sharp contrast to the shorter-term perspective of project approval and predicting the "most likely," versus the most desirable, effects of development. Greater attention needs to be given to exploring alternative futures in CEA; this includes the consideration of hypothetical development scenarios.

Fifth, our assumptions about cumulative effects, as evidenced by practice, are not always consistent with the nature of how environmental systems function. Advancing CEA will require that we rethink our assumptions and, thus, our approach to CEA (Table 11.2). As Ross (1994, 6) points out, "the environmental effects of concern to thinking people are . . . not the effects of a particular project; they are the cumulative effects of everything." In particular, there is a need to think about limits of environmental systems in terms of the types, amounts, and rates of development that can be accommodated.

---

approximately 12 million m³/yr to 595 million m³/yr, of which more than 70 per cent can be attributed to oil sands operations. Between these two time periods, the cumulative annual flow in the Athabasca River decreased by more than 500 m³/s, and temperature increased by 1.4°C; conductivity, turbidity, and phosphorous levels also increased (Squires, Westbrook, and Dubé 2010). Many of these disturbances, such as urban growth and agricultural expansion, have not been subject to assessment. For others, the effects of each project have been deemed unlikely to cause significance adverse environmental effects.

Part of what leads to scenarios like these is that cumulative effects are often ignored or diminished in project assessment, sometimes deliberately and sometimes because the project in question is considered too small to warrant attention. Quite often, individual developments are evaluated independently of other activities and thus deemed "unlikely" to cause significant adverse environmental effects. In other cases, the magnitude of a project's impacts are sometimes erroneously "measured against" or "compared to" the effects of other projects, versus focusing on the overall effects on VEC conditions. When the significance of a project's effects, no matter how small the effect, is evaluated from the perspective of the additional stress placed on VECs that are already stressed by other sources, it is far more likely to be deemed unacceptable, particularly in regions of concentrated development where environmental thresholds may already be exceeded (Gunn and Noble 2012).

Table 11.2 Characteristics of Status Quo CEA versus Requirements for Effective CEA

	Status quo CEA	Required CEA
Assumptions	abundance	limits
Receptors	single media	environmental systems
Spatial context	project	multiple scales
Temporal context	present	past, present, future
Scope	regulated activities	all disturbances
Assessment	stressors *or* effects	stressors *and* effects
Futures	predicted impacts	possible outcomes
Management	mitigation	avoidance
Monitoring	regulatory compliance	thresholds and capacity
Responsibility	individual proponents	multi-stakeholder
Performance	increased efficiency	increased efficacy

A final concern relates to governance, specifically roles and responsibilities for carrying out CEA and ensuring its influence when CEA is conducted outside the scope of the regulatory EIA process—such as regional CEAs or cumulative effects studies. There is a requirement under the Canadian Environmental Assessment Act, 2012, section 19(1), that the EIA of a designated project take into account the results of any relevant regional study conducted by a committee established under the act. However, regional CEAs and cumulative effects studies have had a tradition of being short-term bursts of activity with no long-term support (Kristensen, Noble, and Patrick 2013; Parkins 2011). Notwithstanding considerable advances in the science to support CEA beyond the regulatory and spatial constraints of project EIA, there has been limited attention to the institutional arrangements necessary for implementing and sustaining it. In an analysis of CEA practices in western Canadian watersheds, Sheelanere, Noble, and Patrick (2013) and Kristensen, Noble, and Patrick (2013) argue that among the requirements for implementing, sustaining, and ensuring influential CEA beyond the project scale are:

- an agency with the authority and mandate for CEA, including the means to direct monitoring programs and influence decisions about land use and project development in the region;
- clearly defined stakeholder roles and responsibilities for undertaking the CEA, implementing the results, and monitoring and following-up for continual learning and improvement;
- sharing of monitoring data, both spatial and aspatial and in common data formats, among all stakeholders;
- a means of implementing CEA initiatives, enforcing monitoring programs and compliance, and ensuring influence over development decisions taken at the individual project level;

- sufficient financial and human resources to implement and sustain, over the long term, CEA programs and requirements (e.g., monitoring programs, landscape modelling, reporting, communication and data management, and co-ordination).

The above challenges are not to say that project-based CEAs are not useful; rather, something more is needed to address and manage cumulative environmental change in an effective manner (Cooper 2003; Creasey 2002). CEA should go beyond the evaluation of site-specific direct and indirect project impacts to address broader regional environmental impacts and concerns. Cocklin, Parker, and Hay (1992) identify three main objectives in advancing CEA beyond EIA:

- to develop a broader understanding of the current state of the environment vis-à-vis cumulative change processes;
- to identify, insofar as possible, the extent to which cumulative effects in the past have conditioned the existing environment;
- to consider priorities for future environmental management with respect to general policy objectives and with regard to potential development options.

The underlying notion is that cumulative environmental change is the product of multiple, interacting development actions and that the multiplicity of development decisions in a particular region, while often individually insignificant, cumulatively lead to significant environmental change. Some significant progress has been made in CEA; however, most of this progress has been outside the constraints of the regulatory EIA process.

Key Terms

amplifying effects
analysis scale
cumulative effect
cumulative effects assessment
cumulative effects management
discontinuous effects
effects-based CEA
linear additive effects

non-point-source stress
phenomenon scale
point-source stress
scenario analysis
spatial scale
stressor-based CEA
structural surprises

Review Questions and Exercises

1. Do provisions exist under your provincial, territorial, or state EIA system for cumulative effects assessment?
2. Using the example of multiple reservoir developments in a single watershed, sketch a diagram similar to Figure 11.2, and identify and classify the different

types of cumulative impacts that might result. State the impact "pathways" as illustrated by the example in Box 11.1.

3. What is the difference between effects-based and stressor-based approaches to cumulative effects assessment?

4. Using an example, explain how a proponent might use spatial bounding to its advantage. Given this, should the proponent be solely responsible for determining the spatial boundaries for cumulative effects assessment?

5. It has been said that cumulative effects assessment is simply EIA done right. Do you agree? Given the challenges to and the constraints of EIA in assessing and understanding cumulative effects, should CEA be part of EIA or a separate, independent process? Identify the benefits and limitations of a more integrated versus a more separated CEA process.

6. Cumulative effects often result from multiple and often unrelated project developments in a single region. Should regional cumulative effects assessment be the responsibility of the project proponent?

7. When a project creates environmental damage, it is often the responsibility of the proponent to rectify or compensate for such damage. Assume a region where there are multiple projects and activities, including oil and gas, forestry, highways, recreation, and hydroelectric developments. Individually, each project was approved for development based on the fact that it would not generate significant environmental effects. Cumulatively, however, all of these activities are contributing to overall environmental decline.

    a) Who should be responsible for managing overall cumulative environmental change resulting from the many, unrelated project developments and activities?

    b) How does one determine how much each development or activity is contributing to cumulative change?

    c) Given that each project is "individually insignificant" but that together they are cumulatively damaging, should an additional development be permitted in the region if it too is determined to be individually insignificant? What are the implications of such a decision with regard to equity versus environmental protection?

References

Baxter, W., W.A. Ross, and H. Spaling. 2001. "Improving the practice of cumulative effects assessment in Canada." *Impact Assessment and Project Appraisal* 19 (4): 253–62.

Bonnell, S. 1997. "The cumulative effects of proposed small-scale hydroelectric developments in Newfoundland, Canada." (Memorial University of Newfoundland, MA thesis).

Brydon, J., et al. 2009. "Evaluation of mitigation methods to manage contaminant transfer in urban watersheds." *Canadian Journal of Water Quality Research* 44 (1): 1–15.

Canadian Environmental Assessment Agency. 2013. *Operational Policy Statement. Assessing Cumulative Environmental Effects under the Canadian Environmental Assessment Act, 2012.* Ottawa: Canadian Environmental Assessment Agency.

Canadian Environmental Assessment Research Council. 1988. *The Assessment of Cumulative Effects: A Research Prospectus.* Ottawa: Supply and Services Canada.

Canter, L. 1999. "Cumulative effects assessment." In J. Petts, ed., *Handbook of Environmental Impact Assessment*, vol. 1, *Environmental Impact Assessment: Process, Methods and Potential.* London: Blackwell Science.

———, and B. Ross. 2010. "State of practice of cumulative effects assessment and management: The good, the bad and the ugly." *Impact Assessment and Project Appraisal* 28 (4): 261–8.

Cocklin, C., S. Parker, and J. Hay. 1992. "Notes on cumulative environmental change I: Concepts and issues." *Journal of Environmental Management* 35: 51–67.

Cooper, L. 2003. *Draft Guidance on Cumulative Effects Assessment of Plans.* EPMG Occasional Paper 03/LMC/CEA. London: Imperial College.

Creasey, R. 2002. "Moving from project-based cumulative effects assessment to regional environmental management." In A.J. Kennedy, ed., *Cumulative Environmental Effects Management: Tools and Approaches.* Calgary: Alberta Society of Professional Biologists.

Culp, J., K. Cash, and F. Wrona. 2000. "Cumulative effects assessment for the Northern River Basins Study." *Journal of Aquatic Ecosystem Stress and Recovery* 8: 87–94.

Dubé, M., et al. 2006. "Development of a new approach to cumulative effects assessment: A northern river ecosystem." *Environmental Monitoring and Assessment* 113: 87–115.

Dubé, M., and K. Munkittrick. 2001. "Integration of effects-based and stressor-based approaches into a holistic framework for cumulative effects assessment in aquatic ecosystems." *Human Ecology Risk Assessment* 7 (2): 247–58.

Duinker, P., and L. Greig. 2006. "The impotence of cumulative effects assessment in Canada: Ailments and ideas for redeployment." *Environmental Management* 37 (2): 153–61.

Environment Canada. 2010. "Integrated watershed management." http://www.ec.gc.ca/eau-water/default.asp?lang1/4en&n1/413D23813-1.

———. 2011. "National environmental effects monitoring office." http://www.ec.gc.ca/esee-eem/default.asp?lang1/4En&n1/4453D78FC-1.

European Commission. 1999. *Guidance for the Assessment of Indirect and Cumulative Impacts as well as Impact Interactions.* Luxembourg: Office for Official Publications of the European Communities.

Greig, L., and P. Duinker. 2007. *Scenarios of Future Developments in Cumulative Effects Assessment: Approaches for the Mackenzie Gas Project.* Report prepared by ESSA Technologies Ltd, Richmond Hill, ON, for the Joint Review Panel for the Mackenzie Gas Project.

GSH SAC (Great Sand Hills Scientific Advisory Committee). 2007. *Great Sand Hills Regional Environmental Study.* Regina: Canadian Plains Research Centre.

Gunn, J., and B.F. Noble. 2012. "Critical review of the cumulative effects assessment undertaken by Manitoba Hydro for the Bipole III project." Prepared for the Public Interest Law Centre, Winnipeg. www.cecmanitoba.ca.

———. 2014. "Definitions for cumulative effect, cumulative effects assessment, and cumulative effects management." Winnipeg: Canadian Council of Ministers of the Environment.

Hegmann, G., et al. 1999. *Cumulative Effects Assessment Practitioners' Guide.* Prepared by AXYS Environmental Consulting and CEA Working Group for the Canadian Environmental Assessment Agency, Hull, QC.

Imperial Oil Resources Limited. 1997. *Cold Lake Expansion Project Environmental Impact Statement.* Calgary: Imperial Oil Resources Limited.

Kristensen, S., B.F. Noble, and R. Patrick. 2013. "Capacity for watershed cumulative effects assessment and management: Lessons from the Lower Fraser Basin, Canada." *Environmental Management* doi 10.1007/s00267-013-0075-z.

Munkittrick, K., et al. 2000. *Development of Methods for Effects-Driven Cumulative Effects Assessment Using Fish Populations: Moose River Project*. Pensacola, FL: Society of Environmental Toxicology and Chemistry.

Nasen, L., B.F. Noble, and J. Johnstone. 2011. "Environmental effects assessment of oil and gas lease sites on a grassland ecosystem." *Journal of Environmental Management* 92: 195–204.

Noble, B.F. 2010. "Cumulative environmental effects and the tyranny of small decisions: Towards meaningful cumulative effects assessment and management." Natural Resources and Environmental Studies Institute Occasional Paper no. 8. Prince George: University of Northern British Columbia.

———. 2013. *Development of a Cumulative Effects Monitoring Framework: Review and Options Paper*. Report prepared for the Department of Indian Affairs and Northern Development. Saskatoon: University of Saskatchewan.

Odum, W. 1982. "Environmental degradation and the tyranny of small decisions." *BioScience* 32 (9): 728–9.

Parkins, J.R. 2011. "Deliberative democracy, institution building, and the pragmatics of cumulative effects assessment." *Ecology and Society* 16 (3): 20.

Peterson, E., et al. 1987. *Cumulative Effects Assessment in Canada*. Ottawa: Supply and Services Canada.

RAMP (Regional Aquatic Monitoring Program). 2010. "Regional aquatic monitoring program (RAMP) scientific review." http://www.ramp-alberta.org.

Ross, W. 1994. "Assessing cumulative environmental effects: Both impossible and essential." In A.J. Kennedy, ed., *Cumulative Effects Assessment in Canada: From Concept to Practice*. Papers from the Fifteenth Symposium held by the Alberta Society of Professional Biologists, Calgary.

———. 1998. "Cumulative effects assessment: Learning from Canadian case studies." *Impact Assessment and Project Appraisal* 16 (4): 267–76.

Schindler, D., and W. Donahue. 2006. "An impending water crisis in Canada's western Prairie provinces." *Proceedings of the National Academy of Science of the United States* 103: 7210–16.

Seitz, N.E., C.J. Westbrook, and B.F. Noble. 2011. "Bringing science into river systems cumulative effects assessment practice." *Environmental Impact Assessment Review* 31 (3): 172–9.

Sheelanere, P., B.F. Noble, and R. Patrick. 2013. "Institutional requirements for watershed cumulative effects assessment and management: Lessons from a Canadian trans-boundary watershed." *Land Use Policy* 30: 67–75.

Sonntag, N., et al. 1987. *Cumulative Effects Assessment: A Context for Further Research and Development*. Ottawa: Supply and Services Canada.

Spaling, H. 1997. "Cumulative impacts and EIA: Concepts and approaches." *EIA Newsletter* (University of Manchester) vol. 14.

———, et al. 2000. "Managing regional cumulative effects of oil sands development in Alberta, Canada." *Journal of Environmental Assessment Policy and Management* 2 (4): 501–28.

Spaling, H., and B. Smit. 1994. "Classification and evaluation of methods for cumulative effects assessment." In A.J. Kennedy, ed., *Cumulative Effects Assessment in Canada: From Concept to Practice*, 47–65. Papers from the Fifteenth Symposium held by the Alberta Society of Professional Biologists, Calgary.

Squires, A., and M.G. Dubé. 2012. "Development of an effects-based approach for watershed scale aquatic cumulative effects assessment." *Integrated Environmental Assessment and Management* doi: 10.1002/ieam.1352.

Squires, A., C. Westbrook, and M. Dubé. 2010. "An approach for assessing cumulative effects in a model river, the Athabasca River basin." *Integrated Environmental Assessment and Management* 6 (1): 119–34.

US Council on Environmental Quality. 1978, 1997. *Considering Cumulative Effects under the National Environmental Policy Act.* Washington: Council on Environmental Quality, Executive Office of the President.

12 Strategic Environmental Assessment

Strategically Oriented Environmental Assessment

Strategic environmental assessment (SEA) is about integrating *environment* into higher-order **policy, plan,** and **program (PPP)** development and decision-making processes. In doing so, SEA facilitates an early, overall analysis of the relationships between the environment and PPPs and the potential effects of the projects that might emerge from those PPPs. It is difficult to cover all aspects of SEA in a single chapter; SEA itself deserves its own book. Attention here is limited to SEA principles and practices: basic SEA characteristics, the status of SEA in the Canadian context, and a practical framework for SEA application.

SEA is based on the notion that many of the decisions that affect the environment are made long before project developments are proposed and that environmental, social, economic, and political conditions have significant implications for the success of PPPs. Further, actions taken, or not taken, at the strategic level of PPPs can affect the nature and type of development initiatives that emerge and their impacts. For example, decisions made in a regional land-use management plan can shape the future development of the region and influence the specific types of land uses and development activities undertaken.

Strategic environmental assessment, while variously defined, refers to the environmental assessment of strategic initiatives—typically PPPs and their alternatives. A strategic approach to environmental assessment provides an opportunity to consider a range of PPP options at an early stage when there is greater flexibility with respect to future actions and the decisions to be taken. According to Partidário (2012), SEA is about understanding the context of the PPP or strategy being developed and assessed, identifying problems and potentials, addressing key trends, and assessing environmental and sustainable viable options that will help achieve strategic objectives. As such, SEA is focused on asking such questions as "what are the objectives?," "what are the key drivers?," "what are the strategic options?," and "what are the critical policies to be met?" EIA, in contrast, tends to focus on "what are the characteristics of the project?," "what are the local baseline conditions?," "what are the project alternatives?," and "what are the major impacts and mitigation measures?"

Characteristics of SEA

In essence, SEA extends environmental assessment upstream—but at the same time SEA reflects a different set of characteristics than EIA (Box 12.1). System-based characteristics of SEA refer to the basic provisions and requirements for SEA and the position of SEA in the broader planning and decision-making environment. Procedural characteristics of SEA concern the various methodological and process elements of SEA—in other words, the practice of SEA. Result-based characteristics of SEA capture the overall influence of SEA on decision-making and subsequent actions, including the opportunity for broader system-wide learning and SEA process improvement.

Box 12.1 Characteristics of Strategic Environmental Assessment

System-based characteristics	Description
1. Provisions	• clear provisions, standards, or requirements to undertake the SEA and implement the results
2. Proactive	• application early enough to address deliberation on the root of the problem and opportunities related to PPP priorities and choices, including determining the conditions necessary to achieve desired ends
3. Integrative	• integrative of biophysical, social, economic, cultural, and political knowledge and understanding of issues • integral part of the PPP formulation process
4. Tiered	• assessment undertaken within a tiered system of policy, planning, and decision-making • ability to influence, if not direct, proposals, actions, and decision at lower tiers of planning and development, including project EIA
5. Sustainability	• sustainability/sustainable development a guiding principle and integral concept
Process-based characteristics	**Description**
6. Flexible to PPP context	• can accommodate a range of PPP issues and is sensitive to the particular policy or planning culture or agency or organization undertaking the SEA
7. Responsibility and accountability	• clear delineation of assessment roles and responsibilities • mechanisms to ensure impartiality/independence of assessment review • opportunity for appeal of process or decision output

continued

8. Purpose and objectives	• assessment purpose and objectives clearly stated • centred on a commitment to sustainable development principles
9. Future-oriented	• focused on identifying desirable outcomes and on what is required to achieve those outcomes • about building a desirable future, not attempting to know or predict what the future will be
10. Scoping	• opportunity to develop and apply more or less onerous streams of assessment sensitive to the context and issue • consideration of related strategic initiatives, including other policies, plans, and programs • identification and narrowing of possible valued ecosystem components to focus on those of most importance based on the assessment context
11. Alternatives-based	• comparative evaluation of potentially reasonable alternatives or scenarios
12. Impact evaluation	• identification of potential impacts or outcomes resulting from each option or scenario under consideration • integration or review of sustainability criteria specified for the particular case and context
13. Cumulative effects	• assessment of potential cumulative effects and life-cycle issues associated with alternatives or scenarios
14. Monitoring program	• includes procedures to support monitoring and follow-up of process outcomes and decisions for corrective action
15. Participation and transparency	• opportunity for meaningful participation and deliberations throughout the process • transparency and accountability in assessment process

Result-based characteristics	Description
16. Decision-making	• identification of a "best" option or strategic direction to guide PPP implementation • authoritative decisions, position of the authority of the guidance provided
17. PPP and project influence	• defined linkage with assessment and review or approval of any anticipated lower-tier initiatives, including project EIA • demonstrated PPP influence, modification, or downstream initiative • identification of indicators or objectives for related or subsequent strategic initiatives or activities
18. System-wide	• opportunity for learning and system improvement through learning • regular system or framework review, such as a mandated five-year public review process

Source: Based on Noble 2002; 2003; 2008.

Systems of SEA

The origins of SEA can be linked to the US National Environmental Policy Act (NEPA) of 1969. However, the term "SEA" was not introduced until 1989, first mentioned in a research report to the European Commission:

> The environmental assessments appropriate to policies, plans, and pro-grammes are of a more strategic nature than those applicable to individual projects and are likely to differ from them in several important respects . . . We have adopted the term "strategic environmental assessment" (SEA) to describe this type of assessment (Wood and Djeddour 1989).

It was not until after a number of high-profile international developments—namely, the World Bank's (1999) recommendation for the environmental assessment of policy, the report of the World Commission on Environment and Development, *Our Common Future* (1987), and the United Nations 1992 Earth Summit—that SEA gained international attention. It is only within the past decade that SEA, as a formal process, has been clearly evident in practice.

SEA is currently in place in some 60 countries worldwide (Tetlow and Hanusch 2012). However, SEA still remains far less advanced than EIA, and SEA provisions vary considerably from one jurisdiction to the next (Box 12.2). In the United States, for example, provisions for SEA fall under NEPA, where SEA is broadly interpreted to be

---

Box 12.2 Models of SEA Systems

EIA-based: SEA is implemented under EIA legislation, such as in the Netherlands, or carried out under separately administered procedures, such as in Canada and Hong Kong.

Environmental appraisal: SEA provision is made through a less formalized process of policy and plan appraisal, as in the UK prior to the European SEA directive.

Dual-track system: SEA is differentiated from EIA and implemented as a sep-arate process, such as the Netherlands' "Environmental test" or E-test of legislation and plans.

Integrated policy and planning: SEA is an integrated component of policy and plan development and decision-making, such as in New Zealand or for forest management plans in Saskatchewan.

Sustainability appraisal: SEA elements as separate evaluation and decision tools are replaced by integrated environmental, social, and economic assessment and appraisal of policy and planning issues, such as the former Australian Resource Assessment Commission and current sustainability plans in the UK.

Source: Based on UNEP 2002.

programmatic environmental assessment, or area-wide EIA. Hundreds of programmatic assessments are completed in the United States each year, which essentially involve the direct application of EIA to "programs of development," such as offshore oil and gas infrastructure in the Gulf of Mexico. In the United Kingdom, SEA was initially carried out through a less formalized policy and plan environmental appraisal process. Formal requirements for SEA were adopted by the United Kingdom in 2004 under European SEA Directive 2001/42/EC. In the Czech Republic, SEA was introduced by means of a reform to formal EIA legislation, whereas in Australia SEA was at first adopted informally as part of resource management programs and then in 1999 was introduced through formal legislated requirements under the broad umbrella of environmental assessment that includes planning strategies and programs.

SEA in Canada

Canada has made significant contributions to the development of environmental assessment above the project level (Table 12.1). The underlying principle of SEA in Canada is that in order to make informed decisions in support of sustainable development, decision-makers at all levels must integrate social, economic, and environmental considerations with PPP development. SEA was formally established in Canada in 1990 by means of a federal Cabinet directive and as a separate process from EIA, "making it the first of the new generation of SEA systems that evolved in the 1990s" (Dalal-Clayton and Sadler 2005, 61). Procedural guidance for SEA was first provided in *The Environmental Assessment Process for Policy and Programme Proposals* (FEARO 1993), with implementation subject to oversight by the Federal Environmental Assessment Review Office and later the Canadian Environmental Assessment Agency.

When the Canadian Environmental Assessment Act came into force in 1995, it restricted environmental assessment to "projects" or physical works, meaning that policy decisions would no longer be subject to formal impact assessment. In many respects, the evolution and formalization of SEA as a separate process was a step backward for impact assessment in general insofar as the directive created a non-statutory system for PPP assessment that would remain separate from any legislated environmental assessment process (Noble 2009). In 1999, Canada reinforced its commitment to SEA with release of the *1999 Cabinet Directive on the Environmental Assessment of Policy, Plan and Program Proposals*. It was not until January 2004, however, under an updated SEA directive, that Canadian federal departments and agencies were required to prepare a public statement whenever a full SEA had been completed. Federal guidelines for implementation of the directive were most recently updated in 2010.

The current Cabinet directive states, as a matter of policy, that SEA be conducted when

i)   a proposal is submitted to an individual minister or Cabinet for approval; *and*

ii) implementation of the proposal may result in important environmental effects, either positive or negative.

Table 12.1 Timeline of SEA Development in Canada

1990	A bill is introduced to establish the Canadian Environmental Assessment Act; policies, plans, and programs are not included within the scope of the proposed act.
	Canadian Environmental Assessment Research Council (CEARC) releases guidelines for environmental assessment of policy and program proposals.
	National Round Table on Environment and Economy (NRTEE) and CEARC hold a workshop on the integration of environmental considerations into government policy.
1991	Federal government reform package introduces Canada's first initiative in the development of a system of strategic environmental assessment: *Environmental Assessment in Policy and Program Planning: A Sourcebook.*
1992	Canadian Environmental Assessment Act receives legislative approval. Section 16(2) emphasizes the role and value of regional studies outside the act, but there is no reference to SEA.
1993	FEARO procedural guidelines are released to federal departments on the EI process for policy and program proposals (FEARO 1993).
1995	Amendments to the Auditor General Act require that all federal departments and agencies prepare a sustainable development strategy.
	Federal government releases *Strategic Environmental Assessment: A Guide for Policy and Program Officers.*
	Canadian Environmental Assessment Act comes into force.
1999	Update to the 1990 Cabinet directive is released as well as guidelines for implementation.
	Minister of the environment launches a five-year review of the Canadian Environmental Assessment Act.
2000	Frameworks for regional environmental effects assessment appear in the Canadian Environmental Assessment Agency's research and development priorities for 2000.
2001	Bill C-19 is introduced to amend the Canadian Environmental Assessment Act.
2003	Bill C-9, formerly Bill C-19, receives royal assent, and the amended Canadian Environmental Assessment Act comes into force.
2004	Guidelines on the Cabinet directive on SEA are updated, requiring federal departments and agencies to release a public statement when an SEA has been completed.
2007	Minister of Environment's Regulatory Advisory Committee, Subcommittee on SEA, commissions a report on the state of SEA models, principles, and practices in Canada.
2009	Canadian Council of Ministers of the Environment releases *Regional Strategic Environmental Assessment in Canada: Principles and Guidance.*
2010	Updated federal guidelines for implementing the Cabinet directive on SEA are released.
2012	Federal budget implementation bill, the Jobs, Growth and Long-term Prosperity Act, is introduced.
	National Round Table on Environment and Economy is eliminated.
	Canadian Environmental Assessment Act is repealed and the Canadian Environmental Assessment Act, 2012 enacted, but there is no reference to SEA in the new act.

Sources: Based on Noble 2002; 2003; 2008.

The purpose of the Cabinet directive is to strengthen the role of SEA at the strategic decision-making level by clarifying obligations of departments and agencies and linking environmental assessment to the implementation of sustainable development strategies. Since implementation of the directive, numerous federal departments and agencies have developed their own SEA guidelines. These guidelines focus primarily on procedural support to ensure compliance with the directive; it is assumed that as long as these guidelines are adhered to, the result will be meaningful SEA.

Outside the federal process, at the provincial level SEA is practised largely on an ad hoc basis. SEA is sometimes interpreted as an extension of the regulatory environmental assessment process, such as under the Yukon Environmental and Socio-economic Board, as a separate review process for policy and program proposals, such as Newfoundland's *Strategic Environmental Review Guideline for Policy and Program Proposals*, or as part of project-specific terms of reference, such as for 20-year forest management planning in Saskatchewan. Alberta has adopted the Canadian Council of Ministers of the Environment's (CCME 2009) *Regional Strategic Environmental Assessment in Canada: Principles and Guidance* to help inform regional land-use planning and assessment processes, but there is no formal SEA system. In Quebec, the ministère des ressources naturelles et de la faune recently established a formal SEA program for offshore oil and gas exploration in the Gulf of St Lawrence under which two SEAs have been initiated.

SEA Benefits and Opportunities

SEA is based on the premise that EIA, which in essence reacts to a proposed development, is not sufficient by itself to ensure sustainable development. Consistent with the principles of Agenda 21, SEA attempts to integrate *environment* into higher-order decision-making processes. This early integration of environmental considerations into PPP development provides for a more proactive process whereby strategic options and futures are identified and assessed at an early stage, before irreversible project decisions are taken (Box 12.3). SEA is simply a way of ensuring that downstream project planning and development occurs within the context of the *desirable* outcomes that society wants to achieve. SEA addresses the *sources* rather than the *symptoms* of environmental change.

In their assessment of SEA potential in Canada's western Arctic, Noble et al. (2013) identified a number of perceived benefits of and opportunities for SEA, including:

- the potential to improve regulatory decision-making;
- the opportunity to establish regional environmental baselines and identify gaps in knowledge;
- the ability to better manage cumulative environmental effects;
- the adding of value and certainty to project EIA;
- improved and streamlined local engagement at the more strategic tiers of planning; and
- the provision of a framework for co-ordinating often disparate environmental policy and planning efforts.

In practice, these are often contentious issues that emerge during the EIA process but are beyond the scope and capacity of what EIA and its supporting regulations are equipped to deal with.

Box 12.3 SEA Benefits and Opportunities

- Streamlining project EIA by early identification of potential impacts and cumulative effects
- Allowing more effective analysis of cumulative effects on broader spatial scales
- Facilitating more effective consideration of ancillary or secondary effects and activities
- Addressing the causes of impacts rather than simply treating the symptoms
- Offering a more proactive and systematic approach to decision-making
- Facilitating the examination of alternatives and the effects of alternatives early in the decision process before irreversible decisions are taken
- Facilitating consideration of long-range and delayed impacts
- Providing a suitable framework for assessing overall, sector-wide, and area-wide effects before decisions to carry out specific project developments are made
- Providing focus for project EIA on how proposed actions fit within the broader context of the region
- Verifying that the purpose, goals, and direction of a proposed plan or initiative are environmentally sound and consistent with broader policy, plan, and program objectives for the region
- Saving time and resources by setting the context for subsequent regional and project-based EIAs, making them more focused, effective, and efficient

Sources: Clark 1994; Cooper 2003; Kingsley 1997; Noble 2000; Noble et al. 2013; Sadler and Verheem 1996; Sadler 1998; Wood and Djeddour 1992.

Types of SEA

The nature of SEA will vary depending on the regulatory system and specific assessment context, but in general three types of SEA can be identified (Therivel 1993). **Policy** SEA is perhaps the most significant type of SEA, since large-scale government policies can have more far-reaching effects than plans, programs, or projects. A number of policy SEA applications exist in Canada at the federal level, but their number is limited in comparison to sector SEAs, and many are policy appraisals rather than true SEAs per se. Canadian and international examples of policy SEA include the SEA of the North American Free Trade Agreement (see Hazell and Benevides 2000), Canadian minerals and metals policy SEA (see Noble 2003), and the SEA of two Danish bills under Denmark's EA system, one to amend laws relating to housing and the second a subsidy scheme for private urban renewal (see Elling 1997). There are few publicly available policy SEAs in Canada that go beyond an appraisal of a proposed or existing policy, ensuring compliance with the federal directive. One exception, applied outside the scope of the Cabinet directive, is a 2007 policy SEA funded by the Biosphere Canada

Action Program to assess policy options for greenhouse gas mitigation in prairie agriculture (Box 12.4).

Box 12.4 SEA of GHG Mitigation Policy Options in Prairie Agriculture

The Canadian agriculture and agri-food industries are significant producers of CO_2, NO_2, and CH_4 and responsible for approximately 7 per cent of total CO_2-equivalent emissions in Canada. Agricultural greenhouse gas (GHG) emissions, however, differ from those of other sectors in two respects. First, emissions from agriculture are for the most part not caused by energy production and use but rather through gas discharges from livestock production, soil disturbance, fertilizer use, and cropping practices. Second, the agriculture sector is also part of the solution to mitigating GHG emissions in that agriculture can provide a major carbon sink through enhanced soil management and improved cropping practices. Any single solution for GHG mitigation in the agriculture sector, however, may not necessarily be acceptable in the regulatory context or at the on-farm level. Moreover, a blanket mitigation policy may not be effective across all agricultural regions, creating gainers and losers because of variations in soil management practices, cropping systems, and resulting on-farm impacts.

In 2006, under the Biosphere Canada Action Program, an SEA framework was applied to evaluate competing policy options for GHG mitigation and to identify an appropriate strategy for prairie agriculture (see Noble and Christmas 2008). The assessment considered five alternatives, adopted from Agriculture and Agri-Food Canada's *Opportunities for Reduced Nonrenewable Energy Use in Canadian Prairie Agricultural Production Systems* and from the *Canadian Economic and Emissions Model for Agriculture*, and was applied across the five major soil zones.

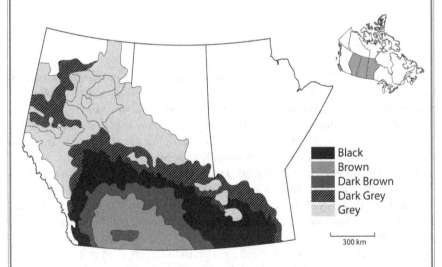

Black
Brown
Dark Brown
Dark Grey
Grey

300 km

The policy alternatives are as follows:

A_1: Enhanced nitrogen-use efficiency

A_2: A 50 per cent increase in zero-tillage over current levels and direct seeding practices

A_3: A 50 per cent reduction in current summer fallow area

A_4: Shifting 10 per cent of the current cropland to forage production

A_5: A 10 per cent increase in the fuel efficiency of farm equipment

A total of 13 VECs were identified against which the alternatives were assessed using a participatory, multi-criteria evaluation process:

Assessment VECs:

crop production quantity	crop production quality
economic risk	economic benefit
economic cost	flexibility of farm operations
institutional support	community support
time requirements	labour requirements
impact on soil resources	impact on water resources
complexity of mitigation	

The impacts of each mitigation policy option were assessed using five standard impact assessment characterization components: magnitude of the potential impact; direction of the expected impact; probability that the VEC would be affected by the proposed alternative; temporal duration of the potential impact; and management potential, or the capacity to offset the impacts of implementing the program given current levels of government or institutional support. In other words, the impact of GHG mitigation option "A_n" on "VEC_i" is determined as a function of the weight (w) of VEC_i and the magnitude (m), direction (d), probability (p), temporal duration (t), and management potential (mp), where the impact of option "n" on VEC_i is defined as $w[d(m \times p \times t) \times mp]$. The total impact of any single policy option is thus defined as:

$$\sum_{i-1}^{n} w\,[d\,(m \times p \times t) \times mp$$

Assessment data were standardized and evaluated using exploratory data analytical techniques to derive the *min-max* solution—the policy option that minimizes potential adverse impacts and maximizes the positive ones. The SEA results indicate considerable differences in the impacts of policy options, with mitigation option A_2 (increased use of zero-tillage cropping systems—an emerging practice in Canadian agriculture) identified as generating the most significant overall positive impacts. Using a concordance analysis, the aggregate results of the SEA indicate a strong preference for alternative A_2, zero-till practices, over all other competing GHG mitigation options: $A_2 \gg A_4 > A_3 \gg A_1 / A_5$, where

continued

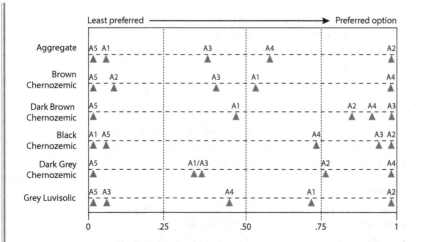

">>," ">," and "/" indicate "strong outranking," "outranking," and "indifference," respectively.

The disaggregate assessment results, however, suggest considerable variation in impacts by soil type. While conservation and zero-tillage are gaining popularity across the prairie region, for example, it is primarily occurring in the moist, dark (black) soil zones. Summer fallow on the other hand—the practice of leaving a normally cultivated field free of vegetation for one growing season so as to conserve soil moisture—is identified by the Canadian Agricultural Census as a *declining* land management practice in the prairie region—a decline that is supported in the relatively moist dark brown and black chernozemic soil zones as a viable GHG mitigation option (A_3). However, increased zero-tillage and decreasing summer fallow are not the preferred mitigation options in other soil zones, in particular the brown chernozemic zone, where fallow is still relied upon as an important soil moisture conservation practice.

Results of the SEA indicate that a blanket policy for GHG mitigation may result in gainers and losers because of variations in soil type and soil management practices, cropping systems, and on-farm impacts. Regional programs, or satisficing solutions, established under a broader national policy and sensitive to soil characteristics and farm-level practices, may be more effective.

Source: Noble and Christmas 2008

Sector SEA applies to sector-based initiatives, plans, and programs, such as forestry plans or oil and gas programs of development. The World Bank (1999) defines sector-based SEA as:

> an instrument that examines issues and impacts associated with a particular strategy, policy, plan, or program for a specific sector; including the evaluation and comparison of impacts against those of alternative options

and recommendation of measures to strengthen environmental management in the sector.

Emphasis is placed on the initiatives of, and alternatives to, particular sector development plans or programs that may lead to environmental change (Box 12.5). Development initiatives and their alternatives are evaluated within the context of sector-based objectives, existing environmental conditions, and current and proposed plans and priorities. Sector SEAs are typically defined by the boundaries of a specific resource sector, such as a forest-harvest management area or oil and gas lease sites.

Box 12.5 Benefits of Sector SEA

- Avoids the limitations of project-specific EIAs in particular sectors by addressing broader environmental, social, and economic concerns
- Prevents or avoids significant environmental effects through the assessment and development of sector-wide plans and programs before individual project decisions are taken
- Provides opportunity for consideration of more effective or efficient sector plans or programs
- Allows for planning and development of sector-wide environmental management strategies
- Facilitates the inclusion of other sector interests in sector-wide planning and development
- Provides a framework for considering the cumulative environmental effects of sector-wide development

Source: World Bank 1999.

A number of sector SEAs have been undertaken in Canada in recent years, including forest management planning in Saskatchewan, electricity sector planning in Ontario, and urban sector planning in Canada's national capital region. The sector that has been subject to the most SEAs is the offshore hydrocarbon sector, particularly in Atlantic Canada, under the Canada–Newfoundland and Labrador and Canada–Nova Scotia offshore petroleum boards (Box 12.6).

Box 12.6 SEA in Atlantic Canada's Offshore Oil and Gas Sector

The Canada–Newfoundland and Labrador Offshore Petroleum Board (C-NLOPB) is responsible for oil and gas activity offshore Newfoundland and Labrador and reports to both the federal and provincial governments. Offshore petroleum activities that require authorization by the C-NLOPB may

continued

also be subject to EIA under federal (and often provincial) legislation. In 2002, the C-NLOPB adopted a policy decision to start conducting SEAs before exploration projects are considered and in conjunction with the call for bids process in offshore areas. The objectives of SEA under the C-NLOPB are to inform licensing in prospective offshore areas and to help streamline issues and considerations for subsequent project EIAs. Six SEAs have been completed by the C-NLOPB, but the three major production facilities currently operating offshore Newfoundland and Labrador all exist in a "non-SEA" region.

Informing offshore planning and rights issuance are the major intents of SEA under the C-NLOPB. SEAs are initiated *only* in areas where no offshore oil and gas operations currently exist. As such, SEA is intended to establish a baseline condition for a potential licensing area. The results are used by the C-NLOPB for licensing decisions and, in principle, by industry to augment baseline assessments in their project EIAs. Using seismic surveying as an example, if there are ecologically sensitive areas identified through an SEA, the intent is that an EIA would focus in detail on these areas as opposed to focusing on areas that may not be regionally significant. The Orphan Basin SEA findings, for example, demonstrate how SEA was designed to inform prospective activity in the study region, whereby special, non-standard, or strict mitigation measures have been identified to be applied to future developments because of the need for special planning around sensitive marine habitats (see LGL 2003).

In practice, however, whether this early baseline assessment is useful to project EIA is unknown (Fidler and Noble 2012). Tiering and influence can only be described in terms of intent, since no projects are operating in areas where SEAs have been completed. There are several operating projects in the Jeanne d'Arc Basin, Canada's most active offshore oil field, but without the guidance of SEA. Doelle, Bankes, and Porta (2012) explain that while the exploration licence is issued after the SEA is completed, the linkage between the SEA and licence issuance is not clear and the expectations of any future oil and gas production in SEA regions are not clearly identified. The SEA reports provide guidance on mitigation measures for exploration activities, but there are no formal decisions that result from the SEA. The SEA process is perhaps best described as an information-gathering exercise. Fidler and Noble (2012), for example, report that industry and regulators in the region perceive there to be limited added value to implementing an SEA in these already-developed regions, suggesting an inherently restrictive view of SEA as simply a high-level information provision tool versus a process for grappling with more strategic issues surrounding the future of offshore energy development. SEA offshore under the C-NLOPB more closely resembles a "regional study" versus a strategic assessment. If SEA has nothing more to offer in areas where development has already occurred, then its strategic influence is questionable. Doelle, Bankes, and Porta (2012) suggest that conducting SEA earlier in the process, before issuing a call for nominations, would enhance the influence of SEA and its role in the decision-making process for offshore planning and development.

Regional SEA is focused on regional plans or programs, such as land-use plans, which may include multiple sectors. In 2009, in an effort to advance SEA beyond the federal Cabinet directive, the Canadian Council of Ministers of the Environment

(CCME) commissioned a study to develop a framework for regional SEA in Canada. The framework, developed by Gunn and Noble (2009), was intended to provide guidance for regionally based applications of SEA and, in particular, to improve the practice of cumulative effects assessment. The CCME (2009) defines regional SEA as:

> a process designed to systematically assess the potential environmental 'effects, including cumulative effects, of alternative strategic initiatives, policies, plans, or programs for a particular region.

Regional SEA is about informing the development of strategic initiatives, policies, plans, or programs for a particular planning or management region, thereby facilitating an opportunity for more informed and efficient EIA and regional environmental management initiatives. The overall objective of regional SEA is to inform the preparation of a development strategy and environmental management framework for a region. Regional SEA is intended to:

- improve the management of cumulative environmental effects;
- increase the effectiveness of project-level environmental impact assessment;
- identify preferred directions, strategies, and priorities for the future use, management, and development of a region.

The outcome of a regional SEA can assist in the identification of preferred, and possible, land-use and development patterns and in setting limits to further development.

Regional SEA is still in its early stages of development in Canada, and the supporting science and tools for its application are still being developed in such areas as Alberta (e.g., land-use planning) (see Johnson et al. 2011) and in Canada's western Arctic (e.g., marine spatial planning for offshore energy development) (see Noble et al. 2013). However, there have been prior attempts at regional SEA in Canada, albeit not under the formal SEA nametag. One such example is the Great Sand Hills Regional Environmental Study, Saskatchewan, which is the focus of this chapter's Environmental Assessment in Action feature.

Environmental Assessment in Action

Regional SEA for Biodiversity Conservation in the Great Sand Hills, Saskatchewan

The Great Sand Hills is located in the southwest region of the province of Saskatchewan, about 375 kilometres west of the city of Regina. The Great Sand Hills comprises approximately 1942 square kilometres of native prairie overlaying a more or less continuous surface deposit of unconsolidated sands, with five sand-dune complexes that total 1500 square kilometres. The northern region of the Great Sand Hills is a designated ecological reserve. Open grasslands and

continued

patches of trees and shrubs characterize the Great Sand Hills, and the region is home to several game species and endangered, threatened, and sensitive species. Natural gas has been exploited in the region since the early 1950s and intensively since the 1980s, with approximately 1500 gas wells now in the area. Livestock grazing has exerted a much longer-term and widespread influence on the landscape, the result of which is a network of permanent trails and vegetative trampling and erosion. In 2004, following decades of growing concern over the effects of natural gas development and ranching activities on the biological diversity and ecological integrity of the region and several land-use planning initiatives, the government of Saskatchewan appointed an independent scientific advisory committee to undertake a regional environmental study of human-induced disturbances in the Great Sand Hills. The terms of reference developed for the regional environmental study directed that the study be based on a strategic environmental assessment framework. The overall objective of the assessment was to ensure that the long-term ecological integrity of the area is maintained while economic benefits are realized.

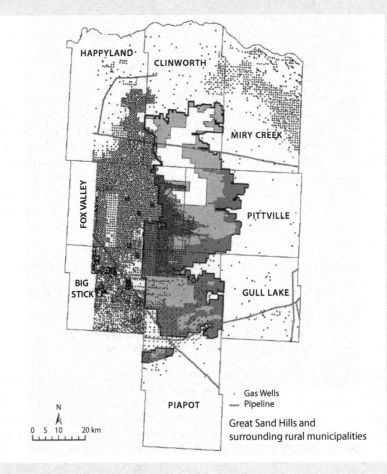

The spatial scale of the assessment was multi-tiered, considering biophysical, socio-economic, and cultural boundaries, as well as the reach of existing PPPs that have the potential to affect any proposed land-use scenario for the region. The biophysical scale of assessment was based primarily on the Great Sand Hills' dunes and grasslands. The socio-economic boundary was based on a larger area of eight regional municipalities that surround the Great Sand Hills. The cultural boundary, capturing First Nations' interests, extended into the neighbouring province of Alberta. The temporal scale of the assessment was from the 1950s, the beginnings of gas development in the region, and projected forward to 2020, at which time gas reserves would be fully tapped.

The assessment consisted of three phases: a baseline assessment, trends identification, and evaluation of alternative scenarios of human-induced surface disturbance. The baseline assessment characterized the biophysical, socio-economic, and cultural environment of the region; identified broad-scale cumulative change; and collected data for identification of stressors, trends projection, and scenarios. The underlying objectives of the baseline assessment were to identify those human activities that have the greatest potential for disturbance and for affecting ecological integrity and VEC sustainability. VECs were identified through an open scoping process involving local stakeholders and First Nations and previous land-use planning initiatives. Data collection was based on field studies, secondary sources, focus groups, participatory mapping, and interviews with more than 250 community members and other interests. The biophysical component of the baseline assessment focused particular attention on biodiversity, specifically delineating concentrated biodiversity using a site selection algorithm, MARXAN, based on collected species and habitat data. MARXAN is a decision-support software that uses an optimization algorithm to aid in the selection of a system of spatially cohesive

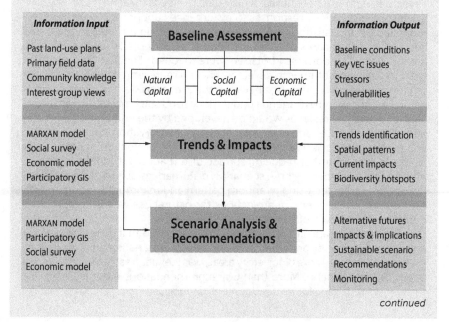

Information Input	Baseline Assessment	Information Output
Past land-use plans Primary field data Community knowledge Interest group views	*Natural Capital* / *Social Capital* / *Economic Capital*	Baseline conditions Key VEC issues Stressors Vulnerabilities
MARXAN model Social survey Economic model Participatory GIS	**Trends & Impacts**	Trends identification Spatial patterns Current impacts Biodiversity hotspots
MARXAN model Participatory GIS Social survey Economic model	**Scenario Analysis & Recommendations**	Alternative futures Impacts & implications Sustainable scenario Recommendations Monitoring

continued

sites to meet biodiversity targets (see Chapter 3, Box 3.8). A total of 37 core biodiversity areas were identified, those with the most to lose if not managed, representing various levels of biological irreplaceability.

In the second phase of the assessment, trends in surface disturbance were identified across the landscape using retrospective analysis of aerial photography and land-use and vegetation/species databases. Rates of change were established for each of the three main sources of stress in the region—roads and trails, gas wells, and cattle watering holes. Significant statistical correlations were found to exist between gas wells and the distribution of roads and trails, and roads and trails were found to be a reasonable surrogate for human-induced surface disturbance in the region. In 1979, for example, there were 76 gas-well surface leases in the region. By 2005, 1391 new wells had been established. Associated with the increase in gas-well development was a growth in roads and trails, which had increased from 2497 kilometres in 1979 to approximately 3175 kilometres by 2005, with new roads and trails 153 times more likely to be built in association with a new gas-well pad than elsewhere in the region. Annual rates of change in development and associated patterns of surface disturbance were used to build statistical models to quantify trends in surface disturbances. Parallel to the biophysical assessment was a regional economic assessment, a telephone-based survey to assess local perceptions of current impacts of ranching and gas activity, and a series of participatory GIS workshops with stakeholders to identify goals for the region, including preferred land-use patterns and designations.

In the third phase of the assessment, alternative land-use scenarios were developed based on possible futures and rates of change in human disturbance, primarily roads and trails, gas wells, and cattle watering holes. A GIS was then used to project future growth rates, based on the statistical models developed, and spatial patterns of disturbance across the landscape. Species, range, and biodiversity responses to disturbances under each scenario were modelled using statistical and spatial relationships determined in the baseline and trends analysis phases, and the patterns of disturbance relative to the locations of biodiversity hotspots were mapped. The first two scenarios were based on past trends in development and represented two variations of the future; the main difference between the scenarios was a more ambitious natural gas development agenda under the second such that all proven, probable, and possible reserves would be developed by the end of the projection period. The third scenario was predicated on a conservation-based approach and designed to conserve biodiversity through the protection of biodiversity hotspots and by further reducing surface disturbance outside the core biodiversity areas. Under all three scenarios, disturbances due to cattle watering holes were projected using a random pattern and minimum spacing criteria. An example of the scenario projections for natural gas well development is depicted in the figure opposite.

The third scenario, a conservation-based scenario that delineated sites of enhanced biodiversity protection and best-practice management for developments outside those core biodiversity areas, was identified as the recommended development scenario. More than 60 recommendations emerged from the assessment regarding future development planning, environmental monitoring,

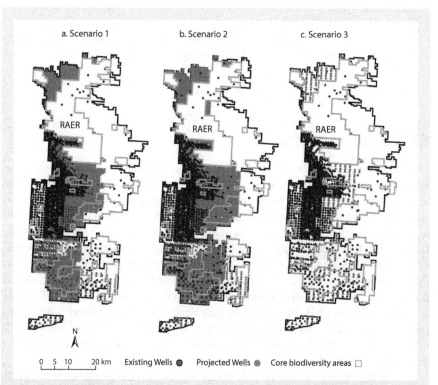

a. Scenario 1   b. Scenario 2   c. Scenario 3

RAER   RAER   RAER

N

0 5 10   20 km   Existing Wells ●   Projected Wells ●   Core biodiversity areas ☐

and biodiversity targets and thresholds. The Great Sand Hills assessment brought to a conclusion nearly 15 years of regional planning to provide a strategic direction for effects management and future land use. What was different about the regional Great Sand Hills assessment, in comparison to past planning and assessment initiatives in the region, was its grounding in an SEA framework. This enabled the assessment to reach beyond the constraints of individual project-based issues, many of which are not subject to any form of impact assessment, in order to address the nature and underlying sources of cumulative change and to identify desirable futures and outcomes.

That said, a number of broader challenges emerged from the Great Sand Hills experience and important lessons for future SEA applications elsewhere. First, the Great Sand Hills assessment, like many regional assessments in Canada, was a "one-off" initiative with no real mechanism to sustain it as an integral part of regional planning and downstream project assessment. Second, the intent in the Great Sand Hills assessment was that it would inform and guide future development activities, land-use zoning, and decision-making. However, as in most other jurisdictions in Canada, there is no formal tiered system of policy, plan, and program assessment to effectively carry regional SEA forward from the strategic to the project scale. Third, although the Ministry of the Environment commissioned the Great Sand Hills assessment, it lacked the capacity and the authority to fully implement the Great Sand Hills plan and recommendations. Many recommendations were beyond the scope of the ministry,

continued

particularly issues that relate to land-use governance and socio-economics. Delineating roles and responsibilities for implementing the results of SEA, and creating an enabling institutional environment, is an important prerequisite for its success.

Sources: Noble 2008; figures and maps adapted from GSH SAC 2007.

SEA Frameworks

There is no single, agreed-upon framework for SEA, but based on recent international experiences a number of common steps can be identified. It is important to note that, as with EIA, specific design requirements are often necessary within each application—including specific methods and techniques. Most of the methods and techniques required for SEA are readily available from impact and policy appraisal. Noble, Gunn, and Martin (2012), for example, examined SEA methods used in a sample of 14 SEAs from Canada and abroad. Their analysis identified 18 different types of methods used, with various forms of expert judgment, public consultation, and ad hoc methods or lessons from elsewhere the most commonly used throughout the SEA process (Table 12.2). The authors suggest that the range of methods used in SEA practice is restrictive and limited to a number of common, qualitative-based methods and that more attention should be given to analytical-based methods, including quantitative approaches.

The following sections present the basic components of a generic SEA framework, including selected examples of methods and techniques. The framework, based on Gunn and Noble (2009), is intended to provide the structure to SEA that allows decision-makers to create and evaluate strategic initiatives and their impacts, without in any way constraining the choice of methods or tools or the level of participation. The framework consists of three broad components (Figure 12.1):

- a pre-assessment or context-setting phase focused on developing a reference framework for the SEA, scoping the baseline and key issues of concern, and identifying trends and stressors of concern;
- an impact-assessment phase, frequently technical in nature, that serves to identify strategic options and assess their potential effects, opportunities, and risks;
- a post-assessment phase focused on developing planning and management measures and moving SEA output forward to PPP implementation and on following up on the results.

Develop a Reference Framework

Context setting is essential to any environmental assessment application. The first phase in the SEA process is to establish the context (decision-making, policy, planning, regulatory) within which the assessment will take place. A number of basic yet fundamental questions should be asked at this stage, including:

Table 12.2 Methods Used in SEA and SEA-Type Applications[1]

■ = used in 50% or more of cases reviewed
● = used in 25–49% of cases reviewed
+ = used in < 25% of cases reviewed

Methods	Stage of SEA process[2]									
	1	2a	2b	2c	2d	2e	3	4	5	6
Expert judgment/workshops	■	■	■	■	■	■	■	■	■	■
Literature/case review	■	■	■	■	■	■	●	●	+	+
Public workshop/consultation	●	●	●	●	●	●	+	●	■	+
Ad hoc (lessons from elsewhere)	+	●	●	●	■	●	+	+	+	+
GIS/mapping applications	+	●	●	+	+	+	+	+	+	
Matrices	●	+	●	+	+	+	+	+		
Trends analysis/extrapolation	+	+	+	+	+	+	+	●		
Risk assessment	+	+	+	+	+	+		+		+
Website/newsletter	+	+	+	+	+	+			+	+
Landscape assessment	+	+	+	+	+	+	+	+		
Systems modelling		+	+	+	+	+	+	+		
Participant funding		+	+	+	+	+		+		
Scenario analysis/modelling		+	+	+	+		+	+		
Weighting/scoring		+	+		+	+		+		
Cost–benefit analysis		+	+		+			+		
Other economic valuation		+	+			+		+		
Sensitivity analysis		+	+	+						
Sustainability assessment							+	+		
Statistical hypothesis testing		+		+						

[1] Some of the methods identified cover a range of methods. For example, there are many different types of matrices, from traditional EIA-type impact assessment matrices to policy compatibility matrices.

[2] Stage of generic SEA process: **1** = Conducting a preliminary scan; **2** = Analyzing environmental effects; **2a** Identifying potential effects; **2b** Characterizing effects; **2c** Predicting outcomes; **2d** Identifying mitigation; **2e** Determining significance; **3** = Identifying PPP alternatives; **4** = Evaluating alternatives; **5** = Identifying stakeholder concern; **6** = Follow-up and monitoring.

Source: Noble, Gunn, and Martin 2012

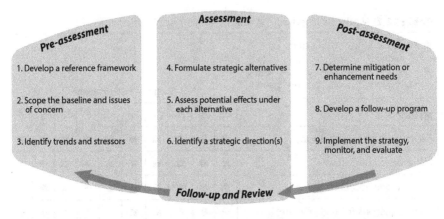

Figure 12.1 ⊚ Framework for Strategic Environmental Assessment
Source: CCME. 2009.

- What are the objectives of the SEA, and what is hoped will be accomplished?
- What are the objectives of the strategic initiative or proposed plan?
- Who is responsible for the SEA, and what stakeholders need to be involved?
- What are the data requirements and information needs for carrying out the assessment?
- What are the needs and opportunities for linking-up with other PPPs and with project EIA?

In sector SEA, such as for offshore oil and gas, the assessment objectives may be more closely related to industry objectives for environmental performance, business initiatives, or industry and industry-related activities within the region, and thus the stakeholders and data requirements can be easily identified. For a broader regional SEA, such as the development of a regional land-use plan, the assessment objectives may focus on identifying an appropriate direction for longer-term sustainable regional development, including the implications of multiple-sector developments, such as housing, oil and gas exploration, and transportation, and interactions between these activities within the region.

Scope the Baseline and Issues of Concern

The current baseline must be established and primary issues of concern relevant to the sector, region, or policy environment identified. This means undertaking an analysis of the environmental and socio-economic setting and the relevant policy and planning environment—conditions that may be affected by the PPP or conditions that may influence the success of the PPP. Depending on the tier of assessment (policy, plan, or program), emphasis may focus on broader policy issues and the policy environment or on the state of the region and on particular environmental VECs. Generally speaking, establishing the baseline involves asking:

- What are the VECs of concern?

- What is the past and current condition of potentially affected VECs?
- What are the observed trends or changes over time?
- What are the thresholds of concern or objectives associated with the VECs?
- What are the appropriate spatial and temporal boundaries for the type of region or sector under consideration?

Only the environmental or socio-economic components likely to be receptors of effects that are important to ecological functioning or valued by society should be considered, including such VECs as air quality, water resources, habitats, spaces of cultural significance, social well-being, or, at the policy level, an existing policy or plan. The objective is to understand the environment in which the PPP will be developed and implemented and will most likely affect, either positively or negatively.

Establish VEC Indicators or Objectives
As with predicting project-based impacts, making useful predictions about effects in SEA requires some "indicator" or specified "objective" against which the incremental effects of the PPP can be assessed. Thus, an indicator should define each VEC that is capable of providing an early warning of the state or health of the VEC of concern. In cases where VECs are not amenable to indicator selection, or when dealing with broader policy-based issues, it is more appropriate to focus on VEC objectives or criteria against which potential changes in VEC conditions, or anticipated outcomes of the PPP, can be appraised. These objectives or criteria can be derived or translated from previously stated goals and objectives or set within the context of a broader environmental vision for the region or sector of concern.

Delineate Assessment Boundaries
Spatial and temporal boundaries provide a frame of reference for SEA and help to establish the resolution of analysis. Spatial and temporal bounding for SEA is no different, in principle, from that for project-based EIA. The primary difference at the strategic level concerns the nature of the activities that must be considered. For example, geographic relationships, common resources, and proposed activities must be viewed not only from the perspective of physical and socio-economic phenomena but also from the perspective of current and proposed policies, plans, or programs that may interact with the proposed plan or program.

Identify Trends and Stressors

This is the retrospective phase of SEA and serves to identify the key trends and driving forces of change in the sector, region, or policy environment of concern. It may require identifying cause–effect relationships or correlations between VEC conditions and various actors or drivers of change. In some cases, particularly at the broader policy level, specific causal relationships may be difficult if not impossible to define. In such cases, attention should focus on identifying past policy initiatives and policy changes and recognized environmental or public responses. The objective is not to spend a great deal of time and resources on identifying ecosystem-wide linkages, for example, but rather to gain an understanding of a limited number of important components and possible interactions such that key trends or observations from the baseline can be carried forward to the future and used as a basis against which alternative options are assessed and appraised. A variety of methods and techniques are available to support trends and stressors analysis, including retrospective modelling, network analysis, and photographic progressions.

Formulate Strategic Alternatives

Unless there is more than one potential and feasible way to proceed, no decision has to be made, and therefore no SEA is required; this includes the option of no plan or action. Alternatives in SEA should include:

- the current baseline condition carried forward;
- the proposed initiative, plan, or program implementation procedure;
- alternatives to the initiative, proposed plan, or plan or program implementation procedure.

This process might include the construction of scenarios for each alternative that emphasize different:

- spatial and temporal attributes or option implementation;
- modes or processes, including technologies or methods;
- means of meeting the plan or strategic initiatives;
- targets or objectives.

Alternatives should be considered only if they are:

- consistent with broader environmental and sustainability goals and objectives;
- not in conflict with existing regional or sector policies, plans, or activities; and/or
- economically, technologically, and institutionally feasible.

The objective is to identify alternatives that are "more sustainable" or "least negative" or that trigger the "least significant" amount of environmental change. If, in the case of a proposed plan or program, a strategic alternative is likely to cause a "more significant adverse effect" than the proposed plan or program itself, then it should not be considered a viable alternative. Thus, for each alternative, the following minimum factors should be considered in order:

- Is the alternative compatible with sustainability goals and objectives?
- Is there a potential interaction (either positive or negative) between activities associated with the proposed alternative and existing VECs?
- Are potentially affected VECs already affected (either positively or

negatively) by existing policies, plans, or activities in the sector or region?
- Will the alternative have an effect on current VECs (either positive or negative) in combination with already existing and proposed policies, plans, and activities?

Assess Potential Effects under Each Alternative

This is the prospective or futures-oriented phase of SEA. Only those options or alternatives that are compatible with the goals and objectives of existing and proposed policies, plans, and initiatives for the sector or region should be considered for detailed assessment. That way, the focus of assessment is on the *desirability* of the initiatives with regard to VEC objectives rather than on trying to resolve conflicting goals and objectives between proposed and existing initiatives. SEA does not deal with predictions of what will happen per se but rather offers contingency statements of possible effects, risks, and opportunities under each alternative

The assessment of options should include consideration of:

- the opportunities and risks created by each strategic option;
- the potential effects on specified VECs;
- the potential for cumulative effects;
- other policies, plans, or actions that may affect the same resources or affect the success of the PPP option.

Determine Environmental Changes Likely to Affect VECs

The analysis of potential effects involves the analysis of the potential effects of each option on the VECs by identifying the primary sources of stress and associated effects pathways. This is a formidable task, and the levels of complexity and uncertainty increase as one moves from projects to plans and from sectors to regions. The objective is to identify potential stressors and VEC responses associated with each alternative. The level of detail depends on the availability of data.

Identify VEC Response

This requires understanding the past and present conditions of VECs and how each might react based on the specified VEC indicators. In short, the objective is to capture, based on the relational networks established, how the VEC receptor might respond or deviate from its current condition and future baseline condition given the multiple interactions and pathways with and without each of the proposed options. Depending on the nature of the VEC receptor and level of information available, it may be possible to model trends over time based on different scenarios of stressors and the range of VEC responses for each option; in other cases, expert-based forecasting, such as the Delphi approach, might be most appropriate. In general, such futures methods can be readily adapted from EIA-driven and planning approaches.

As baseline data and knowledge of cause–effect relationships increase, it is possible to adopt more "science-driven" methods and techniques for effects assessment; in cases where data are lacking or where an understanding of cause–effect relationships is simply not possible, more "judgment-driven" approaches must be replied on. Duinker and Greig (2007) and Cherp, Watt, and Vinichenko (2007) also emphasize the need to consider "external wildcards" (e.g., climate change) and "emergent and external" events (e.g., policy influences, economic change) that may affect VEC responses or the desirability of any given alternative or future scenario.

Identify a Strategic Direction

The outcome of an SEA should lead to the identification of a desirable strategic direction. In other words, when compared to all other proposed options, a preferred strategy for action is identified on the basis of:

- its potential to achieve sustainability goals and objectives;
- acceptable levels of environmental change;
- the potential for cumulative effects on VECs;
- interactions with and the contributions of current and reasonably foreseeable policies, plans, and initiatives within the sector or region, with or without the proposed plan or initiative;
- institutional, financial, social, and political support for implementation.

Evaluate Alternatives

For SEA, the issue is not so much about "predicting" and impacts as much as it is about identifying preferred directions and outcomes. In other words, once environmental effects are identified, they must be evaluated. The question now becomes "what is required to achieve a desired future or VEC outcome, and what are the consequences of different choices?" VEC or plan goals and objectives, such as minimizing the potential for effects and cumulative interactions between plans or programs affecting water quality, for example, and the alternative means of achieving these goals and objectives are evaluated. Thus, emphasis might be placed on assessing the feasibility and desirability of meeting a desired target (e.g., maintaining a specified level of biodiversity) or on selecting alternatives that minimize potential cumulative effects or maximize environmental and socio-economic contributions.

For each VEC objective, one should consider how *desirable* strategic option "i" is compared to option "j" when taken into consideration with the potential effects of current and reasonably foreseeable policies, plans, and initiatives. In the case of SEA for a GHG mitigation policy for prairie agriculture (Box 12.4), for example, alternatives were evaluated on the basis of whether they met a range of VEC objectives simultaneously, including maximizing on-farm flexibility in policy implementation and minimizing economic risk. In the case of the Great Sand Hills assessment in Saskatchewan, the effects of alternative scenarios of development were ultimately evaluated according to their ability to meet specified targets and objectives for biodiversity conservation.

How the options are evaluated against the objectives again depends on the availability of data and desired decision resolution and may include, for example, expert-based judgment or more complex *modelling* and *simulation analysis*. In keeping with the examples and decision-making techniques discussed at various points throughout this text, Box 12.7 presents an example of alternatives evaluation and decision-making based on multi-criteria evaluation and expert judgment.

Box 12.7 Example of Alternatives Evaluation in SEA Based on the Paired Comparison Weighting Technique

How desirable is strategic option "i" compared to option "j" when taken into consideration with the effects of current and reasonably foreseeable policies, plans, and initiatives on the specified VEC?

VEC objective "n"	Option 1	Option 2	Option 3
Option 1	1	1/3	7
Option 2	3	1	7
Option 3	1/7	1/7	1

Where:

9 = option "i" is extremely more desirable in comparison to option "j"

7 = option "i" is strongly more desirable in comparison to option "j"

5 = option "i" is moderately more desirable in comparison to option "j"

3 = option "i" is slightly more desirable in comparison to option "j"

1 = option "i" is equally desirable to option "j"

1/3 = option "i" is slightly less desirable in comparison to option "j"

1/5 = option "i" is moderately less desirable in comparison to option "j"

1/7 = option "i" is strongly less desirable in comparison to option "j"

1/9 = option "i" is extremely less desirable in comparison to option "j"

Compared to option "j", option "i" is more likely to: a) meet the VEC objective and/or b) minimize overall effects on the VEC.

Compared to option "j", option "i" is less likely to: a) meet the VEC objective and/or b) minimize overall effects on the VEC.

Normalized matrix:

VEC objective "n"	Option 1	Option 2	Option 3
Option 1	0.24	0.23	0.47
Option 2	0.72	0.68	0.47
Option 3	0.03	0.10	0.07

Priority vector : Option 1 = 0.31 Option 2 = 0.62 Option 3 = 0.07

continued

$$0.31 \begin{bmatrix} 1 \\ 3 \\ 1/7 \end{bmatrix} + 0.62 \begin{bmatrix} 1/3 \\ 1 \\ 1/7 \end{bmatrix} + 0.07 \begin{bmatrix} 7 \\ 7 \\ 1 \end{bmatrix} = \begin{bmatrix} 1.01 \\ 2.04 \\ 0.21 \end{bmatrix}$$

Rank alternatives based on VEC objective "n" (divide the vector by original priorities):

    Option 1 = 1.01 / 0.31 = (3.26)
    Option 2 = 2.04 / 0.62 = (3.29)
    Option 3 = 0.21 / 0.07 = (3.00)

    Option 2 (3.29) > Option 1 (3.26) > Option 3 (3.00)

Normalize the ranking by $[(i - i_{min}) / (i_{max} - i_{min})]$ to display the relative "magnitude" of the ranking for each VEC objective:

    (Option 1 = 0   Option 2 = 1   Option 3 = 0.10)

Option 1 Option 3   Option 2

    ┼─┼─────────────────────────────────┼──
    0   0.25   0.5   0.75   1.0

Determine "cumulative desirability" of the option:
Repeat the above procedures for each VEC objective. The "cumulative desirability" and final preferred option or initiative is determined by summing the normalized rankings across all VEC objectives. For example:

Objectives	Option 1	Option 2	Option 3
VEC objective 1	0.00	1.00	0.10
VEC objective 2	0.00	1.00	0.24
VEC objective 3	0.36	0.00	1.00
VEC objective 4	0.11	0.00	1.00
"Cumulative desirability"	0.47	2.00	2.34

The net result is a ranking of all options and identification of two competing options (Option 3 and Option 2), of which Option 3 is ultimately the preferred choice based on the (unweighted) objectives across all VECs. To determine the weighted rankings, simply multiply the option scores by VEC weights (importance) prior to normalizing. In this particular example, given the proximity of Option 3 and Option 2, a sensitivity analysis to changing VEC objectives and priorities would be warranted.

Determine Mitigation or Enhancement Needs

Even the "preferred" option may have potentially adverse environmental effects that cannot be avoided. A number of questions can be examined across VECs to determine whether effects associated with strategic initiatives require mitigation, notably:

- Will the proposed initiative or plan option generate effects that will exceed thresholds as defined in regulations or guidance or as established by maximum allowable effects levels?
- Are the effects significant when considered in light of the current conditions of the affected VEC?
- Will the plan or plan option generate effects on VECs that are likely to interact cumulatively with other plans or activities in the region on already stressed VECs?
- Will the effects of the plan or plan option be permanent?
- Does the initiative or plan option conflict with the goals and objectives of existing plans and activities?

Attention should also be given to managing current conditions so that the preferred option can be implemented, such as creating a supporting institutional environment through establishing collaborative planning processes or allocating needed financial and human resources to support PPP implementation. For example, current industry or land-use conflicts may need to be resolved before the preferred option can become viable. Once these questions are addressed for the "preferred" option, its preference should be re-evaluated to determine whether it is still the preferred option based on the mitigation requirements and cost and the feasibility of implementing them. Competing options may need to be considered and a satisfying option pursued.

Develop a Follow-up Program

No assessment process is complete without a follow-up or monitoring component. The post-decision activities of follow-up and monitoring are essential to ensuring that SEA, and the resulting PPP, is delivering its anticipated or desired outcomes and that any mitigation measures prescribed have been implemented and are working. The rationales for follow-up at the strategic tier are generally the same as those at the project level: derived from the notions of uncertainty and risk associated with decision-making and on the need for feedback and learning. However, the complexities and uncertainties of determining post-implementation environmental implications are exacerbated at the strategic level in that:

- PPPs are often formulated in abstract terms, resulting in vague directions for acting;
- whereas significant deviations from the original plan are abnormal at the project level, they are typical of strategic-level processes;
- there are fewer direct linkages between decisions at the strategic level and actual PPP impacts (Cherp, Watt, and Vinichenko 2007).

Depending on the nature and purpose of the SEA—to appraise a PPP or to assess the impacts of a PPP—follow-up should be designed, at a minimum, to track PPP performance on the basis of established goals, objectives, and targets or to track environmental impacts and the effectiveness of management measures, as well as to improve scientific and technical knowledge through feedback.

There are three broad types of follow-up and evaluation in SEA:

1. **Input evaluation** concerns SEA procedures and requirements, such as SEA purpose and objectives, data quality, linkages to PPP initiatives, and ensuring that these requirements and objectives were met.

2. **Process evaluation** concerns the SEA application, such as impact predictions and mitigation measures, and following up to verify the accuracy of these impacts and to determine whether the mitigation measures were implemented and effective.

3. **Output evaluation** concerns SEA effectiveness and involves determining whether the analysis informed the PPP decision process, whether the proposed PPP was modified as a result of the SEA, whether the SEA influenced downstream planning and project impact assessment, and whether what were intended as plan or initiative outcomes were in fact realized.

Implement the Strategy, Monitor, and Evaluate

Regardless of the process efficiency and effectiveness, the output of an SEA is of little benefit if it is not implemented. However, implementation of a new strategy or PPP is easier said than done. The implementation of a strategic initiative requires a degree of collaboration that extends beyond the ability of any single government agency or department. Successful implementation demands a level of commitment and interagency collaboration that is not common in impact assessment.

There is no single best PPP implementation style; rather, implementation depends on the environment within which the PPP is being implemented (Table 12.3) (Noble and Harriman 2008). Generally speaking, however, the more complex the strategy

Table 12.3 Adaptive versus Programmed Approaches to Strategy or PPP implementation

Guidelines	Strategy or PPP environmental context	
	Structured	**Unstructured**
scope of change	incremental	major
certainty in strategy, technology, or management	certain within risk	uncertain
conflict over goals and means	low conflict	high conflict
structure of institutional setting	tightly coupled	loosely coupled
stability of environment	stable	unstable
Preferred implementation style	**Programmed**	**Adaptive**

Source: Based on Berman 1980, from Mitchell 2002.

and the greater the uncertainty or potential conflict involved, the more preferred an adaptive approach to implementation is. It is important to be realistic about the limitations and uncertainties in looking into the future and in developing blueprint implementation and management strategies. Strategies and PPPs must be sufficiently adaptive to system changes, bifurcation, and external and emergent stressors and responsive to new knowledge gained through monitoring and follow-up processes (Cherp, Watt, and Vinichenko 2007).

Directions in SEA

Strategic environmental assessment is still quite new and relatively limited in terms of its adoption and application. Large numbers of SEA applications do not occur under the SEA nametag or under any form of environmental assessment framework. However, based on known experiences to date, Noble (2009) describes Canadian SEA systems and practices as diverse, far from consolidated in scope and function, and encompassing a range of models and practices. As such, there is considerable variability in outcome and expectations.

There is no single model of SEA that can be unequivocally applied; rather, attention must be given to context, to regulatory opportunities and constraints, and to the tier of application. This is not to say that "good-practice" SEA should not be defined but that SEA operates in diverse forms and under a range of institutional arrangements. Common knowledge and understanding of the SEA process and what SEA is supposed to deliver, the influence of SEA over subsequent assessment and decisions, and an institutional commitment and willingness to carry out SEA are among the most significant challenges to its continued development in Canada.

Noble et al. (2013) identify a number of misperceptions about SEA that need to be addressed, either through education and training programs or through good-practice reports, in order to continue to advance SEA update. These include: misperceptions that SEA will mean foregoing anticipated development opportunities; the belief that SEA will result in the loss of flexibility in decision-making; SEA adding an additional layer of bureaucracy to an already cumbersome and expensive regulatory system; and the uncertainties associated with adopting a new approach to decision-making.

Key Terms

input evaluation
output evaluation
plan
policy
policy SEA
process evaluation

program
programmatic environmental assessment
regional SEA
sector SEA
strategic environmental assessment

Review Questions and Exercises

1. What are the potential benefits and challenges of applying environmental assessment to policy?
2. Identify the different types of SEA, and provide an example of the type of problem or situation to which each might apply.
3. What are the provisions, if any, for policy, plan, or program assessment in the political jurisdiction in which you live? What do you see as the main challenges to SEA where you live?
4. Visit the websites for the Canadian Cabinet directive for implementing SEA at www.ceaa.gc.ca and the European directive at http://ec.europa.eu/environment/eia/home.htm. Compare and contrast the directives in terms of their objectives, scope, requirements for reporting, and public involvement.
5. An important part of SEA is the identification of alternatives. One question that emerges from this, however, is who should be involved. Consider two hypothetical plan proposals in your area, one for oil and gas licensing and one for the development of a regional land-use plan. Discuss who should be involved in the identification of alternatives.
6. Obtain a completed SEA from your local library or government registry, or access one online. Using Box 12.1 as a guide, explore the SEA for evidence of "strategic" characteristics. In other words, are the characteristics listed in Box 12.1 present in the SEA? Are there certain characteristics that seem to be missing? Compare your findings to those of others.

References

Berman, P. 1980. "Thinking about programmed and adaptive implementation: Matching strategies to situations." In H. Ingram and D. Mann, eds., *Why Policies Succeed or Fail*. Beverley Hills, CA: Sage.

CCME (Canadian Council of Ministers of the Environment). 2009. *Regional Strategic Environmental Assessment in Canada: Principles and Guidance*. Winnipeg: CCME.

Cherp, A., A Watt, and V. Vinichenko. 2007. "SEA and strategy formation theories: From three Ps to five Ps." *Environmental Impact Assessment Review* 27: 624–44.

Clark, R. 1994. "Cumulative effects assessment: A tool for sustainable development." *Environmental Impact Assessment Review* 12 (3): 319–22.

Cooper, L. 2003. *Draft Guidance on Cumulative Effects Assessment of Plans*. EPMG Occasional Paper 03/LMC/CEA. London: Imperial College.

Dalal-Clayton, B., and B. Sadler. 2005. *Strategic Environmental Assessment: A Sourcebook and Reference Guide to International Experience*. London: Earthscan.

Doelle, M., N. Bankes, and L. Porta. 2012. "Using strategic environmental assessment to guide oil and gas exploration decisions in the Beaufort Sea: Lessons learned from Atlantic Canada." http://ssrn.com/abstract=2142001.

Duinker, P.N., and L. Greig. 2007. 'Scenario analysis in environmental impact assessment: Improving explorations of the future'. *Environmental Impact Assessment Review* 27 (3): 206–19

Elling, B. 1997. "Strategic environmental assessment of national policies: The Danish experience of a full concept assessment." *Project Appraisal* 12 (3): 161–72.

FEARO (Federal Environmental Assessment Review Office). 1993. *The Environmental Assessment Process for Policy and Program Proposals*. Hull, QC: Supply and Services Canada.

Fidler, C., and B.F. Noble. 2012. "Advancing strategic environmental assessment in the offshore oil and gas sector: Lessons from Norway, Canada, and the United Kingdom." *Environmental Impact Assessment Review* 34: 12–21.

GSH SAC (Great Sand Hills Scientific Advisory Committee). 2007. *Great Sand Hills Regional Environmental Study*. Regina: Canadian Plains Research Centre.

Gunn, J., and B.F. Noble. 2009. "A conceptual and methodological framework for regional strategic environmental assessment (R-SEA)." *Impact Assessment and Project Appraisal* 27 (4): 258–70.

Hazell, S., and H. Benevides. 2000. "Toward a legal framework for SEA in Canada." In M.R. Partidário and R. Clark, eds., *Perspectives in Strategic Environmental Assessment*. New York: Lewis Publishers.

Johnson, D., et al. 2011. "Improving cumulative effects assessment in Alberta: Regional strategic assessment." *Environmental Impact Assessment Review* 31: 481–3.

Kingsley, L. 1997. *A Guide to Environmental Assessments: Assessing Cumulative Effects*. Hull, QC: Parks Canada, Heritage Canada.

LGL Ltd. 2003. *Orphan Basin Strategic Environmental Assessment*. St John's: Canada–Newfoundland and Labrador Offshore Petroleum Board.

Mitchell, B. 2002. *Resource and Environmental Management*. 2nd edn. Harlow, UK: Prentice Hall.

Noble, B.F. 2000. "Strategic environmental assessment: What is it and what makes it strategic?" *Journal of Environmental Assessment Policy and Management* 2 (2): 203–24.

——— . 2002. "Strategic environmental assessment of Canadian energy policy." *Impact Assessment and Project Appraisal* 20 (3): 177–88.

——— . 2003. "Auditing strategic environmental assessment in Canada." *Journal of Environmental Assessment Policy and Management* 5 (2): 127–47.

——— . 2008. "Strategic approaches to regional cumulative effects assessment: A case study of the Great Sand Hills, Canada." *Impact Assessment and Project Appraisal* 26 (2): 78–90.

——— . 2009. "Promise and dismay: The state of strategic environmental assessment systems and practices in Canada." *Environmental Impact Assessment Review* 29 (1): 66–75.

——— , et al. 2013. "Strategic environmental assessment opportunities and risks for Arctic offshore energy planning and development." *Marine Policy* 39: 296–302.

Noble, B.F., and L. Christmas. 2008. "Strategic environmental assessment of greenhouse gas mitigation options in the Canadian agricultural sector." *Environmental Management* 41: 64–78.

Noble, B.F., J. Gunn, and J. Martin. 2012. "Survey of current methods and guidance for strategic environmental assessment." *Impact Assessment and Project Appraisal* 30 (3): 139–47.

Noble, B.F., and J. Harriman. 2008. *Regional Strategic Environmental Assessment: Methodological Guidance and Good Practice*. Report prepared for the Canadian Council of Ministers of the Environment. Winnipeg.

Partidário, M.R. 2012. *Strategic Environmental Assessment: Better Practice Guide*. Lisbon: Portuguese Environment Agency and Redes Energeticas Nacionais, SA.

Sadler, B. 1998. "Ex-post evaluation of the effectiveness of environmental assessment." In Alan L. Porter and John J. Fittipaldi, eds., *Environmental Methods Review: Retooling Impact Assessment for the New Century*. Fargo, ND: The Press Club.

———, and R. Verheem. 1996. *Strategic Environmental Assessment: Status, Challenges and Future Directions.* Report 53. Ottawa: Canadian Environmental Assessment Agency.

Tetlow, M.F., and M. Hanusch. 2012. "Strategic environmental assessment: The state of the art." *Impact Assessment and Project Appraisal* 30 (1): 15–24.

Therivel, R. 1993. "Systems of strategic environmental assessment." *Environmental Impact Assessment Review* 13: 145–68.

UNEP (United Nations Environment Programme), Economics and Trade Program. 2002. *Environmental Impact Assessment Training Manual.* 2nd edn. New York: UNEP.

Wood, C., and M. Djeddour. 1989. "Environmental assessment of policies, plans and programs." Interim report to the Commission of European Communities. Manchester: EIA Centre.

———. 1992. "Strategic environmental assessment: EA of policies, plans and programmes." *Impact Assessment Bulletin* 10 (1): 3–23.

World Bank. 1999. *Environmental Assessment.* Operational Policy and Bank Procedures, no. 4.01. Washington: World Bank.

World Commission on Environment and Development. 1987. *Our Common Future.* Oxford: Oxford University Press.

Professional Practice of EIA

Several parties are involved in the EIA process, including the project proponent, regulators and decision-makers, public interest groups, in some cases review panels, and consultants and advisors to each of these parties. As such, there are many roles for the EIA professional. An EIA professional may be an industry employee, responsible for managing the company's EIAs and day-to-day environmental regulatory processes; a government employee responsible for implementing EIA regulations, determining the need for an EIA, and following up on industry commitments; an active member of an environmental non-government organization responsible for EIA awareness and playing a watchdog role during the EIA process; an Aboriginal lands manager, responsible for EIAs and permitting of development projects on lands under Aboriginal jurisdiction; an independent consultant to proponents hired to conduct fieldwork in support of the EIS and assisting with and the preparation of the EIS; or an expert advisor on EIA to industry, governments, and public interest groups, including review panels, on the quality of the EIA, the nature and potential risks associated with a project's impacts, and the quality of the project's overall EIA process.

Perhaps the most common role of the EIA practitioner is that of a consultant. Consultants, working in interdisciplinary teams, usually carry out the preparation of EISs on behalf of project proponents. Boxes 13.1 to 13.4 present samples of actual career postings for environmental assessment professionals in Canada. The samples are limited to the private sector and are certainly not comprehensive of all career fields in environmental assessment. The purpose is to provide some insight into the types of career opportunities in impact assessment and the required experience and skill sets. The career advertisements were selected from the results of a search of career postings on such sites as "simplyhired.ca," "eluta.ca," "indeed.com," and "wowjobs.ca." The company names appearing in the original advertisements and other information that may reveal the company's identity have been removed.

Box 13.1 Senior and Intermediate Socio-economist/ Stakeholder Consultation Specialists

Company ABC requires socio-economist/stakeholder consultation special-ists to support the expansion of socio-environmental consulting services. Knowledge and experience with socio-economic impact assessment, First Nations/indigenous peoples and community consultation, issues manage-ment, traditional use and traditional knowledge studies, and the integration of these studies into Canadian and international regulatory reporting and per-mitting processes is required. The work will involve a strong commitment to business development and client relations and expanding technical excellence, as well as the opportunity to mentor a talented group of socio-environmental scientists. A graduate degree in social sciences or economics, preferably with a focus on community and international development, is also required, as well as a minimum of seven years' experience. Required skills and experience include:

- excellent written and oral communication skills;
- socio-economic and/or consultation experience working on First Nations/indigenous peoples specific studies in Canada and elsewhere;
- proven project management skills and the ability to work with a multi-disciplinary team;
- experience working in Canadian and/or international regulatory and permitting processes;
- strong interpersonal skills, excellent verbal and written communica-tion skills in English (other languages—e.g., Spanish—are an asset);
- ability to travel within Canada and internationally as well as to remote areas.

Box 13.2 Senior Environmental Scientist and Environmental Assessment Lead

As a senior environmental scientist and environmental assessment lead, you will have the opportunity to help lead, develop, and expand our team as well as pursue business development opportunities. This senior role will give you the opportunity to:

- lead baseline and socio-environmental assessment project teams;
- manage projects, budgets, and schedules;
- participate in multi-disciplinary projects and work closely with biol-ogists, ecologists, hydrologists, geoscientists, social scientists, engi-neers, and other project specialists;
- manage project approval process;
- manage client interfaces and maintain key client relationships;
- write high-quality reports, including field reports, permit applications, regulatory documents, environmental impact assessments, manage-ment plans, etc.

- manage, mentor, and train employees;
- provide senior review and quality assurance and quality control;
- travel domestically and internationally for fieldwork.

If you have a bachelor's, or preferably a master's, degree in biology, environmental or related science and experience in environmental work in mining, oil sands, oil and gas, hydroelectric and other electricity generation, transportation, land development, or municipal infrastructure, we want to hear from you. We value your:

- sound working knowledge of provincial and federal environmental legislation, EIA methodology, and permitting (e.g., environmental, mines, water, lands) requirements;
- experience managing, leading, and conducting EIAs;
- proven project management skills and the ability to work with a multi-disciplinary team of professionals;
- previous consulting/practical project experience;
- superior verbal and written communication and presentation skills;
- proven leadership, decision-making, and coaching skills;
- strong commitment to technical excellence;
- strong contract administration and project management skills;
- track record as an effective team player;
- availability to work flexible hours as project requirements dictate.

Box 13.3 Lead Environmental Assessment Specialist

Company XYZ is seeking an environmental assessment specialist with consulting experience and established client relationships to play a key role in growing our business in western Canada. The ideal candidate will have established relationships with oil and gas companies, along with technical and project management experience leading multi-disciplinary EIAs. The main purpose of the lead environmental assessment specialist position includes, but is not limited to, providing senior regulatory and technical support to project managers and clients in support of various projects. In addition, the lead environmental assessment specialist will provide mentoring to staff and participate in business development initiatives. The role may include a field component including project site visits/inspections, safety audits, and liaison with clients and/or regulators. As a senior leader, this individual will be fully accountable for successfully managing and growing a range of oil and gas projects including EIAs. Qualifications include:

- a university degree or diploma in environmental science or a related discipline;
- 10 years of experience focusing on EIA;
- ability to work independently and succeed in a collaborative, multi-disciplinary team environment;
- strong written and verbal communication skills;

continued

- strong knowledge of provincial and federal regulations as well as EIA and regulatory process;
- proven knowledge of environmental permit, approval, and compliance requirements in the oil and gas industry;
- demonstrable knowledge of environmental mitigation and protection planning;
- direct experience in preparing technical and regulatory reports under provincial regulations and the National Energy Board.

Responsibilities include:

- acting as the lead resource on EIA projects;
- assisting in the development and growth of EIA projects with a focus on energy developments;
- supporting the development of the EIA business by developing relationships with existing and future client contacts;
- supporting and preparing regulatory approval and permitting applications;
- liaison with clients, stakeholders, and regulatory agencies to ensure project permitting is advanced;
- support and preparation of environmental protection and mitigation plans to facilitate regulatory approvals;
- preparation and review of technical reports associated with the project EIA, approvals, or compliance requirements;
- establishing and maintaining relationships with government and regulatory agencies.

Box 13.4 Environmental and Regulatory Co-ordinator

As the environmental and regulatory co-ordinator at Company 123, you have a unique opportunity to work in a dynamic environment interfacing with various groups at site to ensure environmental and regulatory compliance across operations. You will act as the liaison between various regulatory bodies and internal and external stakeholders, including engineering, operations, and maintenance and major projects groups and contractors. As a core team member of this group, your ability and desire to work cooperatively with others on a team with the ability to demonstrate interest and skills provide an exceptional opportunity for career progression and development. As a proactive problem-solver, your solution-oriented work style will positively impact operations through increased efficiency, effectiveness, compliance, and overall production and revenue. Key responsibilities include:

- co-ordinate the development and implementation of environmental monitoring initiatives to ensure adherence to legal requirements and standards;
- manage the execution of environmental monitoring programs, including data assimilation, interpretation, and work plan development scheduling;

- contribute to monitoring program development and field-site inspections to assess environmental risks;
- assist in the development of risk mitigation strategies by liaising with business units on potential and realized risks;
- provide advice, resources, and assistance to business units in the interpretation of company environmental systems, policies, and procedures;
- assist in the development delivery and monitoring of personal work plan and setting of objective and targets;
- set clear, measurable goals and expectations and ensure prompt feedback addressing any performance problems or issues;
- develop, maintain, and strengthen partnerships with internal and external stakeholders.

The position requires:

- typically five years of related industry experience with a technical diploma or degree in a related discipline;
- advanced computer skills;
- familiarity with and understanding of relevant provincial environmental regulations and codes.

Skills and Qualifications

EIA professionals come from a diverse range of social science, natural science, and technical backgrounds, including geography, law, business, engineering, public health, biology, political studies, sociology, anthropology, chemistry, geology, and hydrology, to name a few. There is no internationally recognized certification for EIA professionals, though some jurisdictions do have certification requirements for EIA practitioners (e.g., South Africa), whereas most recognize certain professional designations (e.g., professional biologist, geographer, engineer, or geoscientist) but have no EIA certification requirements per se. There are, however, minimum international guidelines for standards for EIA professionals. In 2010, the International Association for Impact Assessment (IAIA) adopted "Guideline standards for IA professionals" (IAIA 2010). The guidelines establish the minimum standards for the profession for both impact assessment practitioners (Table 13.1) and administrators (Table 13.2). The standards for impact assessment administrators are similar to those for practitioners but place much more emphasis on knowledge of relevant environmental and related institutions, legislation, policies, and administrative procedures.

One important criterion that is not addressed in the IAIA guideline standards, likely because it is difficult to assess at the individual practitioner or administrator level, is interdisciplinarity. The EIA process is by nature interdisciplinary—it involves the combining of knowledge from multiple disciplines into a single activity, the development and evaluation of an EIS. The interdisciplinary nature of the teams that carry out EIAs, and subsequently review and evaluate the EIS, is of critical importance to a quality EIA process. The collection of individuals involved in EIA must

Table 13.1 Guideline Standards for Impact Assessment Practitioners

Category	Practitioner	Senior Practitioner	Lead Practitioner
Education and training	Relevant degree from an accredited university or a member in good standing of a relevant professionally accredited organization	Relevant degree from an accredited university or a member in good standing of a relevant professionally accredited organization	Relevant degree from an accredited university or a member in good standing of a relevant professionally accredited organization
Experience	Minimum 2 years' experience in undertaking and reporting on impact assessment studies	Minimum 5 years of progressively senior experience and responsibility in designing, undertaking, and reporting on at least component impact assessment studies, including public participation	Minimum 10 years of progressively senior experience and responsibility in designing, undertaking, and reporting on at least component impact assessment studies, including public participation
Understanding of impact assessment methods	A good understanding of impact assessment methods, including cumulative and strategic assessment	A thorough working knowledge of impact assessment methods, including cumulative and strategic assessment	A thorough working knowledge of impact assessment methods, including cumulative and strategic assessment
Impact assessment study management	Demonstrated, under direction, an ability to effectively plan and carry out specialist impact assessment studies	Demonstrated an ability to effectively lead at least component impact assessment studies and, under direction, some multi-disciplinary studies and to look beyond compliance to develop and promote best practice	Demonstrated an ability to effectively lead and integrate comprehensive, multi-disciplinary impact assessment studies at all scales and to look beyond compliance to develop and promote best practice
Sustainable development	A good understanding of the structure, functioning, and interrelatedness of ecological, socio-economic, health, and political systems that support sustainable development	A good understanding of the structure, functioning, and interrelatedness of ecological, socio-economic, health, and political systems that support sustainable development	A good understanding of the structure, functioning, and interrelatedness of ecological, socio-economic, health, and political systems that support sustainable development and a demonstrated ability to apply this understanding to project planning and impact assessment

Impact assessment administrative systems	Familiar with the relevant impact assessment administrative systems and guidelines	Working knowledge of the relevant impact assessment administrative systems and guidelines and a demonstrated ability to interpret and fulfill their requirements	A broad working knowledge of impact assessment administrative systems and guidelines and a demonstrated ability to interpret and fulfill their requirements
Professional development	Actively engaged in continuing professional development through readings, publications/presentations, and/or training	An active commitment to best practice and continuing professional development through readings, publications/presentations, training, and/or mentoring	An active commitment to best practice and continuing professional development through readings, publications/presentations, training, and/or mentoring
Mentoring	An active commitment to mentoring less experienced practitioners	An active commitment to mentoring less experienced practitioners	An active commitment to mentoring less experienced practitioners

Source: Based on IAIA guidelines, available at www.iaia.org.

Table 13.2 Guideline Standards for Impact Assessment Administrators

Category	Administrator	Senior Administrator	Lead Administrator
Education and training	Relevant degree from an accredited university or a member in good standing of a relevant professionally accredited organization	Relevant degree from an accredited university or a member in good standing of a relevant professionally accredited organization	Relevant degree from an accredited university or a member in good standing of a relevant professionally accredited organization
Experience	Minimum 2 years of impact assessment experience, with an emphasis on the administration of public sector impact assessment processes	Minimum 5 years of progressively senior experience and responsibility, with an emphasis on the administration of public sector impact assessment processes, including some experience with conducting integrated impact assessment studies and related public participation	Minimum 10 years of progressively senior experience and responsibility, with an emphasis on the administration of public sector impact assessment processes, including some experience with conducting integrated impact assessment studies and related public participation
Understanding of impact assessment methods	A good understanding of impact assessment methods, including cumulative and strategic assessment	A thorough working knowledge of impact assessment methods, including cumulative and strategic assessment	A thorough working knowledge of impact assessment methods, including cumulative and strategic assessment
Impact assessment administrative systems	Familiar with the relevant impact assessment administrative systems and guidelines	Good working knowledge of the relevant impact assessment, environmental, and related institutions, legislation, policies, and administrative procedures	Detailed working knowledge of the relevant impact assessment, environmental, and related institutions, legislation, policies, and administrative procedures
Review of impact assessment documents	Capable of drafting, under direction, integrated impact assessment requirements for projects, of evaluating the adequacy of assessment documents, of drafting project approval conditions, and of following up on the implementation of conditions	Demonstrated ability to establish integrated impact assessment requirements for projects in at least a few sectors, to evaluate the adequacy of assessment documents, to draft project approval conditions, and to follow up on the implementation of conditions	Demonstrated ability to establish integrated impact assessment requirements for a full range of project types and scales, to evaluate the adequacy of assessment documents, to draft project approval conditions, and to follow up on the implementation of conditions

Sustainable development	A good understanding of the structure, functioning, and interrelatedness of ecological, socio-economic, health, and political systems that support sustainable development	A good understanding of the structure, functioning, and interrelatedness of ecological, socio-economic, health, and political systems that support sustainable development	A good understanding of the structure, functioning, and interrelatedness of ecological, socio-economic, health, and political systems that support sustainable development and a demonstrated ability to apply this understanding to impact assessment reviews and decision-making
Professional development	Actively engaged in continuing professional development through readings, publications/presentations, and/or training	An active commitment to best-practice and continuing professional development through readings, publications/presentations, training, and/or mentoring	An active commitment to best-practice and continuing professional development through readings, publications/presentations, training, and/or mentoring
Mentoring		An active commitment to mentoring less experienced administrators	An active commitment to mentoring less experienced administrators

Source: Based on IAIA guidelines, available at www.iaia.org.

have access to the complete set of expertise necessary to carry out baseline studies, make predictions of impacts and assess them, propose mitigation and management measures, design follow-up procedures, involve the publics, and evaluate the quality and integrity of the findings. If important skills are lacking, critical impacts or possible mitigation measures can be missed, adverse impacts could occur, and conflict may arise among the affected publics. No individual possesses all of the skills necessary to carry out all aspects of an EIA. However, each individual must appreciate the interdisciplinary nature of the impact assessment process and be willing work outside disciplinary boundaries. This includes a willingness to adapt disciplinary-based methods and find new ways of approaching problems and being open to new methods and different forms of knowledge to assist in understanding the nature and the severity of potential impacts. EIA team leaders, in particular, must have strong teamwork skills, communication skills, and interdisciplinary skills.

Ethics and Practice

EIA is a form of applied research; thus, practitioners must apply ethical research standards (Lawrence 2005). Ethics refers to moral duty or obligation, which typically gives rise to values or governing principles that are used to judge the appropriateness of behaviour or conduct. More specifically, ethics is a branch of philosophy that is focused on the structuring and defending of what constitutes "right" and "wrong." Lawrence (2005) suggests that **normative ethics**, which seeks to arrive at moral conduct standards, and applied or **practical ethics**, which studies specific practical problems and involves a commitment to action, are especially relevant to EIA.

    Professional ethics concerns the conduct of professionals in practice, typically set out in codes of conduct (Box 13.5). These codes of conduct assist members of a professional organization in understanding appropriate conduct (the difference between "right" and "wrong") and in applying a certain standard to their professional behaviour and actions. In many instances, failure to comply with a code of conduct can result in expulsion from the professional organization or withdrawal of professional licences to practice.

Ethical Responsibilities

It is the role of the practitioner to ensure that, to the extent possible, complete, unbiased, and accurate information is available to all parties involved in the EIA process. It is not professional to produce an EIA report solely to meet a legal requirement when such a report must be submitted. Fuggle (2012) explains that the social contract between impact assessment professionals and civil society and decision-makers is such that impact assessments will be conducted with integrity and will be free from misrepresentation or deliberate bias and impact assessments will respect citizen rights to participate in decisions that affect them. Free from bias does not mean that impact assessment is value-free. Values and good judgment are core to the EIA process, from determining the need for an assessment and scoping the valued ecosystem components to interpreting the significance of potential environmental effects and

---

Box 13.5 International Association for Impact Assessment Code of Conduct

---

As a self-ascribed professional member of the International Association for Impact Assessment, the information and services that I provide must be of the highest quality and reliability. I consequently commit myself:

- To conduct my professional activities with integrity, honesty, and free from any misrepresentation or deliberate bias.
- To conduct my professional activities only in subject areas in which I have competence through education, training, or experience. I will engage, or participate with, other professionals in subject areas where I am less competent.
- To take care that my professional activities promote sustainable and equitable actions as well as a holistic approach to impact assessment.
- To check that all policies, plans, activities, or projects with which I am involved are consistent with all applicable laws, regulations, policies, and guidelines.
- To refuse to provide professional services whenever the professional is required to bias the analysis or omit or distort facts in order to arrive at a predetermined finding or result.
- To disclose to employers and clients and in all written reports any personal or financial interest that could reasonably raise concerns as to a possible conflict of interest.
- To strive to continually improve my professional knowledge and skills and to stay current with new developments in impact assessment and my associated fields of competence.
- To acknowledge the sources I have used in my analysis and the preparation of reports.
- To accept that my name will be removed from the list of self-ascribed professional members of IAIA should I be found to be in breach of this code by a disciplinary task-group constituted by the IAIA Board of Directors.

Source: See www.iaia.org/membership/code-of-conduct.aspx.

---

determining whether a project is in the best interest of the environment and society. Ethical responsibilities and how the practitioner reacts to ethical dilemmas is the focus of this chapter's Environmental Assessment in Action feature.

Based on the mission and values statement of the International Association for Impact Assessment (see www.iaia.org), included among the ethical and professional responsibilities of those engaged in environmental assessment is to:

- compile or review impact assessments with integrity and honesty and free from misrepresentation or deliberate bias;
- not condone the use of violence, harassment, coercion, intimidation, or undue force in connection with any aspect of impact assessment or implementation;

- conduct impact assessments in the awareness that different groups in society experience benefits and harm in different ways;
- take gender and other social differences into account and be especially mindful of the concerns of indigenous peoples;
- strive to promote considerations of equity as a fundamental element of impact assessment;
- give due regard to the rights and interests of future generations;
- not advance one's own private interests to the detriment of the public, clients, or employing institutions.

Environmental Assessment in Action

Ethical Dilemmas: What Would You Do?

The stakes are often high in EIA. The fate of projects and, depending on the severity of the potential impacts, the sustainability of environments and communities can hang in the balance. Impact assessment professionals are thus faced with ethical dilemmas on a regular basis. A number of scenarios are presented below for discussion and consideration. There is no "answer" provided—these are ethical challenges. Discuss each of these scenarios with your peers.

Scenario 1

You are a consultant hired by a project proponent to manage an environmental assessment process for a large pulp and paper mill. As the EIS manager, you are responsible for reviewing the technical reports generated by the sub-consultants, presenting the relevant information in the EIS, and concluding whether the predicted impacts will be significant or of concern to environmental quality or public health. Part of the assessment process was based on a predictive model of how chemical X, discharged from pulp mill operations into the river system, might bio-concentrate in sturgeon—a species of considerable importance to a local Aboriginal community. The model used to predict the impacts is an off-the-shelf tool, which itself is known among the scientific community to be imperfect based on its internal assumptions and simplifications of ecological systems. But it is the best available modelling tool. Further, there was limited baseline data available to populate the model, thus introducing considerable uncertainty into the predictive outcome. As a trained expert, you are aware of the limitations and uncertainty, as well as the lack of data (and limited quality of data) used to populate the model and predict the concentration and significance of chemical X. The model output, however, shows that the predicted levels are within the regulatory limits and not of concern.

As a practitioner, hired by the project proponent, what would you do?

- Would you accept the model's results as is and report the impact as insignificant?

- Would you report the model results but disclose the uncertainties and poor data involved in making the prediction, thus making the prediction unreliable from a regulator's perspective?
- What are the possible ethical or career implications of your action?

Scenario 2

The following case was developed based on G. Fred Lee and Anne Jones-Lee's "Practical Environmental Ethics: Is There an Obligation to Tell the Whole Truth?"

Landfill sites can pose a potential threat to local and regional groundwater contamination. To manage the risk, there are two types of standards for land-fill design: first, performance standards that must be achieved by the design, such as the prevention of contamination to groundwater over the life cycle of the project; second, minimum design standards, such as the thickness and permeability of the liner at the bottom of the landfill pit, used to separate soil and groundwater from waste.

Regulations governing disposal of wastes emphasize the need to protect groundwater from use-impairment for as long as the wastes represent a threat. Regulations governing landfill operations identify a "minimum design standard" for the liner thickness to ensure a maximum permeability. The regulations *do not* state that the minimum design standard will meet the performance standard and ensure the long-term protection of groundwater. In previous projects, the minimum design standard model had been breached by leachate in only a few months, posing a risk to groundwater contamination.

  i) A practitioner hired by the project proponent to assess the landfill project claims in her technical report for the EIS that the proposed landfill "will, without question, be protective of groundwater quality over the long term." Note that the practitioner did not claim that the groundwater will be protected from impact forever.

- Is this unethical practice?

  ii) A practitioner hired by the project proponent to assess the landfill project skirts the ethical problem by limiting his evaluation to whether or not the minimum design standard, as set out in regulation, has been met. The practitioner does not say whether the regulations are sufficient to protect health and environmental quality.

- Is this unethical practice?

Scenario 3

Discuss how you would react to each of the following situations:

- The terms of reference you are provided to undertake the assessment unreasonably constrains the study, likely resulting in results that are not representative of the baseline conditions or of the range of potential impacts of the project.

continued

- As a consultant for a project proponent, you are pressured to limit the scope of the assessment or to present only part of your findings.
- You are asked to change your conclusions about the significance of a predicted impact to make it appear less significant than you believe is suggested by your data.
- Communities refuse to meet with you and participate in the study, yet you are required as per the study terms of reference to consult with local communities.
- You are asked not to report on the uncertainties associated with your models or technical studies used to inform impact predictions.
- In reviewing the EIS on behalf of a proponent, you find inaccuracies or errors in the reported information.
- As an independent expert, you are asked to submit a favourable (or unfavourable) review of a proponent's EIS.

The Practitioner as Honest Broker

What is the role of the EIA practitioner? The EIA process is adversarial. This is particularly the case for EIAs for large development projects, especially those subject to review panel assessments. It is often the responsibility of "one side," practitioners working for a project proponent, to present only the strongest possible technical discussion of the project; it is left to the other side, practitioners or experts hired by those wishing to challenge the project, to bring out and discuss the weaknesses in the proponent's technical position. This can create an ethical dilemma for the EIA practitioner.

Practitioners of EIA have choices in the way they interact with project proponents, the public, and decision-makers. These choices have important implications for the decisions taken in EIA and also for the role of EIA practitioners in the EIA process. Pielke (2007) proposes four different roles for experts in decision-making, which are equally applicable to the role of practitioners and other experts in the EIA process. The first role is that of pure scientist, whereby the focus is only on the facts and there is no interaction with the decision-maker. Information is gathered, scientifically analyzed, and the results presented to "speak for themselves." The second role is that of science arbiter, or one who answers specific and factual questions that may be posed by a decision-maker but the information presented is limited to those facts that are relevant to the question at hand. Third is the issue advocate, the individual who seeks to reduce the scope of choice available to the decision-maker by presenting information in a certain way—to try and influence the decision. Finally, the honest broker is one who seeks to expand, or at least clarify, the nature and scope of information and options available to the decision-maker such that the decision-maker is more aware of the potential implications of different decision actions.

As an arena for public debate about the acceptability of a proposed development, the EIA process does benefit from all four roles being represented; however, practitioners must be sure to uphold ethical conduct in EIA. Stakeholders in the EIA process, including project proponents, regulators, and interest groups, can pressure a practitioner to emphasize a particular position—one that either supports or opposes a proposed project by either overemphasizing or underemphasizing certain information. It is improper for an EIA practitioner to advocate private interests to the detriment of the public, clients, or decision-makers. Good impact assessments "enhance the free flow of complete, unbiased, and accurate information to decision makers and affected parties" (Fuggle 2012).

Key Terms

normative ethics   professional ethics
practical ethics

Review Questions and Exercises

1. In addition to the dilemmas presented in Environmental Assessment in Action, what other ethical issues might a professional encounter in the practice of EIA?
2. Are there certain ethical issues that may emerge in the professional practice of EIA that are specific to working with Aboriginal communities? Hint: Consider such matters as traditional knowledge and ownership of information.
3. Using the Internet, search the "codes of conduct" for various professional organizations or associations, such as professional engineers, professional geoscientists, or professional biologists. Are there similarities in the codes of conduct between these professions and the code of conduct presented in Box 13.5 for impact assessment professionals?
4. Select one of the career advertisements in Box 13.1, 13.2, 13.3, or 13.4, or find another advertisement for an EIA professional from a web-based career search engine. Prepare an application for the position. Include in your application a cover letter and a detailed resumé. Are there certain skills sets that you need to further develop?

References

Fuggle, R. 2012. "Ethics." *Fastips* 2. April 2012. www.iaia.org.
IAIA (International Association for Impact Assessment). 2010. "Guidelines standards for IA professionals." Fargo, ND: IAIA.
Lawrence, D.P. 2005. *Environmeal Impact Assessment: Practical Solutions to Recurrent Problems.* Hoboken, NJ: John Wiley & Sons.
Pielke, R.J. 2007. *The Honest Broker: Making Sense of Science in Policy and Politics.* New York: Cambridge University Press.

IV

Environmental Impact Assessment Postscript

The Effectiveness of
Environmental Assessment:
Retrospect and Prospect

Introduction

Conceived initially as a pragmatic tool for ensuring that at least some forethought and foresight are given to the impacts of a proposed development project before that project becomes a reality, EIA is now an accepted part of development decision-making—but has it been a worthwhile part? Would decisions about development have been different and the environmental effects more severe if no EIA had taken place? Has EIA really made a difference? These are challenging questions for the environmental assessment community, and the answers demand a comparison of development outcomes with EIA to those without EIA. In that sense, the answers may perhaps be more hypothetical and argumentative than factual.

Over the years there have been several inquiries into the effectiveness of EIA and whether EIA is making a difference. The majority of these inquiries have focused on the procedural aspects of EIA, from screening and scoping to significance determination and post-decision environmental monitoring. The view is that by improving the process of EIA, its benefits will be more fully realized (Jay et al. 2007). A sound process is necessary for effective EIA procedure, and there is considerable room for EIA process improvement; however, the real measure of EIA is its substantive effectiveness—its contribution to environmental management and longer-term sustainability. This assumes, of course, that there is some consensus that improved environmental management and long-term sustainability are the agreed-upon goals of environmental assessment. This chapter reflects briefly on the substantive effectiveness of environmental assessment in Canada, as illustrated by the current state of project-based EIA, CEA, and SEA systems and outcomes, and makes a case for a more regionalized and integrated approach to impact assessment.

The Substantive Effectiveness of Environmental Assessment

Environmental assessment has evolved considerably since its inception. As Gibson (2002) explains, the concept and practice have moved toward being conducted earlier in the planning process, more open and participative, more comprehensive, more mandatory, and more closely monitored. Environmental assessment has also evolved in a more substantive way, including efforts to "link up" with environmental management, to assess cumulative environmental effects, and to advance application to the strategic tier. At the same time, however, there appears to be a growing dissatisfaction with environmental assessment and increasing pressures on the environmental assessment practitioner and academic community to respond (see Boyden 2007). Surveying the environmental assessment literature, and based on public reaction to the EIA process concerning several recent high-profile development proposals in Canada, it would appear that EIA, CEA, and SEA are falling far short of expectations.

Environmental Impact Assessment

Environmental assessment first appeared on the scene in the early 1970s. Now, thousands of EIAs are completed on an annual basis. The underlying concept of EIA is quite simple—identifying, assessing, and finding ways of mitigating the potential impacts of proposed projects on the human and biophysical environment. The subject of assessment is typically an individual infrastructure project and its potential effects. Procedurally, when a project is proposed and the need for an EIA determined, the most likely impacts of that project are assessed. A decision is then made as to whether the project might cause significant adverse environmental effects and whether such effects can reasonably be mitigated. The philosophy of EIA at the time of its inception, and still very much so today, is one of a rationalist model requiring a technical evaluation of project design and of the local environment in order to provide *sound* advice to proponents and decision-makers. In its early years, much of this advice was about pollution control through engineering and abatement technologies. In recent years, however, this sphere has expanded to include socio-economic and cultural issues, benefits-sharing, sustainability, the assessment of cumulative environmental effects, and an expectation that EIA also contribute to ongoing environmental management.

That said, and as highlighted in Chapter 1, Richard Fuggle, president of the International Association for Impact Assessment (IAIA) at the time, described in an IAIA newsletter a "disillusionment" with EIA "and scepticism that impact assessments are contributing to better decisions" (Fuggle 2005, 1). Noble and Storey (2005) reported on the utility of follow-up in EIA and on the limited attention to social impact management and monitoring post-development decisions (see Chapter 9). Galbraith, Bradshaw, and Rutherford (2007) discussed the rise of privatized environmental and socio-economic impact benefit agreements between communities and private industry—perhaps a response, at least in part, to the failings of the EIA process to adequately consider benefits or build trust and capacity among stakeholders. Whether EIA is delivering on its expectations is at issue.

Since the second edition of this book, global economic conditions have changed considerably—and consequently the climate in which EIA operates. As discussed in Chapter 2, during a period of global economic challenges the Conservative government of Canada introduced its federal budget implementation bill, declaring jobs and growth and to sustain Canada's economy as its top priorities. Perceiving inefficiencies in EIA as a hindrance to economic development, the Canadian Environmental Assessment Act was repealed, and the Canadian Environmental Assessment Act,

2012 was introduced. The new act reduced the scope and number of federal EIAs, enacted tighter timelines for regulatory decisions, and eliminated small project EIAs. For some, this was change long overdue. For example, in a 23 March 2012 commentary by the Canadian Association of Petroleum Producers (CAPP), just prior to the new act, the president of CAPP stated:

> The bottleneck in the regulatory system is about more than delays in project approvals. It's also about potential project cancellations, significant deferrals of projects and a chilling effect on investment because market opportunities pass or competitive alternatives materialize while the regulatory process chugs along. The existing system has grown more complex, inconsistent and uncertain as legislative and administrative layers have been added over time. In our view, this issue must be addressed with some urgency (Collyer 2012).

The views on recent changes to federal EIA in Canada are diverse. According to the president and CEO of the Canadian Chamber of Commerce, for example, "our cumbersome regulatory system [is] one of the top 10 barriers to Canadian competitiveness. The added delays and costs imposed by the over-complicated process dull our competitive edge in global markets and place Canada's standard of living at risk."[1] In sharp contrast, the Sierra Club of Canada has argued that the federal government "is abdicating its responsibility and trying to get out of the protecting-the-environment business."[2]

Neither the economy nor any industry is likely to collapse because of the pressures of EIA (O'Riordan 1982). Regardless of one's view on recent changes to EIA in Canada, the new environment for EIA is one that is occupied with process efficiency. There is, of course, a need to "do things right," and an efficient EIA process is important to timely decision-making. At the same time, however, it is necessary to "do the right things" and ensure that EIA, when applied, is a useful tool for making informed decisions about development initiatives. Delays in EIA are not caused so much by EIA itself as by the failure of the planning and decision-making system to accommodate its findings—the goal is simply to prevent "environmentally idiotic" decisions (O'Riordan 1982).

Cumulative Effects Assessment

The need to better assess and manage cumulative environmental effects really took shape in Canada during the early 1980s, and by the 1990s it had become a requirement under the Canadian Environmental Assessment Act for all project impact assessments. In practice, however, the effects of development continued to be assessed and managed largely on a project-by-project basis, with little regard for the effects that might result in combination with other past, present, and foreseeable development activities. Recent reviews of the state of CEA in Canada have indicated that CEA is simply not working; it remains narrow, reactive, and divorced from the broader planning and decision-making context (e.g., Noble, Sheelanere, and Patrick

2011; Harriman and Noble 2008; Duinker and Greig 2006; Dubé 2003). Impacts are assessed on a stressor-by-stressor basis, and impact statements typically refer to cumulative effects in a separate section, the assumption being that cumulative effects are a different class of effects, derived from summing the total effect of individual stressors. Duinker and Greig (2006) go as far as suggesting that CEA in its current form is doing more harm than good.

In a report on the public hearings for the Wuskwatim Generation and Transmission Projects, Manitoba, for example, the Manitoba Clean Environment Commission (2004) criticized Manitoba Hydro for absorbing the effects of past projects in current and future project baseline conditions, thereby giving inadequate consideration to cumulative environmental effects in their EIA. The proponent was also criticized for the temporal scope of its CEA, considering only 10 years into the future for cumulative effects on caribou. Unfortunately, Gunn and Noble (2012) in their review of Manitoba Hydro's next project, the Bipole III transmission line, note the same practice of ignoring the significance of past cumulative effects and report that the proponent's temporal scope for modelling cumulative effects was only five years into the future.

Part of the challenge is that CEA simply cannot work effectively within the confines of project-based EIA, and properly assessing and managing cumulative environmental effects can be beyond the scope and scale of any individual project proponent (see Ross 1994; Creasey 2002). Under project impact assessment, cumulative effects are considered only in relation to the incremental effects of a proposed activity rather than as the cumulative effects of all human actions on VECs. Perhaps we are asking for all the right things with regard to cumulative effects assessment but looking to the wrong process to achieve them. One cannot expect a truly regional approach to cumulative effects to emerge from within the confines of project-based EIA.

One response to the constraints of doing CEA inside project EIA has been the rise of regional studies in which CEA adopts a more *effects-based* approach and focuses on understanding environmental systems and relationships rather than focusing on individual project stressors. Several regional CEA studies have unfolded in Canada over the past 20 years, and regional frameworks for CEA have been the focus of several Canadian Environmental Assessment Agency research and development programs. The contribution of regional CEA studies to solving the problems of project-based CEA, however, has not materialized. As Harriman and Noble (2008) explain, regional frameworks for CEA have value in their own right, but they have operated at a different spatial scale from that of project EIA, tend to address different VECs and indicators, and occur outside the environmental assessment system as a parallel study with no real requirement for their integration or adoption in development decision-making. According to Spaling et al. (2000), rarely is there authority to implement recommendations or to carry forward regional effects–based findings to EIA, and because regional studies are conducted in many different jurisdictions and for different purposes, it is unlikely that a consistent model will be developed or utilized (Grzybowski and Associates 2001).

CEA requires the participation and cooperation of regulators, stakeholders, and developers to establish environmental objectives and manage development on a regional basis, guided by broader environmental planning and sustainability goals. Problems surface when individual project proponents attempt to address regional environmental management—that is, the effects of activities other than the one(s) they have proposed. Assessing cumulative effects beyond the project is a complex undertaking that requires a level of conceptualization, analysis, and co-ordination that is beyond the knowledge, capacity, and mandate of project proponents (Gunn and Noble 2009); unfortunately, neither is it always within the purview of provincial or regional authorities. The message here is not that proponents should not be required to consider the implications of their projects in the larger, regional context but that the mandates of project-based EIA and of good CEA do not always align.

Strategic Environmental Assessment

Various forms of SEA have been ongoing in Canada for a number of years, both formally and informally and under a variety of labels and institutional models, but reviews of the state of SEA systems and models in Canada conclude that SEA is diverse, founded on a range of principles and frameworks, and not well understood (Noble et al. 2013; Noble 2009). The result is considerable variability in SEA experience and added value, largely because of the institutional and methodological pluralism of SEA, and lack of tiering and integration of strategic policy, plan, program, and project initiatives. Noble (2009) characterizes the current state of SEA as one of "promise and dismay," going on to note that "under the federal system . . . many applications have been disappointing . . . some of the better examples of SEA have neither carried the SEA nametag nor occurred under formal SEA requirements . . . of particular concern is the systematic separation of SEA from downstream decision inputs and assessment activities." The benefits of SEA in facilitating CEA have not been forthcoming.

When SEA was introduced in Canada in the early 1990s, it was touted as the solution to addressing area-wide and cumulative effects problems—a means of helping to streamline EIA and helping to facilitate sustainable development. Yet more than 20 years later, no formal systems of SEA exist in Canada outside the current federal Cabinet directive, and the federal directive itself remains divorced from EIA processes. SEA was intended to address those strategic issues that simply cannot be addressed effectively by project EIA—issues that are a burden to the EIA process and operate on a scale much larger than project development, specifically regional land use, whether development should occur in a region, cumulative effects, and policies that affect resource development. Yet rather than strengthening the role of SEA and its relationship to project EIA, reforms to EIA, particularly at the federal level, have focused on streamlining EIA and the continued separation of SEA from the EIA process. This is ironic, given that one of the purposes of SEA is to ensure that federal departments and agencies are better able to meet sustainable development objectives through their policies, plans, and programs and that project development is often the result of higher-order PPP decisions, or the lack thereof, playing out on the ground. SEA cannot be expected to replace project-based assessment or to deliver the data required by project proponents to get their projects approved; however, SEA can ensure more efficient and more meaningful project-based assessments.

Integrating the Silos of Environmental Assessment

If you have travelled across the Canadian prairies, you will have no doubt seen silos. Silos are tall, cylindrical structures, usually manufactured of steel or concrete, used to store grain. Silos are sealed to exclude air, to the extent possible, so as to prevent moisture from invading and thus preserve the integrity of the grain. In many respects, this is how environmental assessment in Canada has operated—as a series of silos, with each silo containing a preserved process (EIA, CEA, SEA) and with little opportunity for cross-fertilization.

Project proponents have traditionally operated in the silo of stressor-based approaches, focused on identifying and mitigating the stress caused by their project, with little concern for the cumulative effects of their actions in combination with those of other developers. Governments have acted largely as gatekeepers, ensuring that regulatory processes are met and accepting a minimum standard of practices versus ensuring positive contributions to sustainability. The scientific and academic community have tended to operate in the silo of effects-based science, developing models and enhancing the understanding of ecosystem functioning and environmental effects in response to human disturbances but not necessarily providing the type of science needed to support immediate regulatory decisions. Land-use planners and managers have focused on more strategic issues, such as policies and longer-term environmental planning and social goals. Meanwhile, the incremental and day-to-day impacts of project developments have continued to accumulate irrespective of strategic planning and visioning exercises about the future.

Each of these silos is valuable; however, none of them can be expected to deliver the benefits of the others. The real, substantive benefits of impact assessment should be measured in terms of its contribution to environmental management and to better decisions about the sustainability of development. This cannot be realized within any individual silo of environmental assessment but only through a more regional and integrative approach. This is not to say that another layer of environmental assessment is required; rather, there is a need to integrate current knowledge and understanding of environmental assessment systems and practices and to do so at the regional and strategic tier if any of the silos is to have meaning.

Recent initiatives of the Canadian Council of Ministers of the Environment have focused on the advancement of Regional Strategic Environmental Assessment (R-SEA) (see Chapter 12), a means of ensuring a more integrative, regionally based assessment framework, informing the preparation of preferred development strategies and environmental management frameworks at a regional scale, and ensuring that consideration of the effects of future development possibilities trickle down to inform project impact assessment and decision-making. R-SEA is about identifying and assessing future possibilities; it is focused on informing regional development decisions, including the nature and types of development that can or should take place based on regional environmental, social, economic, and sustainability goals and objectives. In this way, R-SEA is a means of identifying preferred development plans, conservation planning initiatives, or resource management programs, which include spatial and temporal considerations of cumulative change, and thus provides direction for project EIA and for decisions about development. Having such a framework of environmental assessment in place is critical to ensuring effective regional environmental planning, cumulative effects management, and sustainability (Creasey 2002).

The success of any regional, integrative approach to environmental assessment, however, depends on good governance, including supporting institutional arrangements. Most regional frameworks for environmental assessment in Canada have been developed external to, and separate from, regulatory-based processes (Dubé 2003; Noble, Sheelanere, and Patrick 2011). Examples include the Northern Rivers Basin Study (Alberta), the Great Sand Hills Regional Environmental Study (Saskatchewan), and the current Elk Valley Cumulative Effects Management Framework (British Columbia). Although advantageous in terms of flexibility in framework design, scope, and application, regional approaches are often met with "limited success because there are no institutional mechanisms to use the results of the assessments—that is, there is no one to tell" (Forest Practices Board 2011). Such initiatives have been described as short-term bursts of activity and short-lived organizational commitments that continue to come up short (Parkins 2011). Good governance and supporting institutional arrangements are necessary for advancing the effectiveness of environmental assessment, yet such matters have received only limited attention.

Prospects

Does this mean that environmental assessment in Canada has not been delivering—that it has not been effective? On the surface, the answer may appear to be an overwhelming yes; however, the question itself is by no means superficial. Environmental assessment has come a long way since first introduced to Canada more than 40 years ago. There have been continual process improvements to project-based EIA, and new tools and frameworks, including CEA and SEA, have emerged. At the same time, however, it must be acknowledged that EIA in the absence of more strategic and regional processes is inherently constrained in its ability to facilitate decisions about development that are consistent with the broader principles of sustainability. EIA is focused on project developments; the aim of the proponent is to get project approval by mitigating, to the point of acceptability, potentially adverse environmental effects. Further, at the strategic tier of SEA, higher-order decisions and sustainability-based principles and objectives are of little benefit if they cannot be operationalized at the regional and project level and inform decisions about the nature and acceptability of development. Each of these processes—EIA, CEA, and SEA—is valuable in its own right, but each is limited in terms of what it can deliver in the absence of a broader, more integrative, more supportive, and more cyclical system of environmental assessment and knowledge translation from the strategic to the regional and the project level.

The knowledge needed to advance the current state-of-the-art of environmental assessment beyond its current limits does exist, but this knowledge is contained within the individual silos of environmental assessment, and the institutional capacity and the will for the types of actions and collaborations necessary for that advancement seem to be lacking. Far more attention has been given to critiquing environmental assessment and achieving greater process efficiencies than to advancing systems and practices and to ensuring substantive outcomes. Perhaps it's just easier to isolate EIA from broader PPP contexts and to continue to streamline the process to make stakeholders "happier" than it is to tackle the tough questions in impact assessment.

References

Boyden, A. 2007. "Environmental assessment under threat." *International Association for Impact Assessment Newsletter* 19 (1): 4.

Collyer, D. 2012. "Regulatory reform recommendations improve jobs, growth, prosperity outlook while continuing to deliver responsible environmental outcomes." Canadian Association of petroleum producers, CAPP Commentary. 23 March 2012. http://www.capp.ca/aboutUs/mediaCentre/Pages/default.aspx.

Creasey, R. 2002. "Moving from project-based cumulative effects assessment to regional environmental management." In A. Kennedy, ed., *Cumulative Environmental Effects Management: Tools and Approaches.* Calgary: Alberta Society of Professional Biologists.

Dubé, M. 2003. "Cumulative effect assessment in Canada: A regional framework for aquatic ecosystems." *Environmental Impact Assessment Review* 23: 723–45.

Duinker, P., and L. Greig. 2006. "The impotence of cumulative effects assessment in Canada: Ailments and ideas for redeployment." *Environmental Management* 37 (2): 153–61.

Forest Practices Board. 2011. "Cumulative effects: from assessment toward management." Special report. FPSB/SR/39.

Fuggle, R. 2005. "Have impact assessments passed their 'sell by' date?" *Newsletter of the International Association for Impact Assessment* 16 (3): 1, 6.

Galbraith, L., B. Bradshaw, and M. Rutherford. 2007. "Towards a supraregulatory approach to environmental assessment in northern Canada." *Impact Assessment and Project Appraisal* 25 (1): 27–41.

Gibson, R.B. 2002. "From Wreck Cove to Voisey's Bay: The evolution of federal environmental assessment in Canada." *Impact Assessment and Project Appraisal* 20 (3): 151–9.

Grzybowski and Associates. 2001. "Regional environmental effects assessment and strategic land use planning in British Columbia." Report prepared for the Canadian Environmental Assessment Agency Research and Development Program. Hull, QC: Canadian Environmental Assessment Agency.

Gunn, J., and B.F. Noble. 2009. "A conceptual and methodological framework for regional strategic environmental assessment (R-SEA)." *Impact Assessment and Project Appraisal* 27 (4): 258–70.

———. 2012. "Critical review of the cumulative effects assessment undertaken by Manitoba Hydro for the Bipole III project." Report prepared for the Public Interest Law Centre, Winnipeg.

Harriman, J., and B. Noble. 2008. "Characterizing project and regional approaches to cumulative effects assessment in Canada." *Journal of Environmental Assessment Policy and Management* 10 (1): 25–50.

Jay, S., et al. 2007. "Environmental impact assessment: Retrospect and prospect." *Environmental Impact Assessment Review* 27: 287–300.

Manitoba Clean Environment Commission 2004. "Report on public hearings, Wuskwatim

Generation and Transmission Projects." www.cecmanitoba.ca.

Noble, B.F. 2009. "Promise and dismay: The state of strategic environmental assessment systems and practices in Canada." *Environmental Impact Assessment Review* 29 (1): 66–75.

——, et al. 2013. "Strategic environmental assessment opportunities and risks for Arctic offshore energy planning and development." *Marine Policy* 39: 296–302.

Noble, B.F., P. Sheelanere, and R. Patrick. 2011. "Advancing watershed cumulative effects assessment and management: Lessons from the South Saskatchewan River watershed." *Journal of Environmental Assessment Policy and Management* 13 (4): 1–23.

Noble, B.F., and K. Storey. 2005. "Toward increasing the utility of follow-up in Canadian EIA." *Environmental Impact Assessment Review* 25 (2): 163–80.

O'Riordan, T. 1982. "Environmental impact assessment in the context of economic recession: Discussion." *The Geographical Journal* 148: 355–61.

Parkins, J. 2011. "Deliberative democracy, institutional building, and the pragmatics of cumulative effects assessment." *Ecology and Society* 16 (3): 20.

Ross, W. 1994. "Assessing cumulative environmental effects: Both impossible and essential." In A. Kennedy, Ed., *Cumulative Effects Assessment in Canada: From Concept to Practice*. Papers from the 15th Symposium held by the Alberta Society of Professional Biologists, Calgary.

Spaling, H., et al. 2000. "Managing regional cumulative effects of oil sands development in Alberta, Canada." *Journal of Environmental Assessment Policy and Management* 2 (4): 501–28.

Notes

[1] See http://actionplan.gc.ca/en/page/r2d-dr2/what-responsible-resource-development.

[2] See http://www.cbc.ca/news/politics/story/2012/04/17/environmental-reviews.html.

Glossary

accuracy The amount of bias applied to the system-wide impact predictions.

acid mine drainage Drainage from surface mining, deep mining, or coal-refuse piles, usually highly acidic with large concentrations of dissolved metals.

active publics Publics that affect assessment decisions.

activity information The effects possibly generated as a result of a project.

adaptive environmental management A multi-step, deliberative process that involves exploring alternative management actions and making explicit forecasts about their outcomes, carefully designing monitoring programs to provide reliable feedback and understanding of the reasons underlying actual outcomes, and then adjusting objectives or management actions based on this new understanding.

additive effects Individual minor actions that may be separate or related and that have a significant overall impact when combined.

adverse effects Direct project effects that have potential to cause long-term, irreversible, undesirable environmental damage or change.

ALCES A Landscape Cumulative Effects Simulator—a system dynamics simulation tool for exploring the behaviour or response of resource systems to disturbances.

alternative means Different ways of carrying out a proposed project—typically alternative locations, timing of activities, or engineering designs.

alternatives to Under the Canadian Environmental Assessment Act, the different ways of addressing the problem at hand or meeting the proposed project objectives; renewable energy, for example, would be considered an "alternative to" a proposed coal-fired generating station.

ambient environmental quality monitoring A pre-project assessment of the surrounding environment inclusive of biophysical and socio-economic factors; collected information is used as a baseline in comparing a project's environment during development, operation, and post-operation against unaffected control sites in order to monitor the impact of a project.

amplifying effects Incremental effect additions whereby each increment has a larger effect than the one preceding it.

analogue approaches The collection of qualitative assessment methods that are based on secondary information, including examination of similar projects, literature and case reviews, document analysis, and synthesis of existing databases.

analysis scale The scale applied to examine VECs and impacts across space.

antagonistic effects Individual adverse effects that have potential to partially cancel each other out when combined.

assessment by responsible authority Under the Canadian Environmental Assessment Act, 2012, an EIA undertaken by the Canadian Environmental Assessment Agency, the Canadian Nuclear Safety Commission, or the National Energy Board.

assessment by review panel Under the Canadian Environmental Assessment Act, 2012, an EIA undertaken by an independent panel of experts appointed by the minister of environment.

assimilative capacity The ability for an environment to incorporate pollutants into its system without adverse effects on its natural state.

auditing An objective examination or comparison of observations with predetermined criteria.

backcasting The process of working backwards from a particular future condition judged to be desirable and then determining the feasibility of achieving that condition given project actions and changing environmental conditions.

balance model Model designed to identify inputs and outputs for specified environmental components; these models are commonly used to predict change in environmental phenomena.

baseline condition A description of the pre-project environment, which is inclusive of the cumulative effects of previous activities and the future environment in the absence of the proposed project.

baseline study Description of the biophysical and socio-economic state of the environment at a given time that can later be used for the comparison of environmental change through time.

benchmark A standard or point of reference against which something can be measured or judged.

biophysical environment Consists of air, water, land, rocks, minerals, plants, animals, and the interacting components of ecosystems.

buffer zone An area of undisturbed environment; a commonly used mitigation practice.

Canadian Environmental Assessment Act The legal basis for federal EIA in Canada from 1995 to 2012; sets out responsibilities and procedures for EIA of projects that involve federal authorities; introduced in Parliament in 1992, proclaimed in 1995, and revised in 2003.

Canadian Environmental Assessment Act, 2012 The current legal basis for federal EIA in Canada, effective July 2012; sets out responsibilities and procedures for EIA of projects that involve federal authorities; introduced and proclaimed in 2012.

Canadian Environmental Assessment Agency The federal agency created in 1994, replacing the Federal Environmental Assessment Review Office, to oversee Canadian federal EIA and implementation of the Canadian Environmental Assessment Act.

case-by-case screening Evaluating project characteristics against a checklist of regulations and guidelines.

cautionary threshold The level of change or set condition at which monitoring efforts should be increased to more closely monitor VEC or indicator conditions and the effectiveness of best management practices verified to prevent any further adverse change.

checklists Method used to create a comprehensive list of effects or indicators of environmental impacts that a project might generate.

chemical antagonism A reaction between two effects, such as in a chemical reaction, whereby the severity of the combined effect is reduced.

closed scoping EIAs with content and scope predetermined by law; modifications can only be made through closed consultation between the proponent and the responsible authority or regulatory agency.

compensation The measures taken by the proponent to make up for adverse environmental impacts of a project that exist after mitigation measures have been implemented.

compliance monitoring Monitoring to ensure that all project regulations, agreements, laws, and specific guidelines have been adhered to.

component interaction matrices Applied to improve understanding of indirect impacts from projects; the matrices identify first-, second-, and higher-order impacts and illustrate the dependencies between environmental components.

comprehensive study EIA An EIA assessment applied on large-scale, complex, environmentally sensitive projects that have a high risk of causing adverse environmental effects.

condition-based indicator Specific indicator (e.g., phosphorus concentrations, benthic invertebrate abundance) that provides direct, measurable information about the condition or state of the VEC.

confirmatory analysis Used to test for uncertainty in impact predictive techniques and to ensure similar predictive outcomes from different types of techniques.

consistency A measure of the quality of an expert's judgment, including a test of the degree of randomness in a set of assessment scores or judgments.

continuous effects Effects that are ongoing over space or time.

control site A reference point where the environment is not affected by the project; used to monitor the nature and extent of project-induced environmental change in areas that are affected by the project.

cost-benefit analysis An assessment method that expresses project impacts in monetary terms.

critical threshold Maximum acceptable change, socially or ecologically, beyond which impacts may be long-term or irreparable.

cumulative effect A change in the environment caused by multiple interactions among human activities and natural processes that accumulate across space and time.

cumulative effects assessment A systematic process of identifying, analyzing, and evaluating cumulative effects.

cumulative effects management The identification and implementation of measures to control, minimize, or prevent the adverse consequences of cumulative effects.

cumulative effects monitoring Monitoring of cumulative effects that emphasizes the accumulated stress of multiple developments within a particular region.

decision-point audit Examination of the role and effectiveness of the EIS based on whether the project is allowed to proceed and under what conditions.

Delphi technique An iterative survey-type questionnaire that solicits the advice of a group of experts, provides feedback to all participants on the statistical summaries of the responses, and gives each expert an opportunity to revise her or his judgments.

designated project A project listed in the Canadian Environmental Assessment Act, 2012 Regulations Designating Physical Activities for which an EIA may be required.

deterministic model A model dependent on fixed relationships between environmental components.

direct effects First-order impacts that have a particular social value.

discontinuous effects Effects that occur in an incremental manner and go unnoticed until some threshold is reached.

dispositional antagonism A type of antagonistic effect in which one effect affects the uptake or transport of another effect, such as ethanol enhancing mercury elimination in mammals.

draft EIS audit Review of the project EIS according to its terms of reference.

duty to consult A formal, legal obligation in Canada for governments to consult with Aboriginal people in cases where Aboriginal rights, claims, or titles are known and may be affected by a development or decision, even in cases where those rights, claims, or titles have not yet been proven in court.

early warning indicators Biological or non-biological indicators that can be measured to detect the possibility of adverse stress on VECs before they are adversely affected.

effects-based CEA Measures multiple environmental responses to stressors; the results are then compared to some control point to determine the actual measure of cumulative change.

environment All environments, including biophysical, social, and economic components.

Environmental Assessment Review Process The first Canadian federal EIA process, formally introduced in 1973 by guidelines order.

environmental baseline The present and likely future state of the environment without the proposed project or activity.

environmental change Measurable change in an environmental parameter over time.

environmental effect The difference in the condition of an environmental parameter with as opposed to without a proposed development activity.

environmental impact assessment (EIA) A systematic process designed to identify, predict, and propose management measures concerning the possible implications that a proposed project's actions may have for the environment.

environmental impacts Environmental effects that have an estimated societal value placed on them.

environmental impact statement (EIS) The formal documentation produced from the EIA process that provides a non-technical summary of major findings, statement of assessment purpose and need, detailed description of the proposed action, impacts, alternatives, and mitigation measures.

environmental management system A voluntary industry-based management system that controls the activities of products and processes that could cause adverse environmental impacts; management systems are a cyclical process of continual improvement in which industries and firms are constantly reviewing and revising their way of doing business to protect the environment.

environmental preview report An early report that is used to determine whether or not an EIA is needed and the level of assessment required; it outlines the project, potential environmental effects, alternatives, and management measures.

environmental protection plans Mandatory management plans that result from project-based EIAs; these management plans are tailored to the project as a result of the identification of key impacts and issues and management measures through the EIA process.

environmental site assessment   An environmental study for the purpose of determining the nature and extent of contamination of a local site and prescribing cleanup and reclamation plans.

environmental systems   Environmental components functioning together as a unit.

equivalency EIA   Applies under the Canadian Environmental Assessment Act, 2012, whereby a provincial or Aboriginal EIA process is substituted as equivalent to the federal process for the purpose of assessing the environmental effects of a designated project when the Canadian Environmental Assessment Agency is the responsible authority.

exclusion list   A screening mechanism listing projects that would be subject to an EIA unless they were *included* in the list; generally, projects excluded involve issues of national defence and emergency or are projects routine in nature.

experimental monitoring   Research into environmental systems and their impacts for the purpose of gathering information and knowledge and testing hypotheses.

ex-post evaluation   Taking action and making decisions based on the result of structure, analysis, and appraisal of information concerning project impacts.

Federal Environmental Assessment Review Office (FEARO)   The federal agency created to oversee implementation of the Federal Environmental Assessment Review Process.

fixed-point scoring   A VEC weighting approach in which a fixed number of values is distributed among all affected environmental components; the higher the point score, the more important the environmental component.

fly-in fly-out   Projects located in remote areas with high amounts of air traffic to and from the site; typically associated with remote mining projects where temporary project work camps are constructed to house workers, who commute by charter air service on a basis, for example, of two weeks at the work site and two weeks off during the lifespan of the project.

functional antagonism   A type of antagonistic effect whereby one effect counterbalances an organism's physiological response to a second effect.

functional scale   Scale relationship based on how different environmental components function across space.

Gaussian dispersion model   A model devised for predicting point-source atmospheric pollution.

Geographic Information System (GIS)   A system of computer hardware and software for working with spatially integrated and geo-referenced data.

gradient-to-background monitoring   Monitoring system that measures the effects caused by the impact source at an increasing distance from the impact origin to the point of background assimilation; an "artificial" control point is established.

gravity model   A deterministic model used to predict population flow or spatial interaction; the model is dependent on a fixed and inverse relationship between mass (population) and distance.

human environment   Aspects of the environment that are non-biophysical components; also referred to as the socio-economic or cultural environment.

hybrid screening   A screening approach that combines the characteristics of case-by-case, list-based, and threshold-based screening.

impact avoidance   A form of impact management whereby impacts are avoided at the outset by way of alternative project designs, timing, or location rather than managed or mitigated after they occur.

impact benefit agreement   Legal agreement between a proponent and a community or group that will potentially be affected by a project; generally applied to ensure that the resources for maximizing the benefits associated with the development are fully capitalized on.

impact mitigation   Minimizing adverse environmental change associated with a project by implementing environmentally sound construction, operating, scheduling, and management principles and practices within project design.

impact significance   The degree of importance of an impact based on the characteristics of the impact, the receiving environment, and societal values.

implementation audit   An evaluation of whether or not the recommendations presented in a project's EIS were actually put into practice.

inactive publics   Publics not normally involved in the environmental planning, decisions, or project issues.

inclusion list   A screening method listing projects that have mandatory or discretionary requirements for an EIA.

incremental effects   Marginal changes in the condition of an environmental component caused by project actions.

initial environmental examination   Information prepared to establish whether or not an EIA is needed and what level of EIA should be implemented.

input evaluation   Follow-up or auditing of SEA procedures and requirements, such as purpose and objectives, data quality, and linkages to policy, planning, or program initiatives.

inspection monitoring   Site-specific monitoring with on-site visits to ensure compliance with procedures and safety standards.

intention surveys   Surveys that attempt to collect the judgment of as many people as possible and record their responses regarding what they intend to do or how they might react, given certain circumstances or situations.

interaction matrices   EIA matrices based on the multiplicative properties of simple matrices to generate a quantitative impact of the proposed project on interacting environmental components.

irreversible impact   An environmental impact that cannot be reversed because of economic, ecological, or technological limitations.

ISO 14001   Internationally recognized industry performance standard for environmental management system certification.

Keynesian multiplier   A basic economic multiplier that notes that an injection of money into a local economy will increase at the local level by some multiple of that initial injection.

Leopold matrix   An EIA matrix for identifying first-order project–environment interactions, consisting of a grid of 100 possible project actions along a horizontal axis and 88 environmental considerations along a vertical axis.

life-cycle assessment   The "cradle-to-grave" assessment of projects from their inception and start-up to post-operation.

linear additive effects   Incremental additions to or from a fixed environment where each additional increment has the same effect.

list-based screening   A checklist of projects that may or may not require an EIA.

Mackenzie Valley Environmental Impact Review Board   A valley-wide public board created as part of the Mackenzie Valley Resource Management Act to undertake EIAs and panel reviews under the jurisdiction of the act.

Mackenzie Valley Resource Management Act   An act implemented by the federal government to give decision-making authority to northerners concerning environment and resource development activities within the Mackenzie Valley region of the Northwest Territories; proclaimed in 1998, the act governs EIA in the region.

magnitude   The amount of change in a measurable parameter relative to baseline conditions (e.g., percentage of habitat change, amount of change in concentration of a contaminant, percentage change in land use) or other target.

magnitude matrices   Matrices that attempt to identify impacts and summarize impact importance, time frame, and magnitude.

MARXAN   Marine Spatially Explicit Annealing, a landscape and marine optimization tool that uses a site-selection algorithm to explore options for conservation and regional biodiversity protection planning.

matrices   Management and assessment tools used for identifying project impacts that typically consist of a two-dimensional checklist with project actions on one axis and environmental components on the other.

maximum allowable effects level   An approach to impact prediction based on specifying certain desired limits or thresholds that a certain impact is not to exceed.

mechanistic model   A type of model based on mathematical equations or flow diagrams that describe cause–effect relationships in a project environment.

methods   The various aspects of an assessment, including organization, identification of impacts, and collection and classification of data.

models   Box-and-arrow or mathematical equations used to simplify real-world environmental systems.

monitoring   A systematic process of data collection or observations used to identify the cause and nature of environmental change.

monitoring for knowledge The monitoring used after impacts occur; data are collected and used for future impact prediction and project management.

monitoring for management Tracking and evaluating changes in a range of environmental, economic, and social variables; usually associated with high-profile projects with uncertain outcomes and with the potential for significant adverse outcomes.

monitoring of agreements Monitoring and auditing of agreements between project proponents and affected groups to ensure compliance.

Monte Carlo analysis A means of statistical evaluation of mathematical functions using random samples.

multi-criteria analysis A structured analytical approach that involves the assessment of competing alternatives or options against multiple criteria.

multiplier A quantitative expression of some initial exogenous change and the expected additional effects caused by interdependences with an endogenous linkage system.

National Environmental Policy Act (NEPA) The US legislation of 1969 that required certain development project proponents to demonstrate that their projects would not cause adverse environmental effects; the beginning of formal EIA.

network diagrams Models based on box-and-arrow diagrams that consist of environmental components linked by arrows indicative of the nature of energy flow or interaction between them.

non-point-source stress Environmental stress from diffuse sources that cannot be traced back to a particular project origin, such as runoff from urban areas or pollution introduced to streams from groundwater.

normative ethics The notion that "right" and "wrong" are found within an individual's behaviour; the focus is on arriving at moral conduct standards.

Nunavut Impact Review Board Established under the Nunavut Land Claims Agreement and the primary authority responsible for EIA in the land-claims area.

Nunavut Land Claims Agreement Canada's largest land claims settlement and land-claims-based EIA process; signed in 1993, giving the Inuit self-governing authority and leading to the establishment of a new territory, Nunavut, in 1999.

off-site impacts Impacts that occur at a distance or removed from the recognized project area.

on-site impacts Impacts that occur directly in the immediate project area.

open scoping A transparent scoping process in which the content and scope of the assessment are determined through consultation with various interests groups and public stakeholders.

output evaluation Monitoring or auditing SEA effectiveness on the basis of the output of the SEA process—namely, whether the process informed policy and whether the intended outcomes were actually realized.

paired comparisons An approach to determining the relative importance of impacts in a hierarchy in ratio form whereby the decision-maker considers trade-offs one at a time for each pair of VECs.

performance audit An assessment of a proponent's capability to respond to environmental incidents and of its management performance.

Peterson matrix A multiplicative EIA matrix consisting of project impacts and causal factors, resultant impacts on the human environment, and relative importance of those human components used to derive an overall project impact score.

phenomenon scale The scale used to determine the spatial extent within which certain environmental components and VECs operate and function.

plan A defined strategy or proposed design to carry out a particular course of action or several actions and various options and means to implement those actions.

point-source stress Environmental stress that can be traced back to a particular project origin, such as discharge or emissions from a development activity.

policy A guiding intent, set of defined goals, objectives, and priorities, either actual or proposed.

policy SEA Strategic environmental assessment applied to policies that have no explicit "on-the-ground" dimension, such as fiscal policies or national energy policies; also referred to as indirect SEA.

practical ethics The notion that "right" and "wrong" can be found within scenarios; the focus is on studying specific practical problems to derive a commitment to action.

precautionary principle The principle that when information is incomplete but there is threat of an adverse effect, the lack of full certainty should not be used as a reason to preclude or postpone actions to prevent harm.

precision The exactness of impact prediction.

predictive technique audit A type of environmental audit in which a project's predicted effects are compared to the actual effects.

probability analysis An analysis that uses quantified probability to classify the likelihood of an impact occurring and under what environmental conditions.

process evaluation Monitoring and auditing SEA based on the actual SEA process, including methods, techniques, openness, and frameworks.

professional ethics The proper conduct of professionals in practice, typically set out in codes of conduct.

program A schedule of proposed commitments or activities to be implemented within or by a particular sector, plan, or area of policy.

programmatic environmental assessment Under the US NEPA, the application of environmental assessment to multiple projects or to programs of development.

programmed-text checklist A type of EIA checklist consisting of a series of filter questions for project screening and impact identification; useful for standard or routine projects.

project evaluation monitoring Performance auditing or productivity measurement, a monitoring program concerned with a project's performance and ability to reach specified goals and objectives.

project impact audit A type of auditing that focuses on determining whether the actual project impacts were predicted in the EIS.

questionnaire checklist An EIA screening method consisting of a set of questions that must be answered when considering the potential effects of a project.

rating An approach to VEC or impact weighting whereby the importance or significance of each is indicated on a numerical scale ranging from, for example, 1 to 5; no direct decision trade-offs are involved.

receptor antagonism A type of antagonistic effect in which one effect blocks the other, such as a toxicant binding to a receptor and blocking the effects of a second toxicant.

receptor information Information pertaining to the processes resulting from project-induced effects, such as habitat fragmentation, that is important to consider when characterizing the environmental setting.

regional SEA An approach to SEA concerned with regional-based environmental planning or development and assessing the impacts of area-specific plans and program initiatives.

Regulations Designating Physical Activities Regulations under the Canadian Environmental Assessment Act, 2012 that identity the activities, undertakings, and projects that may be subject to a federal environmental assessment.

regulatory permit monitoring Site-specific monitoring that includes regular documentation of requirements necessary for permit renewal or maintenance.

relative significance A measurement of the relative importance of one affected environmental component over another.

remediation The process of post-industrial or post-development site cleanup, which typically involves the removal of contaminants or pollution from soil and water.

residual effects Effects that remain after all management and mitigation measures have been implemented.

responsible authority One of three federal authorities responsible for EIA under the Canadian Environmental Assessment Act, 2012: the Canadian Environmental Assessment Agency, the Canadian Nuclear Safety Commission, and the National Energy Board.

reversible impact A change to the environment caused by a project that can be reversed with proper impact management or restoration action.

review panel EIA A level of EIA that is applied to projects with uncertain or potentially significant effects or if public and stakeholder concern warrants an independent review panel.

risk The possibility that an undesired outcome may result from an uncertain situation.

risk assessment Using collected information to identify potential risks.

sanitary landfill A type of waste disposal site where solid waste is contained within an impermeable barrier within the earth's surface and covered.

scenario analysis   An approach used in EIA and SEA to identify hypothetical actions or situations and potential outcomes.

scoping   An early component of the EIA process used to identify important issues and parameters that should be included in the assessment.

screening   The selection process used to determine which projects need to undergo an EIA and to what extent.

screening EIA   A type of project assessment that is an extension of the basic screening process, in which anticipated environmental effects are documented and the need for additional project modification or further assessment is determined.

secondary effects   Effects resulting from a direct impact.

sector SEA   A type of SEA based on initiatives, plans, and programs that are specific to a certain industrial sector.

sensitivity analysis   Examination of the sensitivity of an impact prediction to minor differences in input data, environmental parameters, and assumptions.

socio-economic monitoring   A type of monitoring that looks specifically at socio-economic parameters in the review of a project area.

spatial scale   The actual geographic scale used to define the extent of a project EIA.

spatial model   A type of model used to depict and understand the spatial relationship between phenomena.

statistical model   A type of model used to test relationships between variables and to extrapolate data.

statistical significance   The determination, based on confidence intervals and probabilistic data, as to whether a particular outcome or prediction is significant based on theoretical and empirical findings.

stochastic model   A type of mechanistic model that is probabilistic in nature or gives an indication of the probability of an event occurring within specified spatial and temporal scales.

strategic environmental assessment (SEA)   The environmental assessment of initiatives, policies, plans, and programs and their alternatives.

stress-based indicator   Also referred to as a disturbance-based indicator, an indicator or parameter used to measure stress that

directly affects a VEC (e.g., amount and distribution of surface disturbance or stream crossing density).

stressor-based CEA   Assessment that predicts cumulative effects associated with a particular agent of change.

structural surprises   Cumulative effects that occur in regions with multiple developments; the least understood and most difficult cumulative effects to assess.

substitution EIA   Applies under the Canadian Environmental Assessment Act, 2012, whereby another jurisdiction, other than the federal government, is delegated the responsibility to carry out part of the EIA process for a designated project for which the Canadian Environmental Assessment Agency is the responsible authority.

sustainability assessment   Broadly defined, a process by which the implications of an initiative for sustainability are evaluated to help decision-makers decide what actions to take or not to take to make society more sustainable.

synergistic effects   When the total effects are greater than the sum of the separate, individual effects.

target threshold   Typically, a politically or socially defined limit, a margin of safety, and a mandatory trigger for management action.

techniques   Ways of providing and analyzing data in EIA.

threshold   The point or level at which there is an abrupt change in the condition of an environmental component or where small changes result in large responses.

threshold-based prediction   Basing impact predictions on prior experiences using approaches such as maximum allowable effects levels whereby an impact is capped and not to exceed a certain threshold or level of change.

threshold-based screening   A screening process whereby proposed developments are placed in categories and thresholds are set for each type of development, such as project size, level of emissions generated, or area affected.

traditional environmental knowledge   Local or Aboriginal knowledge acquired from experience, culture, or interaction with land and resources over time.

valued environmental components (VECs)   Components of the human and physical environment that are considered important and therefore require evaluation within EIA.

VEC indicators   Provide a measure of qualitative or quantitative magnitude for an environmental impact and might include, for example, specific parameters of air quality, water quality, or employment rates; allow decision-makers to gauge environmental change efficiently.

weighted impact interaction matrices   An EIA matrix method whereby impacts are multiplied by the relative importance of the affected environmental components and secondary impacts are explicitly incorporated.

weighted magnitude matrices   An EIA matrix method in which degrees of importance, representing the potential impacts of a particular project action on an environmental component, are assigned to the affected environmental components and then multiplied by project impacts.

Index